Springer Series in Advanced Microelectronics

Volume 66

The *Springer Series in Advanced Microelectronics* provides systematic information on all the topics relevant for the design, processing, and manufacturing of microelectronic devices. The books, each prepared by leading researchers or engineers in their fields, cover the basic and advanced aspects of topics such as wafer processing, materials, device design, device technologies, circuit design, VLSI implementation, and sub-system technology. The series forms a bridge between physics and engineering, therefore the volumes will appeal to practicing engineers as well as research scientists.

Sung-Soo Kim

Fundamentals and Principles of Electromagnetic Wave Absorbers

From Theory, Design, and Materials to Measurement

Springer

Sung-Soo Kim
Emeritus Professor
Chungbuk National University
Cheongju, Korea (Republic of)

ISSN 1437-0387 ISSN 2197-6643 (electronic)
Springer Series in Advanced Microelectronics
ISBN 978-981-97-6476-1 ISBN 978-981-97-6477-8 (eBook)
https://doi.org/10.1007/978-981-97-6477-8

This Springer imprint is published by the registered company Springer Nature Singapore Pte Ltd.
The registered company address is: 152 Beach Road, #21-01/04 Gateway East, Singapore 189721, Singapore

To my parents, Gui Taek and Young Sook Kim and to my families, Kyung Sook, Jung Hwan, Hee Jin, Jae Hyung, Hyo Jin, Ha Seong, Jung Woo, and Ga Yoon

Preface

Today, the electromagnetic (EM) radiation from various electronics of high-speed digital and analogue circuits in specific narrow frequency band or wide frequency spectrum is considered a significant challenge with regard to electromagnetic interference (EMI), electromagnetic compatibility (EMC), and reliability of electronic systems. Moreover, the rapid development of the information and communication industries is characterized by high speed of information processing (high frequency) and high integration by chip technology (miniaturization and mobility). High-performance EM wave absorbers are required for electromagnetic wave control to eliminate the interference between electronic devices and systems, particularly with a thin layer thickness at a specific frequency or broad bandwidth from radio frequency even up to a millimeter wave spectrum, depending on their applications. Another area of application is military stealth technology to cope with radar systems by reducing radar cross section (RCS). Radar absorption structures (RAS) and radar absorption materials (RAM) with light weight and wide absorption bandwidth will obviously be needed to achieve RCS reduction.

Entitled *Fundamentals and Principles of Electromagnetic Wave Absorbers: From Theory, Design, Materials to Measurement*, this book consists of three main parts: fundamental theory, design principles and methodology, and potential materials that can be applied to EM wave absorbers. The theoretical part provides the basis of electromagnetism, circuit and transmission line theory, EM wave propagation and reflection, and complex permittivity and permeability by electric polarization and magnetization of materials. The design part describes the design methods for various forms of EM wave absorbers based on equivalent circuit models and simulation techniques. Starting from the traditional resonant absorber, we review the latest metamaterial and frequency selection surface (FSS) absorbers with more advanced design techniques. Recent research results are also included associated with how to design the ultrawide-bandwidth absorbers through multilayering FSSs or shape control of lossy materials. The materials section examines a variety of lossy materials that can be used as EM wave absorbers, including conductive materials, magnetic materials, dielectric materials, core-shell materials, fiber-reinforced composites, metamaterials and metasurfaces. A literature review of electromagnetic properties and EM wave absorption performance is also provided. Finally, the methods and principles for measuring the high-frequency properties (complex permittivity and permeability) and EM wave absorption are described.

This book is suitable for scientists and engineers who are interested in EMI or EMC technology in related fields. By providing a physical and electrical engineering foundation in electromagnetism, it is possible to build up understanding and knowledge of the operating principles and applications of EM wave absorbers. In addition, the included high-frequency physical properties and EM wave absorption characteristics of various absorption materials will be helpful for them to design and select materials suitable for the target application. The book can also be used as a textbook for two semesters of undergraduate or graduate studies. Students will be able to study the physical basis of wave dynamics of EM waves and their interaction with materials. Circuit theory, including transmission lines, is included to model and interpret these physical phenomena as electrical circuits. By introducing high-frequency behavior and loss mechanisms based on electrical conduction, dielectric polarization, and magnetization of materials, the atomic or molecular properties of EM wave absorption materials can be understood. It is hoped that the readers will be able to acquire integrated knowledge and up-to-date information on EM absorber engineering, and this book can be used as a textbook or guidebook for industrial engineers and school students.

Cheongju, Korea (Republic of) Sung-Soo Kim

Acknowledgments

It has been a great blessing and pleasure for me to be able to write this book after retirement through collecting all the lecture materials and research results so far. The main part of this book arose from the research on *electromagnetic wave absorption materials* carried out by the author's research group at Chungbuk National University. I am grateful to all the students who graduated from our lab, and I would not have been able to write this book without their efforts and research products. During the period when I started my research at the Agency for Defense Development, Korea, my collaboration with Dr. Sung-Baek Jo, Dr. Kyung-Il Kwon, and Dr. Kil-Sung Churn became the seed for the publication of this book, and I am very grateful for that. Collaboration and discussions with Dr. Byung-Il Yoon have also been of great help in understanding the physical properties of radar absorption and structural materials. I would like to thank Prof. Young-Cheol Kim for his help in planning and proposing this book. I would like to thank Dr. Kyung-Seop Lee, Mr. Min-Sung Kim, and Ms. Joo-Hee Cho for their help in preparing the manuscript and art production. I would also like to thank the National Research Foundation of Korea for research funding over the long-term period. I am grateful to my parents and my wife for their heartful support. Looking forward to the day when my beloved Ha Seong, Jung Woo, and Ga Yoon will be able to read this booklet.

Cheongju, Korea (Republic of) Sung-Soo Kim

Contents

Chapter 1
Fundamentals of Electromagnetics

This chapter describes the fundamentals of electromagnetism required for the design and application of electromagnetic wave absorbers. Electromagnetic spectrum is introduced, with frequency (wavelength) ranges and applications for each frequency band. This chapter also introduces the physical properties of materials in reaction to the external electric and magnetic fields and the associated material parameters (conductivity, permittivity, permeability) that establish the constitutive equations. SI and decibel units are introduced for representing the electromagnetic physical quantities and the gain or loss of power (or voltage) in electronic devices. Sinusoidal harmonic waves and their vector notation, phasor, are described together with their complex number representation. Vector algebra and vector calculus are included for understanding and mathematical representations of electromagnetic field theory and related phenomena. Maxwell's equations are introduced, including Faraday's law related to electromagnetic induction, Ampere's law to explain the generation of magnetic fields by conduction and displacement currents, and Gauss's law related to electric and magnetic fields. Maxwell's equations in integral and differential forms have been described, expressing the above four electromagnetic phenomena.

1.1 Electromagnetic Wave Spectrum

Electromagnetic waves are synchronized oscillating electric and magnetic fields, transmitting through free space with a speed of light. In general, the wave equations are given by the time-harmonic sine (or cosine) functions:

$$E = A \sin(\omega t - \beta z) \tag{1.1}$$

S. Kim, *Fundamentals and Principles of Electromagnetic Wave Absorbers*, Springer Series in Advanced Microelectronics 66, https://doi.org/10.1007/978-981-97-6477-8_1

where A is the amplitude, $\omega = 2\pi f$ (f is the frequency) is the angular frequency (in radians/second), and β is the phase constant or wave number (in radians/meter). In Eq. (1.1), $\omega t - \beta z$ is the phase (in radians) of the wave and it depends on both time t and space z. At any point, the sinusoidal variation with time is period, given by $T = 2\pi/\omega = 1/f$. At any time, the variation with distance is wavelength $\lambda = 2\pi/\beta$. In response to the variation in both time and location, the speed of the travelling wave is determined by the movement of a point of a constant phase, which is called the phase velocity v:

$$v = f\lambda \tag{1.2}$$

The speed of electromagnetic waves in free space v_0 is

$$v_0 = 2.997925 \times 10^8 (\mathrm{m/s}) \tag{1.3}$$

All electromagnetic waves travel at the speed of light in a vacuum, but have a wide range of frequencies and associated wavelengths, as shown in Fig. 1.1. The entire electromagnetic spectrum, from the lowest frequency to the highest frequency (from the longest to the shortest wavelength), is based on all radio waves (e.g., commercial radio and television, microwaves, radar), infrared, visible light, ultraviolet, X-rays, and gamma rays [1].

Important applications of electromagnetic waves are communications and radar. Applications on the communication side include telephones, TVs, mobile communications, wireless local area networks (WLANs), and satellite communications. Radar is an electronic system based on measuring the distance from the target and the speed of a moving object, and it has various applications such as aircraft takeoff and landing, missile tracking, vehicle collision avoidance, climate prediction, speed measurement, and remote sensing. Table 1.1 shows the frequency bands and their names (originated as code-names during World War II), frequency range, wavelength, and applications [2]. Frequency range is very broad from HF to mm wave, ranging from 3 MHz to 300 GHz. Correspondingly, the wavelength for the bands is reduced from 100 m to 1 mm.

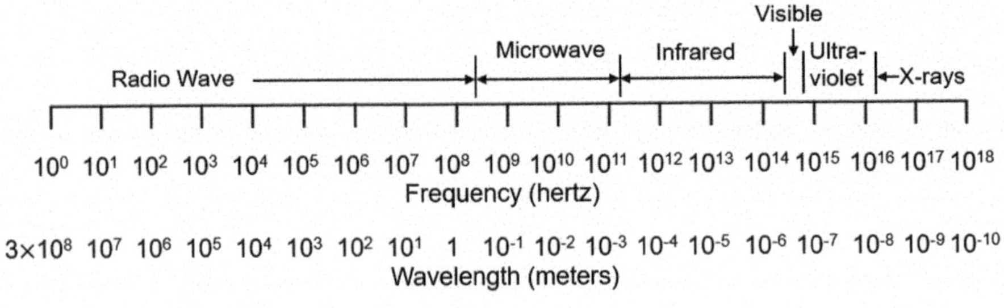

Fig. 1.1 Electromagnetic spectrum and frequency (wavelength)

Table 1.1 Frequency bands for electromagnetic waves and their applications

Frequency band	Band name	Frequency range	Wavelength	Applications
High frequency (HF)	HF	3–30 MHz	10–100 m	Telephone, telegraph, coastal radar systems
Very high frequency (VHF)	P	30–300 MHz	1–10 m	VHF TV, FM radio, very long-range radar
Ultra high frequency (UHF)	UHF	300–1000 MHz	0.3–1 m	UHF TV, satellite, cellular phones, long-range radar
	L	1–2 GHz	15–30 cm	
Super high frequency (SHF)	S	2–4 GHz	7.5–15 cm	Airplane radar, missile guidance, marine and weather radar, speed detector, mapping, airport surveillance
	C	4–8 GHz	3.75–7.5 cm	
	X	8–12 GHz	2.5–3.75 cm	
	Ku	12–18 GHz	1.67–2.5 cm	
	K	18–27 GHz	1.11–1.67 cm	
Extreme high frequency (EHF)	Ka	27–40 GHz	0.75–1.11 cm	Various radar applications, military communications, visual sensor for autonomous vehicles
	mm	40–300 GHz	1–7.5 mm	
	Q	40–60 GHz	5–7.5 mm	
	V	50–75 GHz	4–6 mm	
	W	75–110 GHz	2.7–4 mm	

1.2 Material Parameters and Constitutive Relation

When an electric field E (V/m) and a magnetic field H (A/m) are applied to a material or substance, it exhibits the inherent electromagnetic properties of the material. When an electric field is applied to metals and conductors, the phenomenon of electrical conduction occurs due to the movement of free electrons or mobile ions, which is expressed as the conduction current density J (A/m^2). When electric field is applied to an ionic bonding material such as ceramics, the phenomenon of electrical polarization appears due to the displacement of ions, which is expressed as the electric flux density or displacement D (C/m^2). When electric currents flow through a coil, magnetic field is generated. When the magnetic field is applied to a material, magnetic polarization (magnetization) occurs due to the orbital motion of electrons or the alignment of spins. This is denoted by magnetic flux density B (Wb/m^2). These are summarized in Table 1.2.

The conductivity is a material constant that indicates the electrical conduction, and is denoted by the Greek σ (sigma), and the unit is (S/m). Permittivity is a material constant that indicates the ability of a material to generate the electrical polarization, expressed in the Greek ε (epsilon), and its unit is (F/m). Permeability refers to the magnetization ability of a material to an external magnetic field, written in the Greek μ (mu), and the unit is (H/m).

In a linear medium, the following constitutive relations are established between the applied field and the physical quantities of electromagnetic response that is specific

Table 1.2 Classification of electromagnetic properties of materials in response to applied fields

Properties	Applied field	Electromagnetic response	Material constants
Electric conduction	Electric field	Migration of Coulomb charges (electrons, ions)	Conductivity σ
Electric polarization	Electric field	Displacement of Coulomb charges (electrons, ions)	Permittivity ε
Magnetization	Magnetic field	Alignment of spins or orbital magnetic moment	Permeability μ

to a material or substance:

$$D = \varepsilon E \tag{1.4}$$

$$B = \mu H \tag{1.5}$$

$$J = \sigma E \tag{1.6}$$

If the constitutive parameters ε, μ, σ are different in position, the medium is said to be inhomogeneous. If different in applied strength of electric and magnetic field, the medium is non-linear. If different in direction, the medium is anisotropic. If different in frequency, the parameter is said to be dispersive.

For free space, the permittivity and permeability, denoted by ε_0 and μ_0, have the following values:

$$\varepsilon_0 = \frac{1}{36\pi} \times 10^{-9} \simeq 8.854 \times 10^{-12} (\text{F/m}) \tag{1.7}$$

$$\mu_0 = 4\pi \times 10^{-7} (\text{H/m}) \tag{1.8}$$

The permittivity and permeability of the medium are usually expressed as relative values to free space, $\varepsilon_r = \varepsilon/\varepsilon_0$ and $\mu_r = \mu/\mu_0$, which are called the relative permittivity and relative permeability, respectively.

1.3 Units and Decibels

1.3.1 SI Units

SI unit is an acronym that comes from the French word '*Le Systeme International d'Unites*' and refers to the 'International System of Units' that are now officially adopted and used in most countries of the world. The basic units of SI are MKS

Table 1.3 SI derived units with special names and symbols used in electromagnetics [3]

Quantity	Name	Symbol	in SI units
Phase angle	radian	rad	m/m
Frequency	hertz	Hz	s^{-1}
Force	newton	N	$kg{\cdot}m/s^2$
Energy, work, heat	joule	J	N·m
Power	watt	W	J/s
Electric charge	coulomb	C	A·s
Voltage	volt	V	J/C, W/A
Capacitance	farad	F	C/V
Resistance, impedance	ohm	Ω	V/A
Electrical conductance	siemens	S	A/V
Magnetic flux	weber	Wb	V·S
Magnetic flux density	tesla	T	Wb/m^2
Inductance	henry	H	Wb/A

units based on meter (m, length), kilogram (kg, mass), and second (s, time), and with the use of ampere (A, electric current), the MKSA unit is formed. To this were added kelvin (K, thermodynamic temperature), candela (cd, luminous intensity), and mole (mol, amount of substance), making it the current seven base units of SI. The fundamental derived units in electromagnetics can be expressed in the SI base units and represented in Table 1.3 [3]. The SI system uses a mixture of basic units and various prefixes. These prefixes represent multiples or submultiples of original unit (for example, THz $= 10^{12}$ Hz and fF $= 10^{-15}$ F, etc.). A list of commonly-used prefixes for the SI units and their abbreviations and etymologies is presented in Table 1.4 [4].

1.3.2 Decibels

An electric system has inputs and outputs, and their ratio is used as one of the indexes that indicate the performance of the system. For the convenience of comparing the relative ratios, the unit decibel (dB) is often used [5]. Decibel is originated from the words of Deci and Bel (in short, B) and means 1/10 of Bel, in which Bel is named after Alexander Graham Bell, the inventor of telephone. dB is a unit that expresses the relative ratio of two power values (e.g., electric power and sound energy) on a logarithmic scale for magnitude comparison. Particularly, dB is useful in electrical and electronic engineering, because the handling range of basic quantities of power, voltage, and current is very wide (e.g., 1 μV to 200 V for voltages).

Taking the amplification circuit as an example (Fig. 1.2a), when the power of the input signal is referred to as P_{in} and the output power P_{out}, the power gain G_p is defined as

Table 1.4 Prefixes for the international system of units (SI)

Base 10	Symbol	Name	Etymology
10^{18}	E	exa	Greek ex, "beyond"
10^{15}	P	peta	Greek petasos, "outreach"
10^{12}	T	tera	Greek teras, "monster"
10^{9}	G	giga	Latin gigas, "giant"
10^{6}	M	mega	Greek megas, "great"
10^{3}	k	kilo	Greek chilioi, "thousand"
10^{2}	h	hecto	Greek hekaton, "hundred"
10^{1}	da	deca	Greek deka, "ten"
10^{-1}	d	deci	Latin decimus, "tenth"
10^{-2}	c	centi	Latin centum, "hundred"
10^{-3}	m	milli	Latin mille, "thousand"
10^{-6}	μ	micro	Greek mikros, "small"
10^{-9}	n	nano	Greek nanos, "dwarf"
10^{-12}	p	pico	Spanish pico, "small quantity"
10^{-15}	f	femto	Danish femten, "fifteen"
10^{-18}	a	atto	Danish atten, "eighteen"

$$G_p = \frac{P_{\text{out}}}{P_{\text{in}}} \tag{1.9}$$

Expressing this in decibels gives

$$G_{p\text{dB}} = 10 \log\left(\frac{P_{\text{out}}}{P_{\text{in}}}\right) \tag{1.10}$$

If the power is lost ($P_{\text{out}} < P_{\text{in}}$), we call it power loss which has a negative value in (dB). For example, if input power is 2 (W) and output power is 4 (W), the power gain is 10 log(4/2) = 3 (dB). Conversely, if the output power is reduced to 0.2 (W), it has a value of 10 log(0.2/2) = −10 (dB).

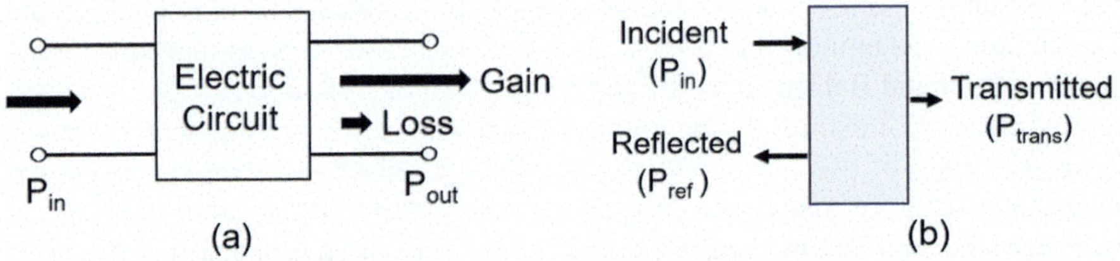

Fig. 1.2 Schematic representation of **a** power gain or power loss in electric circuit and **b** reflection and transmission loss for electromagnetic waves

Reflection and transmission of electromagnetic waves can also be expressed in decibels, as shown in Fig. 1.2b. For the plane wave of incidence with power P_{in}, when the reflected power is P_{ref} and transmitted power is P_{trans}, the reflection loss (RL) and transmission loss (TL) in (dB) are expressed as

$$RL = 10 \log\left(\frac{P_{\text{ref}}}{P_{\text{in}}}\right) \tag{1.11}$$

$$TL = 10 \log\left(\frac{P_{\text{trans}}}{P_{\text{in}}}\right) \tag{1.12}$$

The voltage gain G_v is defined as

$$G_v = \frac{V_{\text{out}}}{V_{\text{in}}} \tag{1.13}$$

Expressed in decibels, we get

$$G_{v\text{dB}} = 20 \log\left(\frac{V_{\text{out}}}{V_{\text{in}}}\right) \tag{1.14}$$

Decibels are also used to compare the absolute values of voltage, current, and power with respect to a reference value. For voltage, dBμV is based on 1 μV as a reference value:

$$\text{dB}\ \mu\text{V} = 20 \log(\text{volts}/1\ \mu\text{V}) \tag{1.15}$$

Likewise, with a reference value of 1 μA, we use the relative units of dBμA. For power, dBm is often used, indicating the power measurement relative to 1 mW:

$$\text{dBm} = 10 \log(\text{watts}/1\ \text{mW}) \tag{1.16}$$

The hallmark of decibels is that they can be expressed simply by compressing very large or small values.

1.4 Sinusoidal Harmonic Waves and Phasor

A current that is constant in magnitude and direction regardless of a change in time is called the direct current (DC), and a current that periodically changes in magnitude and direction is called the alternating current (AC). Among alternating currents, the current with a waveform of sine curve is called the sinusoidal wave, which is represented in Fig. 1.3 and expressed as

$$i = I_m \sin(\omega t + \theta) \tag{1.17}$$

where I_m is the amplitude, ω is the angular frequency, and θ is the initial phase.

In the case of AC circuits consisting of resistance R, inductance L, and capacitance C, the voltage and current always have the same frequency. Therefore, only the magnitude and the relative phase of voltage and current are the main subject in circuit analysis. A vector that expresses the voltage and current of a sinusoidal wave in magnitude and initial phase is called the phasor, which can be represented by the position vector in Fig. 1.3 and expressed as

$$I = I_m \angle \theta \tag{1.18}$$

The phasor greatly facilitates the analysis of alternating current and voltage of sinusoidal wave. Because the phasors are quantities with magnitudes and angles, they can be expressed in complex numbers. Complex numbers are functions consisting of real and imaginary parts, expressed as

$$z = a + jb \tag{1.19}$$

where $j = \sqrt{-1}$, and a is the real part and b is the imaginary part, which can be denoted

$$a = \mathrm{Re}(z), b = \mathrm{Im}(z) \tag{1.20}$$

$z^* = a - jb$ is called the conjugated complex for $z = a + jb$. Thus, $z + z^* = 2a$, $z - z^* = j2b$, and $zz^* = a^2 + b^2$.

Complex number is represented by a point in the complex plane of Cartesian coordinate in which x-axis represents real part and y-axis represents the imaginary part, as shown in Fig. 1.4a. The polar form of complex numbers in terms of r and θ is shown in Fig. 1.4b, where r is the absolute value and θ is the argument or phase angle. Thus, the following relations are established:

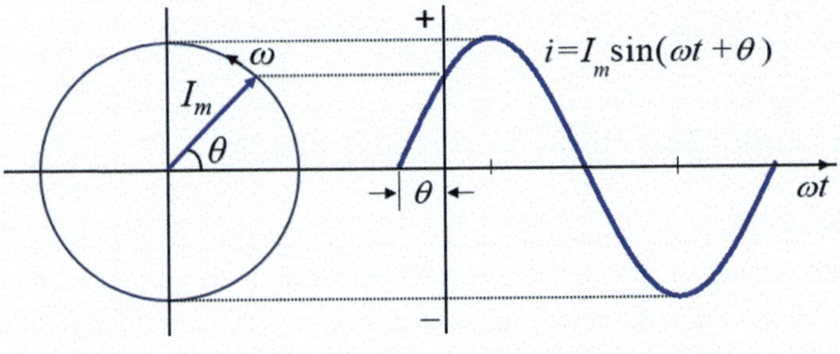

Fig. 1.3 Sinusoidal harmonic wave of current and its vector notation (phasor)

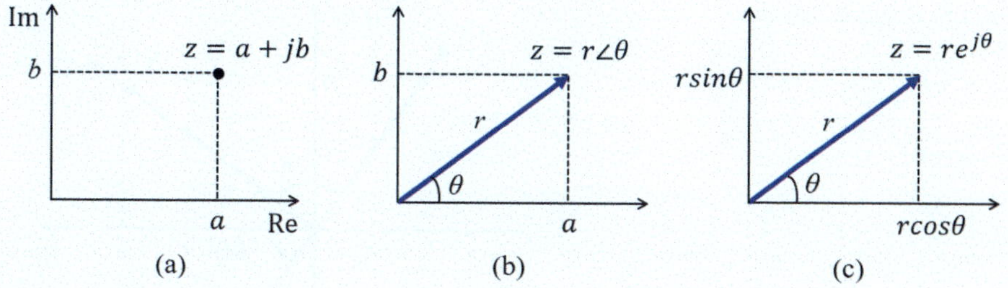

Fig. 1.4 Various representations of complex numbers: **a** rectangular form, **b** polar form, **c** exponential form

$$r = |z| = \sqrt{a^2 + b^2} \tag{1.21}$$

$$\theta = \arg z = \tan^{-1}\left(\frac{b}{a}\right) \tag{1.22}$$

The expression in phasor form is

$$z = r\angle\theta \tag{1.23}$$

A phasor can be expressed in exponential form using the following Euler's formula:

$$e^{j\theta} = \cos\theta + j\sin\theta \tag{1.24}$$

Since $e^{j\theta}$ is a phasor with a magnitude of 1 and a phase angle θ, the phasor can be expressed as an exponential function, as shown in Fig. 1.4c:

$$z = re^{j\theta} = r(\cos\theta + j\sin\theta) \tag{1.25}$$

Complex conjugate $z^* = a - jb$ can be represented in polar and exponential form as $z^* = r\angle - \theta = re^{-j\theta}$.

The argument $e^{j\theta}$ has the meaning of rotating the phasor by θ counterclockwise. Thus, multiplying the imaginary $j(= 1\angle 90°)$, the phasor rotates by $\pi/2$ (rad) counterclockwise, as shown in Fig. 1.5. Likewise, the multiplication of the phasor by $-j$ $(= 1/j = 1\angle - 90°)$ results in the phase shift by $-\pi/2$ (rad), a clockwise rotation.

The addition or subtraction of complex numbers can be done in rectangular form. For two complex numbers, $z_1 = a + jb$ and $z_2 = c + jd$, the addition and subtraction are given by

$$z_1 + z_2 = (a + c) + j(b + d) \tag{1.26}$$

$$z_1 - z_2 = (a - c) + j(b - d) \tag{1.27}$$

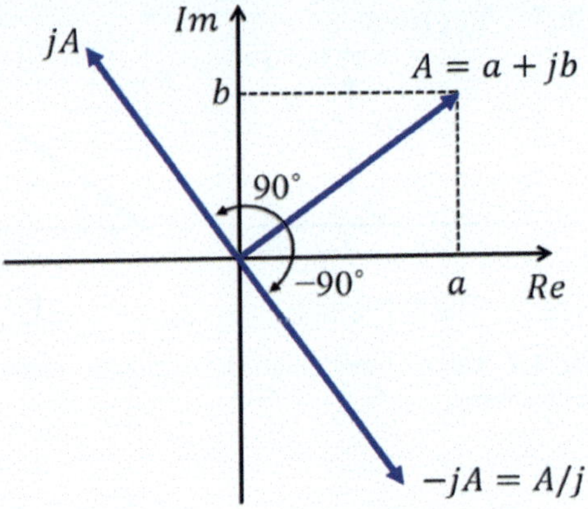

Fig. 1.5 Rotation operator *j*

The multiplication and division of complex numbers are better carried out using polar form. For two complex numbers, $z_1 = r_1\angle\theta_1$ and $z_2 = r_2\angle\theta_2$, the multiplication and division are given by

$$z_1 z_2 = r_1 r_2 \angle(\theta_1 + \theta_2) \tag{1.28}$$

$$\frac{z_1}{z_2} = \frac{r_1}{r_2}\angle(\theta_1 - \theta_2) \tag{1.29}$$

Power, root, and logarithm of complex numbers are also better carried out using polar form. For a complex number, $z = r\angle\theta = re^{j\theta}$,

$$z^n = r^n \angle n\theta = r^n e^{jn\theta} \tag{1.30}$$

$$z^{\frac{1}{n}} = r^{\frac{1}{n}} \angle \frac{\theta + 2\pi k}{n} = r^{\frac{1}{n}} e^{j\frac{\theta+2\pi k}{n}} \quad (k = 0, 1, 2, \cdots, (n-1)) \tag{1.31}$$

$$\ln z = \ln r + j(\theta + 2\pi k) \quad (k = 0, 1, 2, \cdots) \tag{1.32}$$

For example, for a given complex number $z = 2\angle\frac{\pi}{3} = 2e^{j\frac{\pi}{3}}$, (a) $z^2 = 4\angle\frac{2\pi}{3} = 4e^{j\frac{2\pi}{3}}$, (b) $\sqrt{z} = \sqrt{2}\angle\frac{\pi}{6} = \sqrt{2}e^{j\frac{\pi}{6}}$ (for $k = 0$) and $\sqrt{z} = \sqrt{2}\angle\frac{7\pi}{6} = \sqrt{2}e^{j\frac{7\pi}{6}}$ (for $k = 1$), (c) $\ln z = \ln 2 + j\left(\frac{\pi}{3}\right)$ (for $k = 0$) and $\ln z = \ln 2 + j\left(\frac{7\pi}{3}\right)$ (for $k = 1$), $\cdots$ for higher values of k.

1.5 Vectors

1.5.1 Vector Operations

A vector is a physical quantity with magnitude and direction. Figure 1.6 shows a vector of $\vec{A}$ of which component is A_x, A_y, A_z in x, y, z Cartesian coordinate, and can be expressed as

$$\vec{A} = A_x a_x + A_y a_y + A_z a_z \tag{1.33}$$

where a_x, a_y, a_z denote the unit vectors along each axis.

Addition and subtraction of two vectors $\vec{A}$ and $\vec{B}$ are carried out component by component,

$$\vec{A} + \vec{B} = (A_x + B_x)a_x + \left(A_y + B_y\right)a_y + (A_z + B_z)a_z \tag{1.34}$$

$$\vec{A} - \vec{B} = (A_x - B_x)a_x + \left(A_y - B_y\right)a_y + (A_z - B_z)a_z \tag{1.35}$$

The dot (scalar) product $\vec{A} \cdot \vec{B}$ of two vectors is expressed as

$$\vec{A} \cdot \vec{B} = AB\ \cos\theta_{AB} = A_x B_x + A_y B_y + A_z B_z \tag{1.36}$$

where θ_{AB} is the angle between two vectors $(0 \le \theta \le \pi)$, as shown in Fig. 1.7a. The cross (vector) product of two vectors $\vec{A}$ and $\vec{B}$ is expressed as

$$\vec{A} \times \vec{B} = AB\ \sin\theta_{AB} a_n \tag{1.37}$$

or

$$\begin{aligned} \vec{A} \times \vec{B} &= \begin{vmatrix} a_x & a_y & a_z \\ A_x & A_y & A_z \\ B_x & B_y & B_z \end{vmatrix} \\ &= \left(A_y B_z - A_z B_y\right)a_x + (A_z B_x - A_x B_z)a_y \\ &\quad + \left(A_x B_y - A_y B_x\right)a_z \end{aligned} \tag{1.38}$$

where a_n is the unit vector normal to the plane containing $\vec{A}$ and $\vec{B}$, as shown in Fig. 1.7b. The direction of a_n is that of the advance of right-hand screw from $\vec{A}$ to $\vec{B}$. Thus, $\vec{A} \times \vec{B} = -\vec{B} \times \vec{A}$.

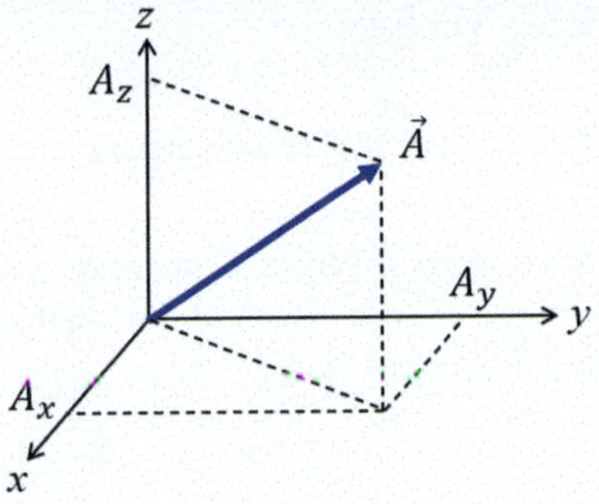

Fig. 1.6 Vector notation in Cartesian coordinate

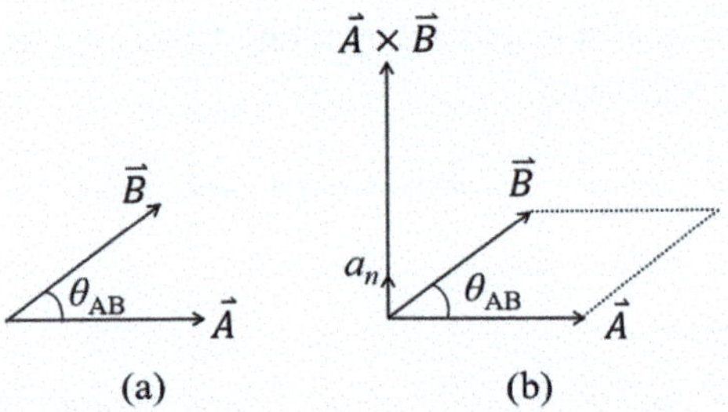

Fig. 1.7 Geometrical representation of vector multiplication: **a** dot product, **b** cross product

1.5.2 Line, Surface, and Volume Integrals

Given a vector field $\vec{A}$ and a path L, as shown in Fig. 1.8, the line integral of vector $\vec{A}$ along the path L is defined as

$$\int_L \vec{A} \cdot d\vec{l} = \int_L A \cos\theta \, dl \tag{1.39}$$

where θ is the angle between $\vec{A}$ vector and tangential component of differential curve $d\vec{l}$. In the case of integrating along a closed curve, which is called the circulation of $\vec{A}$ around path L, Eq. (1.39) is expressed as $\oint_L \vec{A} \cdot d\vec{l}$.

Given a vector field $\vec{A}$ passing through the curved surface S, as shown in Fig. 1.9, the flux Ψ through S is obtained by the surface integral defined as

$$\Psi = \int_S \vec{A} \cdot d\vec{S} = \int_S A \cos\theta \, dS \tag{1.40}$$

where θ is the angle between $\vec{A}$ and unit normal to S. For a closed surface, Eq. (1.40) becomes

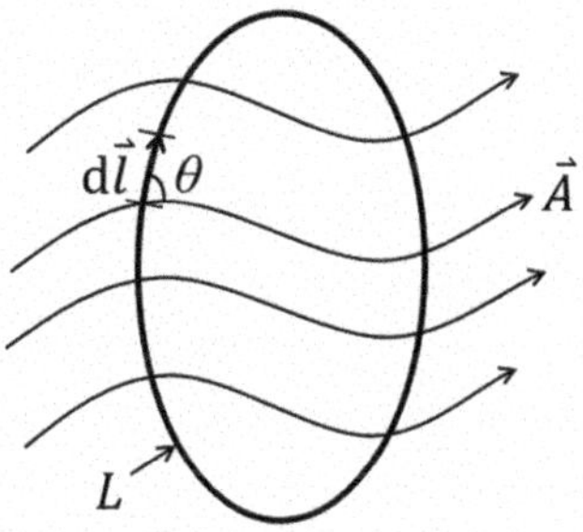

Fig. 1.8 Line integration of vector field $\vec{A}$ along the path L

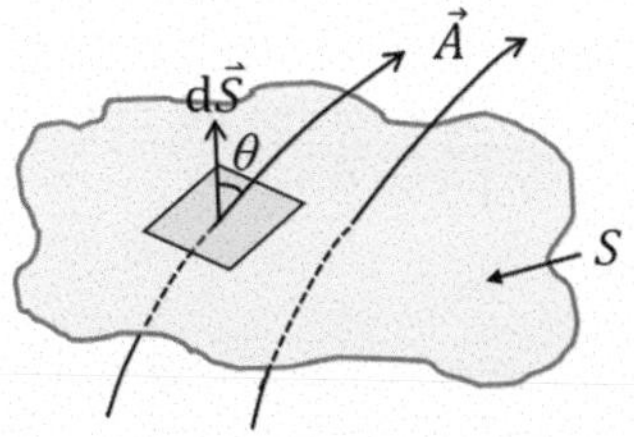

Fig. 1.9 Surface integration of vector $\vec{A}$ through S

$$\Psi = \oint_S \vec{A} \cdot \mathrm{d}\vec{S} \tag{1.41}$$

The integration of scalar amount per unit volume (e.g., the charge density ρ_v) over the volume is called the volume integral, and the total charge Q is obtained by the following equation:

$$Q = \int_v \rho_v \mathrm{d}v \tag{1.42}$$

1.5.3 Del Operator

Del operator, denoted by ∇, has the meaning of a vector differential operator and is represented in rectangular coordinate:

$$\nabla = \frac{\delta}{\delta x} a_x + \frac{\delta}{\delta y} a_y + \frac{\delta}{\delta z} a_z \tag{1.43}$$

This operator is used as follows:

(a) ∇V means the gradient of the scalar V and is expressed as

$$\nabla V = \left(\frac{\delta}{\delta x}a_x + \frac{\delta}{\delta y}a_y + \frac{\delta}{\delta z}a_z\right)V = \frac{\delta V}{\delta x}a_x + \frac{\delta V}{\delta y}a_y + \frac{\delta V}{\delta z}a_z \tag{1.44}$$

(b) $\nabla \cdot \vec{A}$ means the divergence of the vector $\vec{A}$, that is the maximum spread, and is expressed as

$$\begin{aligned}\nabla \cdot \vec{A} &= \left(\frac{\delta}{\delta x}a_x + \frac{\delta}{\delta y}a_y + \frac{\delta}{\delta z}a_z\right) \cdot (A_x a_x + A_y a_y + A_z a_z) \\ &= \frac{\delta A_x}{\delta x} + \frac{\delta A_y}{\delta y} + \frac{\delta A_z}{\delta z}\end{aligned} \tag{1.45}$$

(c) $\nabla \times \vec{A}$ means the curl of the vector $\vec{A}$, that is the maximum circulation, and is expressed as

$$\begin{aligned}\nabla \times \vec{A} &= \begin{vmatrix} a_x & a_y & a_z \\ \frac{\delta}{\delta x} & \frac{\delta}{\delta y} & \frac{\delta}{\delta z} \\ A_x & A_y & A_z \end{vmatrix} \\ &= \left[\frac{\delta A_z}{\delta y} - \frac{\delta A_y}{\delta z}\right]a_x + \left[\frac{\delta A_x}{\delta z} - \frac{\delta A_z}{\delta x}\right]a_y \\ &+ \left[\frac{\delta A_y}{\delta x} - \frac{\delta A_x}{\delta y}\right]a_z\end{aligned} \tag{1.46}$$

(d) $\nabla^2 V$ is the Laplacian of scalar V, meaning the divergence of gradient V, and is expressed as

$$\nabla^2 V = \nabla \cdot \nabla V = \frac{\delta^2 V}{\delta x^2} + \frac{\delta^2 V}{\delta y^2} + \frac{\delta^2 V}{\delta z^2} \tag{1.47}$$

1.6 Maxwell's Equations

1.6.1 Faraday's Law

Faraday's law of electromagnetic induction is that when the magnetic flux changes with time, an electric field is induced around it. This is formulated as

$$\int_C E \cdot \mathrm{d}l = -\frac{\mathrm{d}}{\mathrm{d}t}\int_S B \cdot \mathrm{d}S \tag{1.48}$$

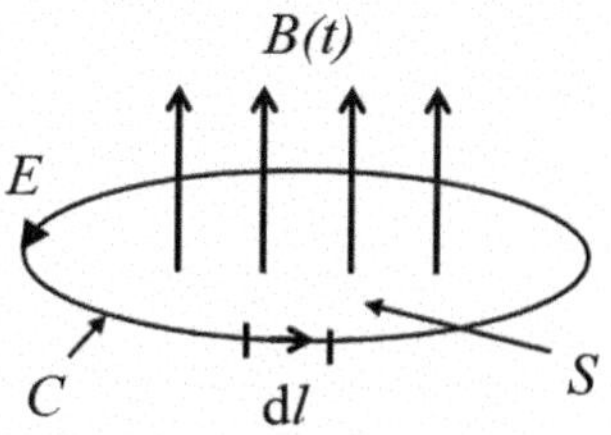

Fig. 1.10 Faraday's law of electromagnetic induction

where E (V/m) is the electric field intensity and B (Wb/m^2) is the magnetic flux density. dl is a unit length vector along the closed contour C and dS is a unit area vector in the curved surface that the contour C contains, as shown in Fig. 1.10.

The left-hand side of Eq. (1.48) is the electromotive force (*emf* (V)) induced along the contour C, and the right-hand side is the time rate of change of the magnetic flux Φ passing through the curved surface S, and expressed as

$$emf = -\frac{\mathrm{d}\Phi}{\mathrm{d}t} \tag{1.49}$$

Using the following Stokes' theorem, which states a mathematical procedure of converting the line integral of E field over a loop to the surface integral (*flux*) of its curl through the enclosed surface,

$$\int_C E \cdot \mathrm{d}l = \int_S (\nabla \times E) \cdot \mathrm{d}S \tag{1.50}$$

the differential form is obtained:

$$\nabla \times E = -\frac{\delta B}{\delta t} \tag{1.51}$$

1.6.2 Ampere's Law

In contrast to Faraday's law, Ampere's law is that the changes in the electric field with time induce the magnetic field. As shown in Fig. 1.11, in a curved surface S enclosed by the contour C, the following equation is established:

$$\oint_C H \cdot \mathrm{d}l = \int_S \left(J + \frac{\delta D}{\delta t}\right) \cdot \mathrm{d}S \tag{1.52}$$

where H (A/m) is the magnetic field strength, J (A/m^2) is the conduction current density, and D (C/m^2) is the electric flux density. The line integral of a magnetic field

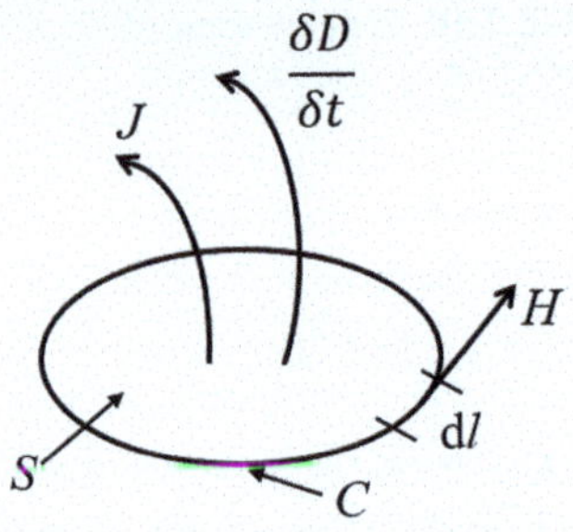

Fig. 1.11 Ampere's circuital law

along the curve C is equal to the sum of the current J and the rate of time change of electric flux density $\delta D/\delta t$ passing through the curved surface S.

$\delta D/\delta t$ was introduced by Maxwell and is defined as the displacement current density, which can be treated as an AC current arising in connection with capacitors, as illustrated in Fig. 1.12. In an AC circuit of capacitor, applying the voltage V increases the charge Q in the capacitor, which can be expressed as

$$V = \frac{Q}{C} = \frac{1}{C}\int i\,\mathrm{d}t \tag{1.53}$$

where C is the capacitance. The current flow in the circuit is given by

$$i = C \cdot \frac{\mathrm{d}V}{\mathrm{d}t} = \frac{\varepsilon A}{d} \cdot \frac{\mathrm{d}V}{\mathrm{d}t} \tag{1.54}$$

where A and d are the area and spacing of parallel plate capacitor, and ε is the permittivity of a dielectric between two conducting plates. Using the relation $D = \varepsilon E$, the current density is given by

$$J_d = \frac{i}{A} = \varepsilon\frac{\mathrm{d}V/d}{\mathrm{d}t} = \varepsilon\frac{\mathrm{d}E}{\mathrm{d}t} = \frac{\mathrm{d}D}{\mathrm{d}t} \tag{1.55}$$

Here, the process of increasing the electric field E has the effect of current flow in the circuit, and this is called the displacement current.

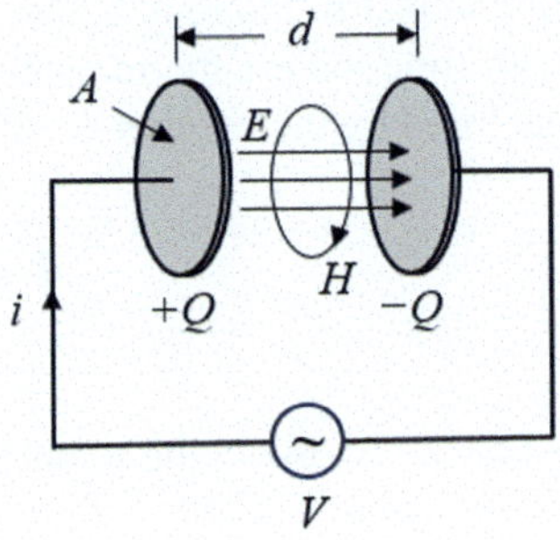

Fig. 1.12 Illustration of displacement current in the circuit of parallel plate capacitor

Maxwell extended the Ampere's circuital law to understand that a magnetic field is created when the intensity of an electric field changes. This proved that magnetic field is created not only in wires but everywhere the strength of the electric field changes, and the propagation of electromagnetic waves could be explained by introduction of the displacement current.

With the help of following Stokes' theorem, which means that the line integral of a H field over a loop is equal to the flux of its curl through the enclosed surface,

$$\int_C H \cdot \mathrm{d}l = -\int_S (\nabla \times H) \cdot \mathrm{d}S \tag{1.56}$$

the differential form is obtained:

$$\nabla \times H = J + \frac{\delta D}{\delta t} \tag{1.57}$$

1.6.3 *Gauss's Law*

Gauss's law describes the magnitude of electric field generated by the charge. Gaussian law of electric fields is represented by a volume integral of the charge density ρ_v (C/m^3):

$$\int_S D \cdot \mathrm{d}S = \int_v \rho_v \mathrm{d}v \tag{1.58}$$

Total amount of electrical flux passing through the closed surface S is equal to the total amount of charge in the enclosed volume. The electric flux lines cannot intersect each other and are perpendicular to the surface of conductor.

Using the following divergence theorem, which means that the electric flux through the surface is equal to the volume integral of its divergence (charge density) over the region inside the surface,

$$\int_S D \cdot \mathrm{d}S = \int_v \nabla \cdot D \mathrm{d}v \tag{1.59}$$

the differential equation is obtained:

$$\nabla \cdot D = \rho_v \tag{1.60}$$

Gauss law of magnetic field is expressed as

Table 1.5 Maxwell's equations for time-varying electromagnetic field

Integral form	Differential form	Related law
$\int_S D \cdot \mathrm{d}S = \int_v \rho_v \mathrm{d}v$	$\nabla \cdot D = \rho_v$	Gauss's Law
$\int_S B \cdot \mathrm{d}S = 0$	$\nabla \cdot B = 0$	No magnetic single pole
$\oint_C E \cdot \mathrm{d}l = -\frac{\mathrm{d}}{\mathrm{d}t} \int_S B \cdot \mathrm{d}S$	$\nabla \times E = -\frac{\delta B}{\delta t}$	Faraday's Law
$\oint_C H \cdot \mathrm{d}l = \int_S \left(J + \frac{\delta D}{\delta t}\right) \cdot \mathrm{d}S$	$\nabla \times H = J + \frac{\delta D}{\delta t}$	Ampere's Law

$$\int_S B \cdot \mathrm{d}S = 0 \tag{1.61}$$

Total magnetic flux through a closed surface is zero. That means, unlike electricity, a magnet has no single pole, and N and S poles are always together. Because of this nature of magnetism, the magnitude of magnetic flux entering a certain space and that going out are always the same. Using the following divergence theorem,

$$\int_S B \cdot \mathrm{d}S = \int_v \nabla \cdot B \, \mathrm{d}v \tag{1.62}$$

the differential form is obtained:

$$\nabla \cdot B = 0 \tag{1.63}$$

Maxwell's equations for time-varying electromagnetic field are summarized in Table 1.5.

1.7 Further Reading

The topics in this chapter, including sinusoidal harmonic wave, vector algebra and vector calculus, and Maxwell's equations, are presented in greater detail in advanced textbooks on electromagnetics [6, 7] and in the public domain [8].

References

1. Electromagnetic spectrum on Wikipedia. https://en.wikipedia.org/wiki/Electromagnetic_spectrum
2. Radio spectrum on Wikipedia. https://en.wikipedia.org/wiki/Radio_spectrum
3. D. Newell, E. Tiesinga, *The International System of Units (SI)*, 2019 Edition, Special Publication 330 (National Institute of Standards and Technology, Gaithersburg, MD, 2019), p. 15
4. International System of Units on Wikipedia. https://en.wikipedia.org/wiki/International_System_of_Units
5. Decibel on Wikipedia. https://en.wikipedia.org/wiki/Decibel
6. M.N.O. Sadiku, *Elements of Electromagnetics*, 2nd edn. (Oxford University Press, New York, 1995), pp.3–110
7. W.H. Hayt Jr., *Engineering Electromagnetics*, 5th edn. (McGraw-Hill, New York, 1989), pp.311–335
8. Maxwell's Equations on Wikipedia. https://en.wikipedia.org/wiki/Maxwell%27s_equations

Chapter 2
Circuit Theory

Electromagnetic (EM) wave absorbers are based on the fundamental principles in wave dynamics of EM waves and their interaction with materials. This chapter provides the circuit theory, including transmission lines, to gain basic knowledge for modeling and interpreting these physical phenomena with electrical circuits. The electrical behavior of DC and AC circuits of passive devices (resistor R, inductor L, capacitor C) is described through analysis of magnitude and phase between voltage and current. Subsequently, series and parallel circuits of R, L, C are described by circuit parameters of impedance and admittance. Transmission lines are useful for expressing the propagation of electromagnetic waves as equivalent electrical circuits. Based on the distributed element circuit theory, the transmission line equation is derived in terms of the characteristic impedance and the propagation constant. The input impedance is defined in the load-connected transmission line, which provides a circuit theory that connects to the design of the EM wave absorber.

2.1 Passive Devices

Circuit elements include an active element (power supply) that delivers power to a circuit and a passive element that performs its functions by utilizing power or energy supplied into a circuit [1]. Active components include diodes, transistors, DC generators, current and voltage sources, etc. Passive components include resistors, inductors, capacitors, transformers, motors, etc.

S. Kim, *Fundamentals and Principles of Electromagnetic Wave Absorbers*, Springer Series in Advanced Microelectronics 66, https://doi.org/10.1007/978-981-97-6477-8_2

2.1.1 *Resistance, Inductance, Capacitance*

(1) **Resistance**

A resistor is a circuit element that consumes the energy supplied from a power source, and a proportional relationship is established between the voltage and current. The proportional constant is called the resistance, and the following Ohm's law is established: when DC current I passes through to the element of resistance R, the voltage V between the terminals is expressed as

$$V = RI \tag{2.1}$$

If the voltage $v(t)$ across a device has the following relationship to the change in current $i(t)$ with time t, the device is called a resistor,

$$v(t) = Ri(t) \tag{2.2}$$

The resistance of a conductor is proportional to its length l, inversely proportional to its cross-sectional area A. The proportional constant ρ is called the resistivity, and a unit (Ωm) or (Ωcm) is commonly used,

$$R = \rho\frac{l}{A} \tag{2.3}$$

(2) **Conductance**

Another form of Ohm's law can be expressed as

$$i = \frac{1}{R}v = Gv \tag{2.4}$$

where G is defined as the conductance, the reciprocal of resistance R. The unit of conductance is siemens (S) or mho (℧). G is inversely proportional to the length l and proportional to the cross-sectional area A of the conductor, expressed as

$$G = \sigma\frac{A}{l} \tag{2.5}$$

The proportional constant σ is called the conductivity, and a unit (S/m) or (S/cm) is used. Another unit is (℧/m) or (℧/cm), reciprocal of resistivity:

$$\sigma = \frac{1}{\rho} \tag{2.6}$$

(3) **Inductance**

When electric current I flows along an electrical conductor or coil, a magnetic flux Φ is created. The magnetic flux is proportional to the magnitude of current, and the proportional constant L is defined as the inductance [2]. For a coil having N turns, total magnetic flux is $N\Phi$, and the inductance is defined as the ratio of the total flux linkages to the current which they link, that is

$$L = \frac{N\Phi}{I} \quad (2.7)$$

Inductance is defined as the ability of coil to convert an electric current to magnetic flux. The unit of inductance uses henry (H). The inductance of 1 (H) means that when a current of 1 (A) is applied to the coil, a magnetic flux of 1 (Wb) is produced. Any change in magnetic field produces an electromotive force by Faraday's induction law that is proportional to the time rate of current change, establishing the following relationship:

$$v(t) = L\frac{\mathrm{d}i(t)}{\mathrm{d}t} \quad (2.8)$$

The voltage drop $v(t)$ in the coil is proportional to the time differential of current $i(t)$, and the proportional constant is the induction coefficient (inductance) L of coil.

(4) **Capacitance**

A capacitor is a two-terminal device for storing electrical energy in an electric field [3]. The symbol C, the unit of which is farad (F), represents the ratio of the amount of accumulated charge to a voltage applied. The capacitance of 1 (F) means that when a voltage of 1 [V] is applied to a capacitor, a charge of 1 (C) can be accumulated on each parallel plate. The relationship between voltage V, charge Q, and capacitance C is expressed as

$$C = \frac{Q}{V} \quad (2.9)$$

The capacitance of capacitors may typically be in the range of a microfarad (μF $= 10^{-6}$ F), a nanofarad (nF $= 10^{-9}$ F), and a picofarad (pF $= 10^{-12}$ F).

Capacitance is determined by the shape of conductor and the dielectric property of the insulator between the conductors. The capacitance of a parallel plate capacitor is given by

$$C = \varepsilon_0 \varepsilon_r \frac{A}{d} \quad (2.10)$$

where A: area of one conductor plate, d: distance between the conductor plates, ε_0: permittivity of vacuum, ε_r: relative dielectric permittivity of the insulating material. The capacitance C is proportional to the dielectric constant ε_r and the area of the electrode A and inversely proportional to the distance between the electrodes d.

The following relationship is established between the time-varying voltage $v(t)$ and current $i(t)$ between the terminals of capacitor:

$$v(t) = \frac{1}{C}\int i(t)dt \tag{2.11}$$

The voltage across a capacitor $v(t)$ is proportional to the time integral of the current $i(t)$, and the proportional constant is the inverse of capacitance $1/C$.

2.1.2 Time-Dependent Voltage and Current

When treating time-varying voltage and current in AC circuits, they are expressed as a harmonic wave of sine or cosine,

$$v \text{ or } i = A\sin(wt + \theta) = B\cos(\omega t + \theta) \tag{2.12}$$

where A, B: amplitude, ω: angular frequency ($= 2\pi f$), and θ: initial phase. $wt + \theta$ is the overall phase. Either a sine or a cosine can be used.

(1) **Resistor**

As shown in Fig. 2.1a, when the voltage $v(t) = V_m\sin\,\omega t$ is applied to a resistor R, the current $i(t)$ is expressed as

$$i(t) = \frac{1}{R}V_m\sin\,\omega t = I_m\sin\,\omega t \tag{2.13}$$

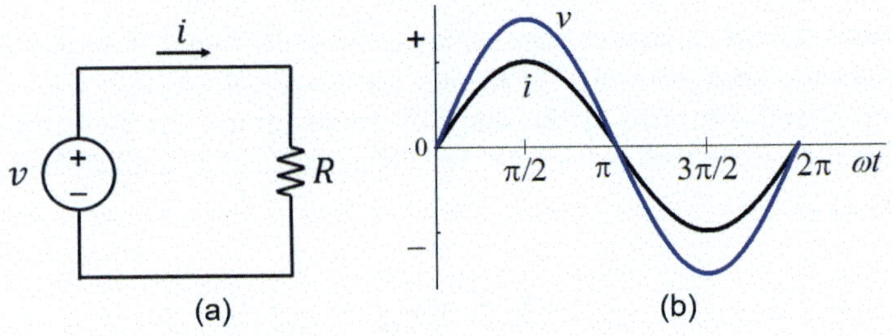

Fig. 2.1 **a** A circuit of resistor and **b** its voltage-current relationship

It is seen that the voltage and current are in phase (Fig. 2.1b). Between the current and voltage maximums, Ohm's law is established ($I_m = V_m/R$).

(2) **Inductor**

The voltage drop $v(t)$ across an inductor caused by the flow of alternating current, $i(t) = I_m \sin \omega t$, is given by

$$v = L\frac{\mathrm{d}i(t)}{\mathrm{d}t} = L\frac{\mathrm{d}}{\mathrm{d}t}(I_m \sin \omega t) = \omega L I_m \cos \omega t = V_m \sin\left(\omega t + \frac{\pi}{2}\right) \tag{2.14}$$

The voltage and current are out of phase. It is to be noted that the voltage leads the current by $\pi/2$ (rad) (i.e., the current lags the voltage by $\pi/2$ (rad)), that is shown in Fig. 2.2b. From the relationship between the amplitude of voltage and current, $V_m = \omega L I_m$, we obtain

$$I_m = \frac{V_m}{\omega L} = \frac{V_m}{X_L} \tag{2.15}$$

Here $X_L (= \omega L)$ is called the inductive reactance, and has the meaning of AC resistance of an inductor [2]. The X_L is proportional to its own inductance L and frequency f. Thus, the inductor plays as a low-pass filter, passing the direct current and blocking the high-frequency currents.

(3) **Capacitor**

When an AC voltage, $v(t) = V_m \sin \omega t$, is applied to a capacitor, the current through the capacitor is given by

$$i(t) = C\frac{\mathrm{d}v(t)}{\mathrm{d}t} = C\frac{\mathrm{d}}{\mathrm{d}t}(V_m \sin\omega t) = \omega C\, V_m \cos\omega t = I_m \sin\left(\omega t + \frac{\pi}{2}\right) \tag{2.16}$$

Equation (2.16) states that the current leads the voltage by $\pi/2$ (rad) (i.e., the voltage lags the current by $\pi/2$ (rad)), as shown in Fig. 2.3b.

It exhibits the opposite phase characteristics to the inductor element. From the relationship between the amplitude of current and voltage, we obtain

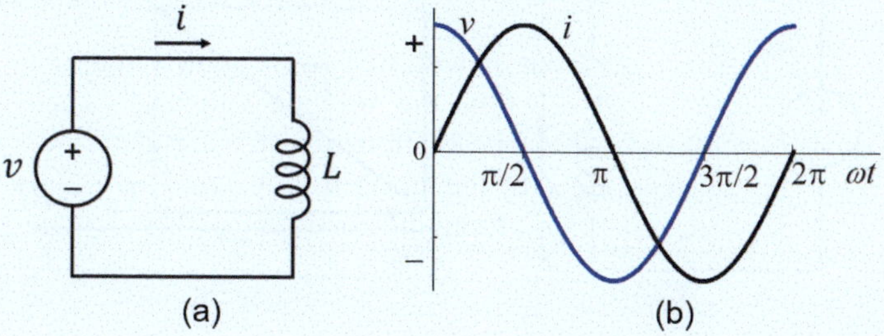

Fig. 2.2 **a** A circuit of inductor and **b** its voltage-current relationship

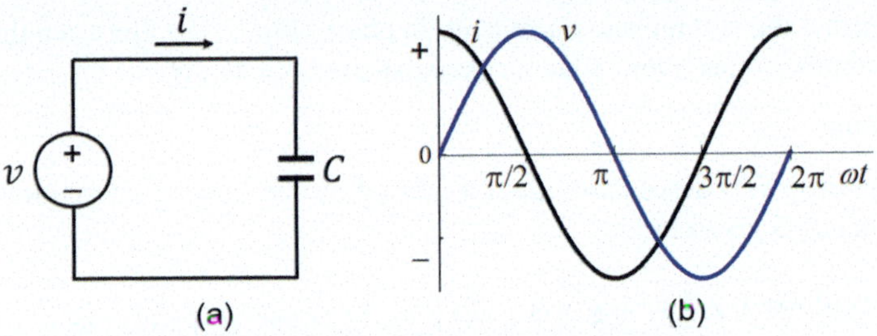

Fig. 2.3 **a** A circuit of capacitor and **b** its voltage-current relationship

$$I_m = \frac{V_m}{1/\omega C} = \frac{V_m}{X_C} \tag{2.17}$$

Here $X_C(= 1/\omega C)$ is called the capacitive reactance and has the meaning of AC resistance of a capacitor [3]. X_C is inversely proportional to capacitance C and frequency f. Thus, the capacitor exhibits a frequency characteristic of high-pass filter that blocks the direct current and passes the high-frequency currents.

2.1.3 *RLC Series Circuits*

(1) ***RL* Series Circuit**

As shown in Fig. 2.4a, when the current I flows through the series circuit of resistor R and inductor L, the voltage drops across R and L elements are as follows:

(a) $V_R = RI$, and V_R and I are in phase,
(b) $V_L = X_L I = \omega L I$, and V_L leads I by $\pi/2$ (rad).

The vector diagram for voltage and current of the *RL* series circuit is shown in Fig. 2.4b. The total voltage and current are given by

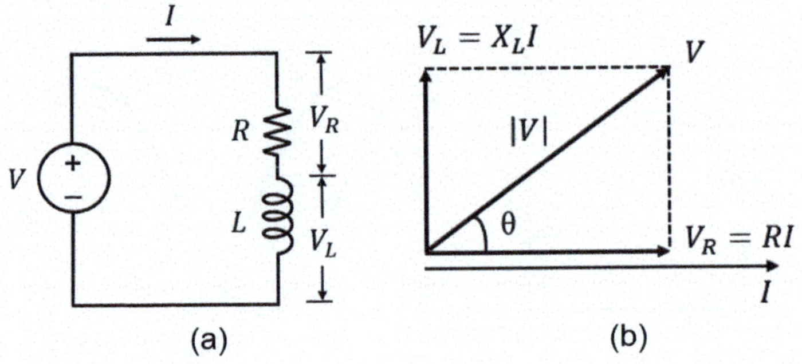

Fig. 2.4 **a** A series circuit of resistor and inductor and **b** its voltage-current relationship

$$V = \sqrt{V_R^2 + V_L^2} = I\sqrt{R^2 + X_L^2} = I\sqrt{R^2 + (\omega L)^2} \tag{2.18}$$

$$I = \frac{V}{\sqrt{R^2 + X_L^2}} \tag{2.19}$$

The ratio of voltage and current is called the impedance Z and is expressed as the total resistance of R and X_L, that is

$$Z = \sqrt{R^2 + X_L^2} = \sqrt{R^2 + (\omega L)^2} \tag{2.20}$$

The phase difference θ between total voltage and current is given by

$$\theta = \tan^{-1}\frac{X_L}{R} = \tan^{-1}\frac{\omega L}{R} \tag{2.21}$$

The voltage leads current by θ (rad) and the range of θ is $0 \leq \theta \leq \pi/2$. If $\theta = 0$, it will be a circuit of only resistor R, and when $\theta = \pi/2$, it will be a circuit of only inductor L.

(2) ***RC* Series Circuit**

As shown in Fig. 2.5a, when current I flows through the series circuit of resistor R and capacitor C, the voltage drops across R and C elements are as follows:

(a) $V_R = RI$, and V_R and I are in phase,
(b) $V_C = X_C I = \frac{1}{\omega C} I$, and V_C lags I by $\pi/2$ (rad).

The vector diagram for voltage and current of the *RC* series circuit is shown in Fig. 2.5b. The total voltage and current are given by

$$V = \sqrt{V_R^2 + V_C^2} = I\sqrt{R^2 + X_C^2} = I\sqrt{R^2 + (\frac{1}{\omega C})^2} \tag{2.22}$$

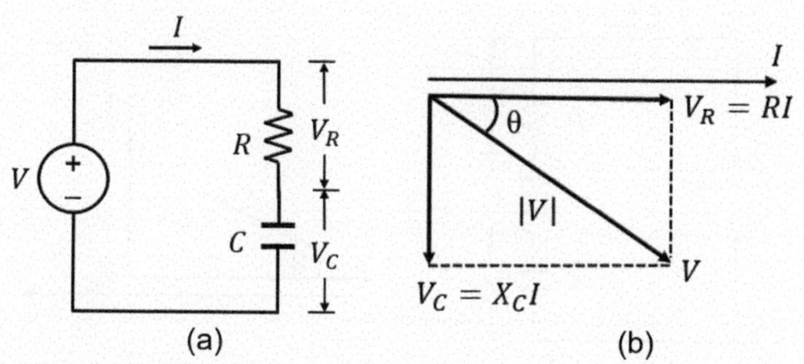

Fig. 2.5 **a** A series circuit of resistor and capacitor and **b** its voltage-current relationship

$$I = \frac{V}{\sqrt{R^2 + X_C^2}} \tag{2.23}$$

Then the impedance Z of the series RC circuit is expressed as

$$Z = \sqrt{R^2 + X_C^2} = \sqrt{R^2 + (\frac{1}{\omega C})^2} \tag{2.24}$$

The phase difference θ between total voltage and current is given by

$$\theta = \tan^{-1}\frac{X_C}{R} = \tan^{-1}\frac{1}{\omega CR} \tag{2.25}$$

The voltage lags the current by θ (rad), and θ has values in the range of $-\pi/2 \leq \theta \leq 0$. In the case of $\theta = -\pi/2$, it becomes a circuit of only capacitor C, and when $\theta = 0$, it becomes the circuit of only resistor R.

(3) ***RLC* Series Circuit**

As shown in Fig. 2.6a, when current I flows through the series circuit of resistor R, inductor L, and capacitor C, the voltage drops across R, L, and C elements are as follows:

(a) $V_R = RI$, and V_R and I are in phase,
(b) $V_L = X_L I = \omega LI$, and V_L leads I by $\pi/2$ (rad),
(c) $V_C = X_C I = \frac{1}{\omega C}I$, and V_C lags I by $\pi/2$ (rad).

When $X_L > X_C$, the vector diagram for voltage and current of the RLC series circuit is shown in Fig. 2.6b. The total voltage and current are given by

$$V = \sqrt{V_R^2 + (V_L - V_C)^2} = I\sqrt{R^2 + (\omega L - \frac{1}{\omega C})^2} \tag{2.26}$$

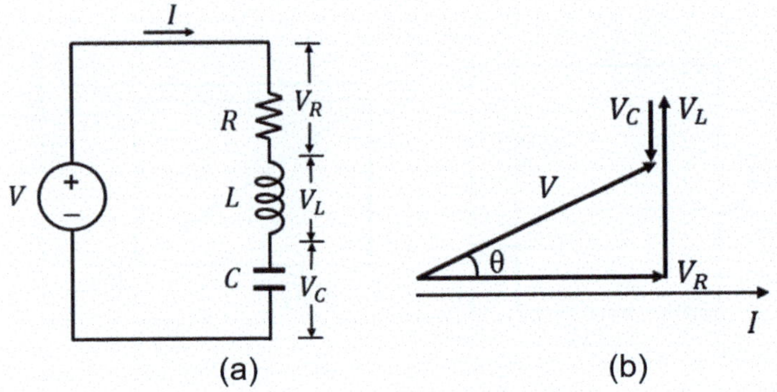

Fig. 2.6 **a** A *RLC* series circuit and **b** its voltage-current relationship

$$I = \frac{V}{\sqrt{R^2 + (X_L - X_C)^2}} \tag{2.27}$$

The impedance Z of the series *RLC* circuit is expressed as

$$Z = \sqrt{R^2 + (X_L - X_C)^2} = \sqrt{R^2 + (\omega L - \frac{1}{\omega C})^2} \tag{2.28}$$

The phase angle θ is given by

$$\theta = \tan^{-1}\frac{X_L - X_C}{R} = \tan^{-1}\frac{\omega L - (1/\omega C)}{R} \tag{2.29}$$

The inductive and capacitive nature of the *RLC* series circuit depends on the magnitude of $X_L - X_C$:

(a) if $X_L > X_C$, the circuit is inductive and the current lags behind the voltage,
(b) if $X_L < X_C$, the circuit is capacitive and the current leads ahead the voltage,
(c) if $X_L = X_C$ ($\omega L = \frac{1}{\omega C}$), the current and voltage have the same phase and the series *RLC* circuit becomes a resonance circuit.

In the case of a series resonance circuit, the impedance becomes $Z = \sqrt{R^2 + (X_L - X_C)^2} = R$, which makes the circuit be purely resistive. Thus, at the resonance frequency f_0, which is given by Eq. (2.30), the impedance becomes minimum and equal to the resistance R, and the circuit current is at its maximum I_0, as shown in Fig. 2.7:

$$f_0 = \frac{1}{2\pi\sqrt{LC}} \tag{2.30}$$

$$I_0 = \frac{V}{R} \tag{2.31}$$

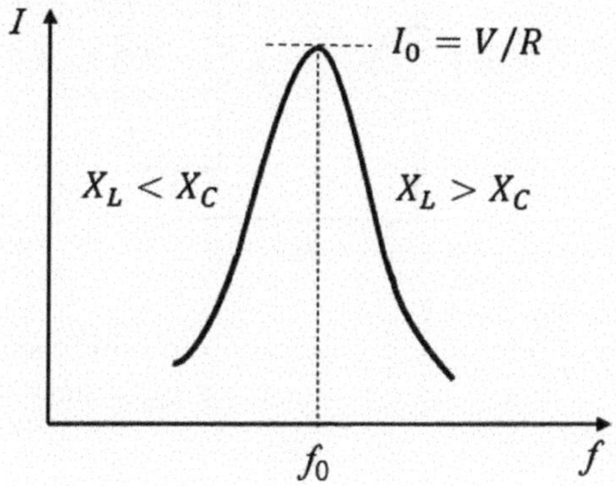

Fig. 2.7 Frequency response of current in a series *RLC* resonance circuit

The circuit is inductive above the resonance frequency ($X_L > X_C$) and capacitive below the resonance frequency ($X_L < X_C$).

2.1.4 *RLC* Parallel Circuits

(1) *RL* Parallel Circuit

As shown in Fig. 2.8a, applying voltage V to a parallel circuit of resistor R and inductor L, the currents flowing through R and L elements are as follows:

(a) $I_R = \frac{V}{R}$, and I_R and V are in phase,
(b) $I_L = \frac{V}{\omega L}$, and I_L lags V by $\pi/2$ (rad).

The vector diagram for voltage and current of the *RL* parallel circuit is shown in Fig. 2.8b. The total current is given by

$$I = \sqrt{I_R^2 + I_L^2} = V\sqrt{(\frac{1}{R})^2 + (\frac{1}{\omega L})^2} \tag{2.32}$$

The impedance Z and phase angle θ of the parallel *RL* circuit are given by

$$Z = \frac{1}{\sqrt{(\frac{1}{R})^2 + (\frac{1}{\omega L})^2}} \tag{2.33}$$

$$\theta = \tan^{-1}\frac{I_L}{I_R} = \tan^{-1}\frac{V/\omega L}{V/R} = \tan^{-1}\frac{R}{\omega L} \tag{2.34}$$

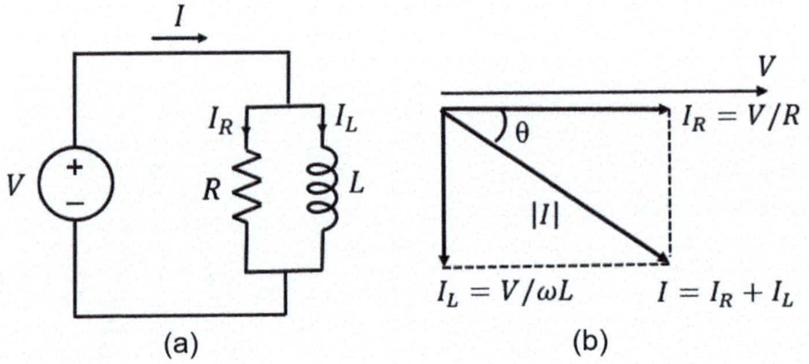

Fig. 2.8 **a** A parallel circuit of resistor and inductor and **b** its voltage-current relationship

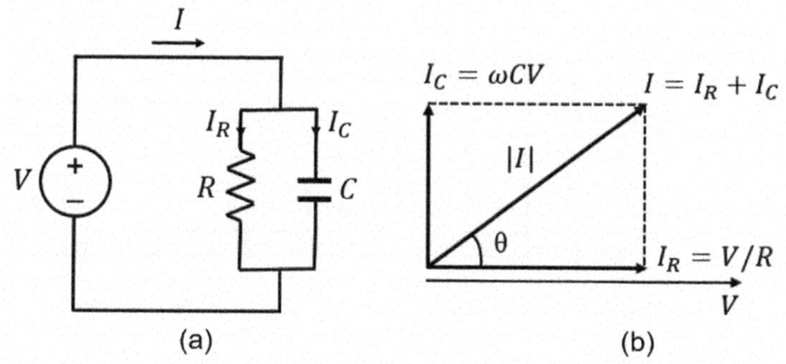

Fig. 2.9 **a** A parallel circuit of resistor and capacitor and **b** its voltage-current relationship

The current lags the voltage by θ (rad), and θ has values in the range of $-\pi/2 \le \theta \le 0$. If $\theta = 0$, it will be a circuit of only resistor R, and when $\theta = -\pi/2$, it will be the circuit of only inductor L.

(2) ***RC* Parallel Circuit**

As shown in Fig. 2.9a, applying voltage V to the parallel circuit of resistor R and capacitor C, the currents flowing through R and C elements are as follows:

(a) $I_R = \frac{V}{R}$, and I_R and V are in phase,
(b) $I_C = \frac{V}{1/\omega C} = \omega CV$, and I_C leads V by $\pi/2$ (rad).

The vector diagram for voltage and current of the RC parallel circuit is shown in Fig. 2.9b. The total current is given by

$$I = \sqrt{I_R^2 + I_C^2} = V\sqrt{(\frac{1}{R})^2 + (\omega C)^2} \tag{2.35}$$

The impedance Z and phase angle θ of the parallel RC circuit are given by

$$Z = \frac{1}{\sqrt{(\frac{1}{R})^2 + (\omega C)^2}} \tag{2.36}$$

$$\theta = \tan^{-1}\frac{I_C}{I_R} = \tan^{-1}\frac{\omega CV}{V/R} = \tan^{-1}\omega CR \tag{2.37}$$

The current leads the voltage by θ (rad), and θ has values in the range of $0 \le \theta \le \pi/2$. If $\theta = 0$, it will be a circuit of only resistor R, and when $\theta = \pi/2$, it will be a circuit of only capacitor C.

(3) ***RLC* Parallel Circuit**

As shown in Fig. 2.10a, applying voltage V to the parallel circuit of resistor R, inductor L, and capacitor C, the currents flowing through R, L, and C elements are as follows:

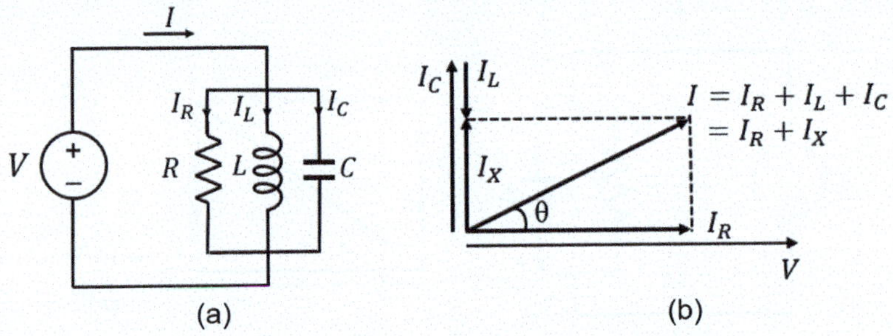

Fig. 2.10 **a** A *RLC* parallel circuit and **b** its voltage-current relationship

(a) $I_R = \frac{V}{R}$, and I_R and V are in phase,
(b) $I_L = \frac{V}{\omega L}$, and I_L lags V by $\pi/2$ (rad),
(c) $I_C = \frac{V}{1/\omega C} = \omega CV$, and I_C leads V by $\pi/2$ (rad).

When $I_L < I_C$, the vector diagram for voltage and current of the *RLC* parallel circuit is shown in Fig. 2.10b. The total current is given by

$$I = \sqrt{I_R^2 + (I_C - I_L)^2} = V\sqrt{(\frac{1}{R})^2 + (\omega C - \frac{1}{\omega L})^2} \tag{2.38}$$

From this, the impedance Z and phase angle θ are expressed as

$$Z = \frac{1}{\sqrt{(\frac{1}{R})^2 + (\omega C - \frac{1}{\omega L})^2}} \tag{2.39}$$

$$\theta = \tan^{-1}\frac{I_C - I_L}{I_R} = \tan^{-1}\frac{\omega CV - \frac{V}{\omega L}}{V/R} = \tan^{-1}\left(\omega C - \frac{1}{\omega L}\right)R \tag{2.40}$$

The characteristics of the parallel *RLC* circuit depend on $\left(\omega C - \frac{1}{\omega L}\right)$:

(a) if $\omega C > \frac{1}{\omega L}$, the current leads the voltage and the circuit is capacitive,
(b) if $\omega C < \frac{1}{\omega L}$, the current lags the voltage and the circuit is inductive,
(c) if $\omega C = \frac{1}{\omega L}$, the current and voltage have the same phase and the parallel *RLC* circuit becomes a resonance circuit.

In the case of parallel *RLC* resonance circuit, the total impedance is equal to the $Z = R\ (\Omega)$. At the resonance frequency f_0, the impedance becomes maximum and the circuit current is at its minimum, as shown in Fig. 2.11. The parallel resonance circuit exhibits the band-stop filter characteristics, opposite to the series resonance circuit showing the band-pass filter characteristics.

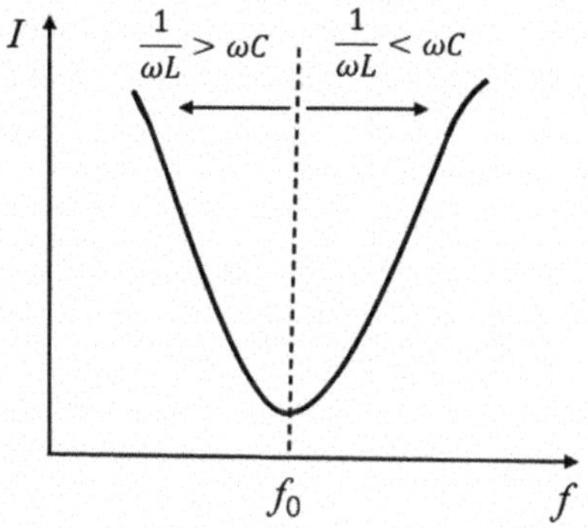

Fig. 2.11 Frequency response of current in a parallel *RLC* resonance circuit

2.1.5 *Complex Impedance and Admittance*

Time-dependent voltage and current in trigonometric functions ($V_m\cos\ \omega t$ and $I_m\cos\ \omega t$) can be represented by the complex-valued functions of time, with the help of Euler's formula ($e^{j\omega t} = \cos\ \omega t + j\ \sin\ \omega t$):

$$V(t) = V_m e^{j\omega t} \tag{2.41}$$

$$I(t) = I_m e^{j\omega t} \tag{2.42}$$

In the case of a resistor, the current flow for the applied voltage, $V(t) = V_m e^{j\omega t}$, is given by

$$I(t) = \frac{V_m}{R} e^{j\omega t} = \frac{1}{R} V(t) \tag{2.43}$$

For an inductor, the voltage drop across it for the circuit current, $I(t) = I_m e^{j\omega t}$, is given by

$$V(t) = L\frac{\mathrm{d}I(t)}{\mathrm{d}t} = j\omega LI(t) \tag{2.44}$$

where the rotational operator $j(= 1\angle 90^\circ)$ means that the voltage leads the current by $\pi/2$ (rad). In the case of a capacitor, the current flow in the circuit for $V(t) = V_m e^{j\omega t}$ is given by

$$I(t) = C\frac{\mathrm{d}V(t)}{\mathrm{d}t} = j\omega CV(t) \tag{2.45}$$

As opposed to an inductance circuit, the current leads the voltage by $\pi/2$ (rad).

The current–voltage relationships for resistor, inductor, and capacitor show that $V(t)$ and $I(t)$ are in proportional relationship in all cases. Thus $V(t)/I(t)$ or $I(t)/V(t)$

is independent of time t. The former is called the impedance and the latter is called the admittance, and denoted by Z and Y, respectively. The units of Z and Y are ohms (Ω) and siemens (S), respectively. The Z and Y of resistor, inductor, and capacitor are represented in Table 2.1.

When the current and voltage are expressed by complex numbers (phasors), $I = I_m\angle\alpha$ and $V = V_m\angle\beta$, respectively, the impedance and admittance are also represented by complex numbers. These are called the complex impedance Z and complex admittance Y, and expressed as

$$Z = \frac{V}{I} = \frac{V_m}{I_m}\angle(\beta - \alpha) = |Z|\angle\theta \tag{2.46}$$

$$Y = \frac{I}{V} = \frac{I_m}{V_m}\angle(\alpha - \beta) = |Y|\angle\theta' \tag{2.47}$$

where $|Z|$ and $|Y|$ are magnitudes, θ and θ' are phases of impedance and admittance, respectively, and have the relationship, $|Y| = 1/|Z|$ and $\theta' = -\theta$. In rectangular form, the complex impedance is expressed as

$$Z = |Z|\angle\theta = |Z|(\cos\theta + j\sin\theta) = R + jX \tag{2.48}$$

where R is the real or resistive component and X is the imaginary or reactive component, as shown in Fig. 2.12.

The complex admittance is the reciprocal of complex impedance, which is written in rectangular form:

$$Y = |Y|\angle\theta' = \frac{1}{Z} = \frac{1}{R + jX} = G + jB \tag{2.49}$$

Table 2.1 Impedance Z and admittance Y for resistor, inductor, and capacitor

	Resistor	Inductor	Capacitor
Impedance $Z(=\frac{V}{I})$	$\frac{RI}{I} = R$	$\frac{j\omega LI}{I} = j\omega L$	$\frac{I/j\omega C}{I} = -j(1/\omega C)$
Admittance $Y(=\frac{I}{V})$	$\frac{V/R}{V} = 1/R$	$\frac{V/j\omega L}{V} = -j(1/\omega L)$	$\frac{j\omega CV}{V} = j\omega C$

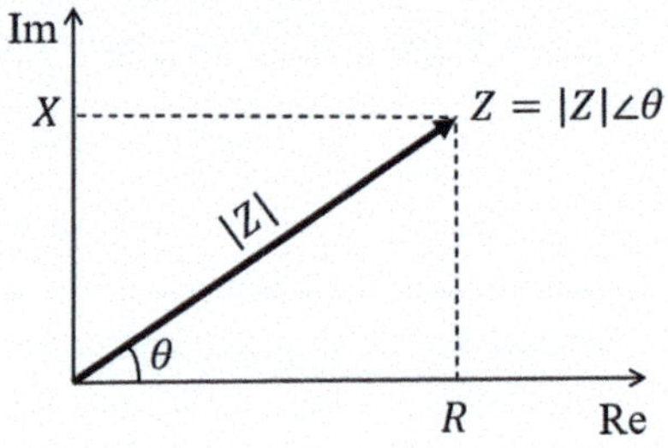

Fig. 2.12 Complex impedance diagram

The real part of the complex admittance is called the conductance G, and the imaginary part is called the susceptance B. Y, G, and B have the unit siemens (S) or mho (℧), and the following relationship is established:

$$|Y| = \sqrt{G^2 + B^2} \tag{2.50}$$

From Eq. (2.49), we can show that

$$G = \frac{R}{R^2 + X^2} \tag{2.51}$$

$$B = \frac{-X}{R^2 + X^2} \tag{2.52}$$

For purely resistive circuit ($X = 0$),

$$G = \frac{1}{R} \tag{2.53}$$

For purely reactive circuit ($R = 0$),

$$B = -\frac{1}{X} \tag{2.54}$$

The susceptance for inductor and capacitor is called inductive susceptance B_L and capacitive susceptance B_C, respectively, and they are expressed as

$$B_L = -\frac{1}{X_L} = -\frac{1}{\omega L} \tag{2.55}$$

$$B_C = -\frac{1}{X_C} = \omega C \tag{2.56}$$

The impedance and admittance of the passive elements are summarized in Table 2.2.

Impedances can be combined using the same rules that we establish for resistor combinations. If $Z_1, Z_2, \cdots, Z_n$ are connected in series, the total impedance Z is the sum of the component impedances:

Table 2.2 Impedance and admittance for passive elements

Element	Impedance (Ω)	Admittance (S)
Resistor	R (resistance)	$G = \frac{1}{R}$ (conductance)
Inductor	$X_L = \omega L$ (inductive reactance)	$B_L = -\frac{1}{\omega L}$ (inductive susceptance)
Capacitor	$X_C = -\frac{1}{\omega C}$ (capacitive reactance)	$B_C = \omega C$ (capacitive susceptance)

$$Z = Z_1 + Z_2 + \cdots + Z_n \tag{2.57}$$

and if $Z_1, Z_2, \cdots, Z_n$ are connected in parallel, the total impedance is given by

$$\frac{1}{Z} = \frac{1}{Z_1} + \frac{1}{Z_2} + \cdots + \frac{1}{Z_n} \tag{2.58}$$

For example, for a series circuit of resistor and inductor, the total impedance becomes $Z = Z_R + Z_L = R + j\omega L$.

Since admittance is a similar concept to conductance, the rules for combining admittances are the same as those for combining conductance. If $Y_1, Y_2, \cdots, Y_n$ are connected in series, the total admittance Y is given by

$$\frac{1}{Y} = \frac{1}{Y_1} + \frac{1}{Y_2} + \cdots + \frac{1}{Y_n} \tag{2.59}$$

If $Y_1, Y_2, \cdots, Y_n$ are connected in parallel, the total admittance is given by

$$Y = Y_1 + Y_2 + \cdots + Y_n \tag{2.60}$$

For example, for a parallel circuit of resistor and capacitor, the total admittance becomes $Y = Y_R + Y_C = G + j\omega C$.

2.2 Transmission Line

2.2.1 Distributed Element Circuit

Conventional circuit theory has interpreted only circuits that operate at sufficiently long wavelengths (sufficiently low frequencies) compared to physical dimension of lumped elements. These circuits are called the lumped element circuits which assume all electrical components are concentrated at a single point and this model is valid when circuit's characteristic length is much smaller than circuit's operating wavelength. To the contrary, a circuit that uses a short wavelength (very high frequency) signal compared to a physical circuit length cannot be interpreted by general circuit theory, that is called the distributed element circuits. In a distributed element circuit, the physical quantities associated with the electrical components are infinitesimally small and distributed in space. Even the wires connecting the devices are not assumed to be perfect conductors and subject to consideration that they have distributed characteristics of passive elements (resistor, inductor, capacitor). This is because the wavelength is as short as a partial portion of the wire, so the magnitude and phase of the voltage and current change depending on the position along the wire. A connector created in consideration of this characteristics is called the transmission line, which

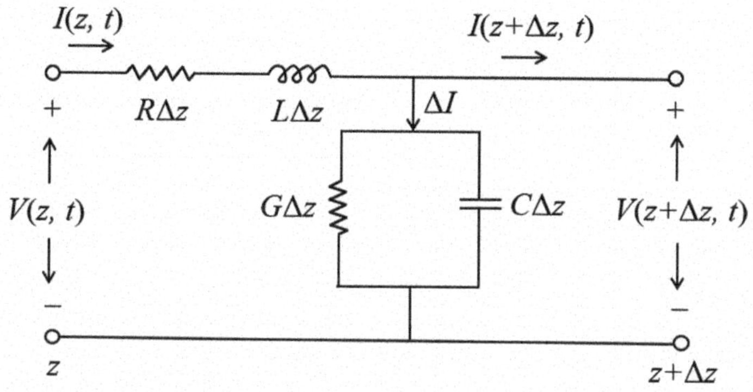

Fig. 2.13 Equivalent circuit of transmission line

is a special cable or structure created to carry a signal (voltage, current) of radio frequency (RF) [4–6]. The current and voltage in the distributed element circuit have characteristics similar to the electric field or magnetic field derived from Maxwell's equations, so it is easy to understand the relationship between circuit theory and electromagnetic field theory.

To interpret a transmission line operating in RF as a lumped element circuit, the line is divided as short as wavelength. This is shown in Fig. 2.13 with the line parameters of capacitance between conductors C (F/m), inductance of conductor loop L (H/m), resistance of conductors (conductor loss) R (Ω/m), and shunt conductance (dielectric loss) G (S/m). R is the resistance of line conductors. The current along the conductors forms a magnetic field and generates an inductance L. Capacitance C is formed in the space (insulator) between two conductors. Since there is no perfect insulator, shunt conductance G is created across the line. Since R, L, C, G are values per unit length, total values can be obtained by multiplying the finite line length Δz.

2.2.2 *Transmission Line Equations*

The amplitude of voltage $V(z)$ and current $I(z)$ varies in place z. Consider the two points z and $z + \Delta z$, as shown in Fig. 2.13. Δz is small enough, and finally $\Delta z \to 0$. The voltage and current at the starting point z are $V(z, t)$ and $I(z, t)$, and those at the end point $z + \Delta z$ are $V(z + \Delta z, t)$ and $I(z + \Delta z, t)$, respectively. When Δz is a small value, this part can be treated like a lumped element circuit: resistance $R\Delta z$, inductance $L\Delta z$, conductance $G\Delta z$, and capacitance $C\Delta z$. Using Kirchhoff's law, the voltage and current are expressed as

$$V(z) = \left[R\Delta z + j\omega(L\Delta z)\right]I(z) + V(z + \Delta z) \tag{2.61}$$

$$I(z) = \left[G\Delta z + j\omega(C\Delta z)\right]V(z) + I(z + \Delta z) \tag{2.62}$$

Taking the limit as $\Delta z \to 0$ leads to

$$\lim_{\Delta z \to 0} \frac{\Delta V(z)}{\Delta z} = \frac{\mathrm{d}V}{\mathrm{d}z} = -(R + j\omega L)I \tag{2.63}$$

$$\lim_{\Delta z \to 0} \frac{\Delta I(z)}{\Delta z} = \frac{\mathrm{d}I}{\mathrm{d}z} = -(G + j\omega C)V \tag{2.64}$$

Equations (2.63) and (2.64) are called the basic equations of the distributed element circuit [4–6]. At the transmission line, the distributed series impedance Z and parallel admittance Y per unit length are expressed as

$$Z = R + j\omega L \tag{2.65}$$

$$Y = G + j\omega C \tag{2.66}$$

If we differentiate Eq. (2.63) once more and employing Eq. (2.64), we obtain

$$\frac{\mathrm{d}^2 V(z)}{\mathrm{d}z^2} = (R + j\omega L)(G + j\omega C)V(z) = \gamma^2 V(z) \tag{2.67}$$

If Eq. (2.64) is differentiated once more and employing Eq. (2.63), we obtain

$$\frac{\mathrm{d}^2 I(z)}{\mathrm{d}z^2} = (R + j\omega L)(G + j\omega C)I(z) = \gamma^2 I(z) \tag{2.68}$$

where γ is defined as the propagation constant, given by

$$\gamma = \sqrt{(R + j\omega L)(G + j\omega C)} = \alpha + j\beta \tag{2.69}$$

The real part α and imaginary part β of the propagation constant are called the attenuation constant and phase constant, respectively, and expressed as

$$\alpha = \frac{1}{\sqrt{2}}\sqrt{\sqrt{\left(R^2 + \omega^2 L^2\right)\left(G^2 + \omega^2 C^2\right)} + \left(RG - \omega^2 LC\right)} \tag{2.70}$$

$$\beta = \frac{1}{\sqrt{2}}\sqrt{\sqrt{\left(R^2 + \omega^2 L^2\right)\left(G^2 + \omega^2 C^2\right)} - \left(RG - \omega^2 LC\right)} \tag{2.71}$$

The general solutions of the basic Eqs. (2.67) and (2.68) are given by

$$V(z) = V_0 e^{-\gamma z}(+z \text{ direction}) \text{ or } V(z) = V_0 e^{\gamma z}(-z \text{ direction}) \tag{2.72}$$

$$I(z) = I_0 e^{-\gamma z}(+z \text{ direction}) \text{ or } I(z) = I_0 e^{\gamma z}(-z \text{ direction}) \tag{2.73}$$

where V_0 and I_0 are wave amplitudes. Substituting Eqs. (2.72) and (2.73) into Eq. (2.63), we obtain

$$\frac{dV(z)}{dz} = -\gamma V_0 e^{-\gamma z} = -(R + j\omega L)I(z) = -(R + j\omega L)I_0 e^{-\gamma z} \tag{2.74}$$

$$Z_c = \frac{V_0}{I_0} = \frac{R + j\omega L}{\gamma} = \sqrt{\frac{R + j\omega L}{G + j\omega C}} = R_c + jX_c \tag{2.75}$$

Z_c is defined as the characteristic impedance, which means the ratio of voltage to current at transmission line. The real part of Z_c becomes the resistive component R_c, and the imaginary part becomes the reactive component X_c.

(1) **Lossless Line**

If the resistance of the conductor is $R = 0$ and the dielectric conductance between the conductors is $G = 0$, the line is called lossless. For such a line, Eq. (2.69) becomes

$$\alpha = 0 \tag{2.76}$$

$$\beta = \omega\sqrt{LC} \tag{2.77}$$

and Eq. (2.75) becomes

$$Z_c = \sqrt{\frac{L}{C}} = R_c \tag{2.78}$$

The propagation constant γ becomes pure imaginary and wave attenuation does not occur. The speed of movement of sinusoidal waves along the line is defined as the phase velocity v_p, which is given by

$$v_p = \frac{\omega}{\beta} = \frac{1}{\sqrt{LC}} \tag{2.79}$$

(2) **Coaxial Line**

The formulas for calculating the values of R, L, G, and C for coaxial line consisting of inner radius a and outer radius b, as shown in Fig. 2.14, are given by [5]

$$R = \frac{1}{2\pi\delta\sigma_c}\left[\frac{1}{a} + \frac{1}{b}\right] (\Omega/\text{m}) \tag{2.80}$$

$$L = \frac{\mu}{2\pi}\ln\frac{b}{a}\ (\text{H/m}) \tag{2.81}$$

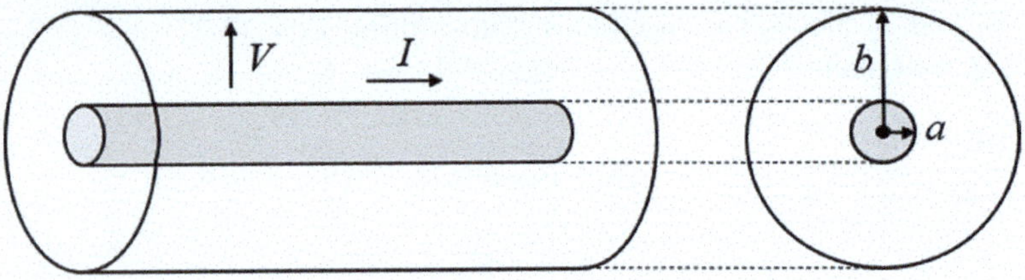

Fig. 2.14 Coaxial transmission line

$$G = \frac{2\pi\sigma}{\ln\frac{b}{a}}\ (\mathrm{V/m}) \tag{2.82}$$

$$C = \frac{2\pi\varepsilon}{\ln\frac{b}{a}}\ (\mathrm{F/m}) \tag{2.83}$$

where σ_c: electrical conductivity of conductor, δ: skin depth of conductor ($\delta \ll a$), μ: permeability of medium, σ: electrical conductivity of medium, ε: permittivity of medium. If the medium is air ($\mu = 4\pi \times 10^{-7}$ H/m, $\sigma = 0$, $\varepsilon = 8.854 \times 10^{-12}$ F/m) and $\sigma_c = \infty$, $R = 0$ and $G = 0$, which leads to

$$Z_c = 0.159 \times 377 \times \ln\frac{b}{a}\ (\Omega) \tag{2.84}$$

For a commercial coaxial line of $2a = 3$ mm and $2b = 7$ mm (such as APC-7 connector), $Z_c = 50\ (\Omega)$.

2.2.3 *Load-Connected Transmission Line*

When a transmission line with a characteristic impedance Z_c is connected to a load with an impedance Z_L which is located at a distance $z = l$, as illustrated in Fig. 2.15, a wave is reflected at the load due to the impedance mismatch between the load and line. If we $V^+ = V_0^+ e^{-\gamma z}$ refer to the incident voltage and $V^- = V_0^- e^{\gamma z}$ the reflected voltage, the total voltage and current waves in the line are given by

$$V(z) = V_0^+ e^{-\gamma z} + V_0^- e^{\gamma z} \tag{2.85}$$

$$I(z) = \frac{V_0^+}{Z_c} e^{-\gamma z} - \frac{V_0^-}{Z_c} e^{\gamma z} \tag{2.86}$$

In Fig. 2.15, V_0 and I_0 are input voltage and input current at $z = 0$, and V_L and I_L are load voltage and load current at $z = l$, respectively. Substituting these into Eqs. (2.85) and (2.86), we obtain

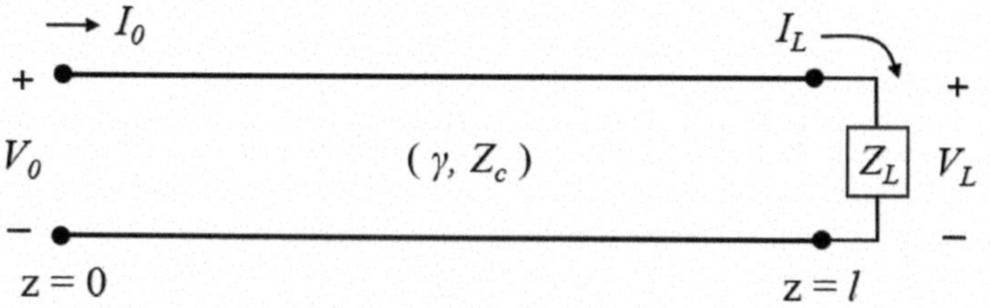

Fig. 2.15 Transmission line to which the load is connected

$$V_0^+ = \frac{1}{2}(V_L + Z_c I_L)e^{\gamma l} \tag{2.87}$$

$$V_0^- = \frac{1}{2}(V_L - Z_c I_L)e^{-\gamma l} \tag{2.88}$$

Since the input impedance is defined as $Z_{in} = V(z)/I(z)$ at any point on the line, the input impedance at $z = 0$ is given by

$$Z_{in} = \frac{Z_c\left(V_0^+ + V_0^-\right)}{V_0^+ - V_0^-} \tag{2.89}$$

Substituting Eqs. (2.87) and (2.88) into Eq. (2.89) and using the formula,

$$\tanh \gamma l = \frac{\sinh \gamma l}{\cosh \gamma l} = \frac{e^{\gamma l} - e^{-\gamma l}}{e^{\gamma l} + e^{-\gamma l}} \tag{2.90}$$

we obtain the input impedance expressed as

$$Z_{in} = Z_c\left[\frac{Z_L + Z_c \tanh \gamma l}{Z_c + Z_L \tanh \gamma l}\right] \tag{2.91}$$

For a lossless line ($\alpha = 0$), $\gamma = j\beta$ and $\tanh j\beta l = j \tan \beta l$, then Eq. (2.91) becomes

$$Z_{in} = Z_c\left[\frac{Z_L + jZ_c \tan \beta l}{Z_c + jZ_L \tan \beta l}\right] \tag{2.92}$$

The voltage reflection coefficient Γ is defined as the ratio of the magnitude of reflected voltage to that of incident voltage, given by

$$\Gamma = \frac{V_0^- e^{\gamma l}}{V_0^+ e^{-\gamma l}} \tag{2.93}$$

Substituting Eq. (2.87) and Eq. (2.88) into Eq. (2.93) and using $V_L = Z_L I_L$, the reflection coefficient at the load Γ_L is expressed as

$$\Gamma_L = \frac{Z_L - Z_c}{Z_L + Z_c} \tag{2.94}$$

Γ_L is proportional to the difference in impedance between the load and the line. The power reflection coefficient Γ_p is the ratio of the reflected power to the incident power, given by

$$\Gamma_p = |\frac{V^-}{V+}|^2 \tag{2.95}$$

Then we get

$$\Gamma_p = |\Gamma|^2 \tag{2.96}$$

A reflection of 10% in voltage is equivalent to a reflection of 1% in power.

If the transmission line is not fully matched to the load, the incident wave and the reflected wave coexist in the line, and a standing wave is formed. The standing wave ratio s of a voltage or current is defined as the ratio of the maximum value to the minimum value of a voltage or current, given by

$$s = \frac{V_{max}}{V_{min}} = \frac{I_{max}}{I_{min}} = \frac{1 + |\Gamma_L|}{1 - |\Gamma_L|} \tag{2.97}$$

Since $|\Gamma_L| \leq 1$, s has a positive value greater than 1. In the absence of reflection ($|\Gamma_L| = 0$), $s = 1$, and it is said that the line and load are matched. Conversely, in the case of the complete reflection ($|\Gamma_L| = 1$), s becomes infinite.

If the impedance of the load is 0 ($Z_L = 0$), we call it the short-circuited line, and Eq. (2.92) becomes

$$Z_{in} = Z_{sc} = jZ_c \tan \beta l \tag{2.98}$$

From Eq. (2.94) and Eq. (2.97), $\Gamma_L = -1$ and $s = \infty$. The input impedance of the shorted line Z_{sc} is a pure reactance (inductive or capacitive), and Fig. 2.16 shows the variation of Z_{sc} with the length of line l [5]. When $l = (2n + 1)\lambda/4$ (where $n = 0, 1, 2, \ldots$), $Z_{sc} = \pm\infty$, resulting in a high-impedance surface. If $l = n\lambda/2$, $Z_{sc} = 0$, resulting in a zero-impedance surface.

For a lossless material of $\lambda/4$ thickness and terminated by conductor, it forms the high-impedance surface ($Z_{in} = \infty$). Granting a resistive film of free-space impedance ($R = 377\ \Omega$) on this surface, incident wave is fully absorbed through impedance matching, which is the operating principle of the Salisbury screen. Figure 2.17 shows the schematic representation of Salisbury screen and its equivalent transmission line. The equivalent impedance Z at the surface of the dielectric spacer with a thickness $\lambda/4$ is determined by the parallel connection of input impedance of spacer Z_{in} and film resistance R, which is expressed as

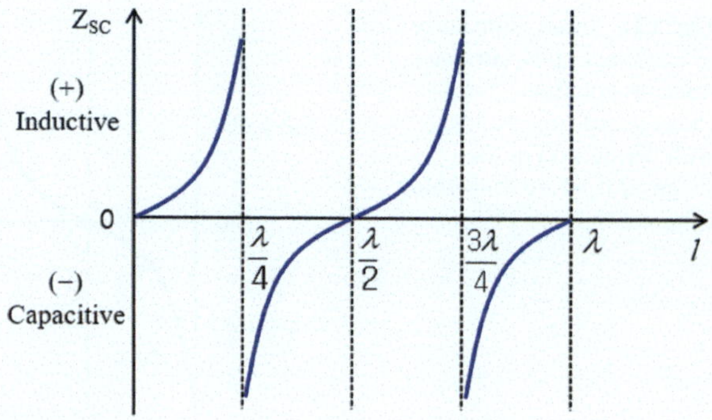

Fig. 2.16 Input impedance of a lossless shorted transmission line (reproduced from Ref. [5] with permission of the Licensor (Oxford Publishing Limited) through PLS clear)

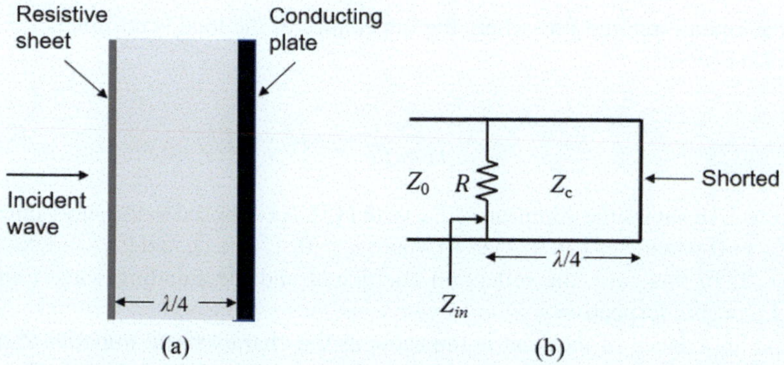

Fig. 2.17 **a** Schematic representation of Salisbury screen and **b** its equivalent transmission line

$$Z = \frac{Z_{in}R}{Z_{in} + R} \tag{2.99}$$

When $Z_{in} = \infty$, the equivalent impedance is equal to the film resistance ($Z = R$). If R is controlled to be the free-space impedance ($Z_0 = 377\ \Omega$), it becomes a non-reflected absorber through impedance matching.

With a similar design concept, when using a lossy material, the loss component can match to the free-space impedance, which enables it to be a non-reflected absorber with a layer thickness of $\lambda/4$. In this case of lossy material shorted by conductor ($Z_L = 0$), the input impedance of Eq. (2.91) becomes

$$Z_{\text{in}} = Z_c \tanh \gamma l \tag{2.100}$$

If we control Z_{in} to be the free-space impedance, impedance matching can be obtained, leading to zero wave reflection. That is called the Dallenbach layer (1/4 wavelength thickness absorber), which will be explained in detail in Chap. 5.

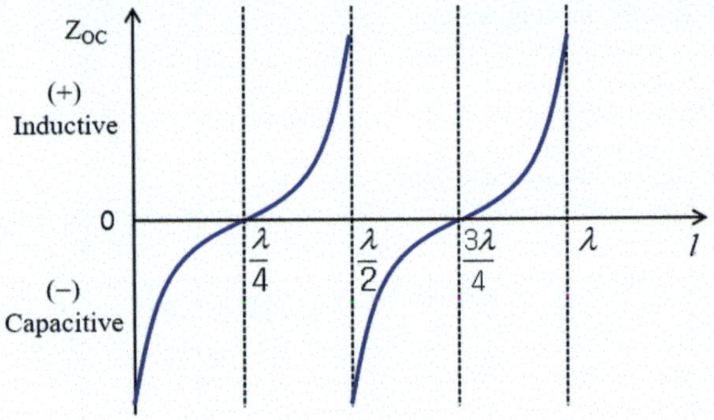

Fig. 2.18 Input impedance of a lossless open-circuited transmission line (reproduced from Ref. [5] with permission of the Licensor (Oxford Publishing Limited) through PLS clear)

For an open-circuited line where the impedance of the load is infinite ($Z_L = \infty$), Eq. (2.92) becomes

$$Z_{in} = Z_{oc} = \frac{Z_c}{j\tan \beta l} = -jZ_c \cot \beta l \tag{2.101}$$

Figure 2.18 shows the variation of Z_{oc} with l [5]. As opposed to the short-circuited line, $Z_{in} = 0$ when $l = (2n + 1)\lambda/4$ (where $n = 0, 1, 2, \ldots$), and $Z_{in} = \pm\infty$ when $l = n\lambda/2$. In this case, the reflection coefficient and the standing wave ratio are $\Gamma_L = 1$, $s = \infty$, respectively.

If the impedance of the load is the same as the characteristic impedance of the line ($Z_L = Z_c$), we call it the matched line. In this case, Eq. (2.92) becomes $Z_{in} = Z_c$, and $\Gamma_L = 0$, $s = 1$. There is no wave reflection at the load and the maximum power is transferred to the load.

2.2.4 *Microstrip Line*

A microstrip line is one of electrical transmission lines. Because it is easily integrated with other passive and active high-frequency devices, it is commonly used to carry the high-frequency signals [7]. The structure of a typical microstrip line is shown in Fig. 2.19.

It consists of a dielectric substrate, a signal line at the top, and a ground conductor at the bottom. The signal is preserved and transmitted in the form of electromagnetic field between the signal line and the ground plane. In Fig. 2.19, w is the width of signal line, t is the thickness of signal line, h is the height of dielectric substrate, and ε_r is the dielectric constant of substrate. These are important parameters in the design of microstrip lines. In addition, the loss tangent of substrate tanδ affects the loss of the line.

For strip lines where the signal line is centered inside the dielectric substrate, the electromagnetic field is uniformly distributed within the dielectric substrate and the

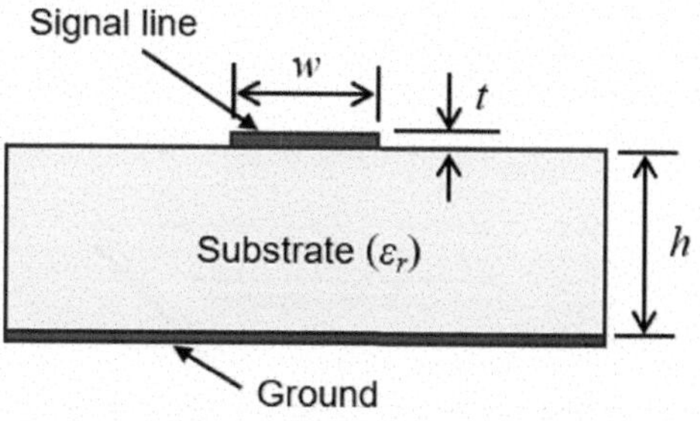

Fig. 2.19 Structure of microstrip line

wave propagates in TEM mode. In the microstrip line, however, the electromagnetic field is asymmetric (partly in the dielectric substrate and partly in the air) and the wave propagates in quasi-TEM mode. For this reason, effective dielectric constant ε_{eff} is introduced with an approximation of homogeneous medium [8]:

$$\varepsilon_{\text{eff}} = \frac{\varepsilon_r + 1}{2} + \frac{\varepsilon_r - 1}{2} \frac{1}{\sqrt{1 + 12(h/w)}} \tag{2.102}$$

The phase velocity v_p, propagation constant β, and wave number k_0 in the microstrip line are given by

$$v_p = \frac{c}{\sqrt{\varepsilon_{\text{eff}}}} \tag{2.103}$$

$$\beta = \frac{k_o}{\sqrt{\varepsilon_{\text{eff}}}} \tag{2.104}$$

$$k_o = \frac{\omega}{c} \tag{2.105}$$

where ω is the angular velocity and c is the speed of light in free space.

Once the dimension of microstrip line is given, we can calculate the characteristic impedance Z_c using Eq. (2.106) [8, 9]:

$$Z_c = \begin{cases} \frac{60}{\sqrt{\varepsilon_{\text{eff}}}} \ln\left(\frac{8h}{w} + \frac{w}{4h}\right) & \text{for } \frac{w}{h} \leq 1 \\ \frac{120\pi}{\sqrt{\varepsilon_{\text{eff}}}\left[\frac{w}{h} + 1.393 + 0.667 \ln\left(\frac{w}{h} + 1.444\right)\right]} & \text{for } \frac{w}{h} \geq 1 \end{cases} \tag{2.106}$$

Conversely, given the characteristic impedance Z_c of the line and the dielectric constant ε_r, w/h ratio can be calculated as follows [8]:

$$\frac{w}{h} = \begin{cases} \frac{8e^A}{e^{2A} - 2} & \text{for } \frac{w}{h} \leq 2 \\ \frac{2}{\pi}\left[B - 1 - \ln(2B - 1) + \frac{\varepsilon_r - 1}{2\varepsilon_r}\left\{\ln(B - 1) + 0.39 - \frac{0.61}{\varepsilon_r}\right\}\right] & \text{for } \frac{w}{h} \geq 2 \end{cases} \tag{2.107}$$

where A and B are given by

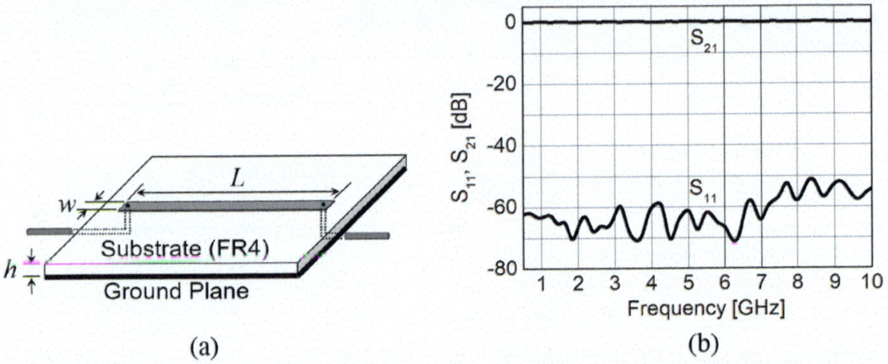

Fig. 2.20 Microstrip line: **a** schematic structure and **b** reflection and transmission characteristics (S_{11}, S_{21})

$$A = \frac{Z_c}{60}\sqrt{\frac{\varepsilon_r + 1}{2}} + \frac{\varepsilon_r - 1}{\varepsilon_r + 1}\left(0.23 + \frac{0.11}{\varepsilon_r}\right) \tag{2.108}$$

$$B = \frac{377\pi}{2Z_c\sqrt{\varepsilon_r}} \tag{2.109}$$

Figure 2.20a shows a schematic structure of a microstrip line designed for a characteristic impedance of 50 Ω at a center frequency $f_0 = 3$ GHz. FR4 (with dielectric constant $\varepsilon_r = 4.8$ and thickness $h = 1.26$ mm) is used as the substrate, and the width of the signal line calculated using Eq. (2.107) is $w = 2.24$ mm. S parameters measurement with the designed microstrip line with a substrate size of 100×100 mm^2 and a length of signal line $L = 79.54$ mm shows the ideal transmission line characteristics (reflection loss $S_{11} \approx -60$ dB and transmission loss $S_{21} \approx 0$ dB), as shown in Fig. 2.20b. The results indicate that the good impedance matching is satisfied with 50 Ω connector (APC-7) and the insertion loss (ohmic loss from conductor and dielectric loss from substrate) is very small.

2.3 Further Reading

The topics in this chapter, including *RLC* circuits and transmission line theory, are presented in greater detail in the public domain [10, 11] and in advanced textbooks on electromagnetics [5, 6, 8].

References

1. Electronic component on Wikipedia. https://en.wikipedia.org/wiki/Electronic_component
2. Inductance on Wikipedia. https://en.wikipedia.org/wiki/Inductance
3. Capacitance on Wikipedia. https://en.wikipedia.org/wiki/Capacitance
4. Transmission Line on Wikipedia. https://en.wikipedia.org/wiki/Transmission_line
5. M.N.O. Sadiku, *Elements of Electromagnetics*, 2nd edn. (Oxford University Press, New York, 1995), pp.518–538
6. Y. Naito, *Radio Wave Absorbers* (Ohm, Japan, 1987), pp. 21–35. (in Japanese)
7. Microstrip on Wikipedia. https://en.wikipedia.org/wiki/Microstrip
8. D.M. Pozar, *Microwave Engineering*, 4th edn. (Wiley, USA, 2011), pp. 48–63 and pp. 147–153
9. I.J. Bahl, D.K. Trivedi, A designer's guide to microstrip line. Microwaves **16**, 174–182 (1977)
10. Electrical resistance and conductance on Wikipedia. https://en.wikipedia.org/wiki/Electrical_resistance_and_conductance
11. *RLC* circuit on Wikipedia. https://en.wikipedia.org/wiki/RLC_circuit

Chapter 3
Electromagnetic Waves

This chapter discusses plane waves that propagate in an unbounded medium, starting from Maxwell's equations. Applying the initial and boundary conditions that characterize the medium to Maxwell's equations gives a solution for the electric and magnetic fields of plane waves. The propagation characteristics are examined by dividing the medium into free space, dielectrics, and lossy media including good conductors. This chapter also examines the mathematical solutions for the reflection and transmission of plane waves for normal and oblique incidence depending on the polarization. The formation of standing waves at complete reflection provides information about the distribution of electric and magnetic fields, and their ratio leading to the input impedance provides a physical basis for the resonant 1/4 wavelength absorbers.

3.1 Maxwell's Equations

When dealing with electromagnetic waves, Maxwell's equations are usually used [1]. These equations include a time-varying electric and magnetic fields, which are related to two important electrical phenomena [2]. The first is that the electric field is produced by changing the magnetic field, as explained by Faraday's induction law. The second is the displacement current introduced by *J. C. Maxwell*. Displacement current results from time-varying electric flux density and provides the theory that predicts the existence and propagation of electromagnetic waves.

In a space without power source, time-varying electric field $E(t)$, electric flux density $D(t)$, magnetic field $H(t)$, magnetic flux density $B(t)$, and conduction current density $J(t)$ satisfy the following equations:

$$\nabla \times E(t) = -\frac{\partial B(t)}{\partial t} \tag{3.1}$$

S. Kim, *Fundamentals and Principles of Electromagnetic Wave Absorbers*, Springer Series in Advanced Microelectronics 66, https://doi.org/10.1007/978-981-97-6477-8_3

$$\nabla \times H(t) = J(t) + \frac{\partial D(t)}{\partial t} \tag{3.2}$$

Equation (3.1) is based on the Faraday's law of electromagnetic induction, which states that when the magnetic flux changes, an electric field is produced around it. Equation (3.2) is based on the Ampere's circuital law, which states that when current flows through a wire (conduction current), a magnetic field is produced around it. Maxwell demonstrated that magnetic field is produced not only in wires but in empty space where the intensity of electric field changes with time, and he explained this by introducing the concept of displacement current density J_d:

$$J_d = \frac{\delta D(t)}{\delta t} \tag{3.3}$$

Replacing the time-varying sinusoidal waves $E(t) = Ee^{j\omega t}, D(t) = De^{j\omega t}, H(t) = He^{j\omega t}$, $B(t) = Be^{j\omega t}$, and $J(t) = Je^{j\omega t}$ by phasors E, D, H, B, and J, Eq. (3.1) and Eq. (3.2) are expressed as

$$\nabla \times E = -j\omega B \tag{3.4}$$

$$\nabla \times H = J + j\omega D \tag{3.5}$$

In a linear medium, the following constitutive relations are established:

$$D = \varepsilon E \tag{3.6}$$

$$B = \mu H \tag{3.7}$$

$$J = \sigma E \tag{3.8}$$

where ε represents the dielectric permittivity in unit (F/m), μ magnetic permeability (H/m), and σ electrical conductivity (S/m) of the medium, respectively. Substituting the constitutive relations into Eq. (3.4) and Eq. (3.5), we obtain

$$\nabla \times E = -j\omega\mu H \tag{3.9}$$

$$\nabla \times H = (\sigma + j\omega\varepsilon)E \tag{3.10}$$

3.2 Plane Waves

3.2.1 Wave Equations

A plane wave is the case in which the physical quantities of the electric and magnetic fields remain constant at any moment in all planes perpendicular to the wave vector which means the direction of wave propagation. As shown in Fig. 3.1, when the electric field and magnetic field are polarized in x-axis and y-axis, respectively, and the wave propagates in z direction in the rectangular coordinate, the field vectors are expressed as

$$\vec{E} = E_x(z, t)a_x \tag{3.11}$$

$$\vec{H} = H_y(z, t)a_y \tag{3.12}$$

where a_x and a_y are unit vectors to x-axis and y-axis, respectively. The direction of propagation energy is indicated by a pointing vector $\vec{P}$, which is given by cross product $\vec{P} = \vec{E} \times \vec{H}$ [3].

Since the electric and magnetic fields are constant in x–y plane, the following equations are established:

$$\frac{\partial E_x}{\partial x} = \frac{\partial E_x}{\partial y} = 0 \tag{3.13}$$

$$\frac{\partial H_y}{\partial x} = \frac{\partial H_y}{\partial y} = 0 \tag{3.14}$$

For these polarized electric and magnetic fields, applying the Maxwell's equations expressed by phasors (Eq. (3.9) and Eq. (3.10), respectively) yields

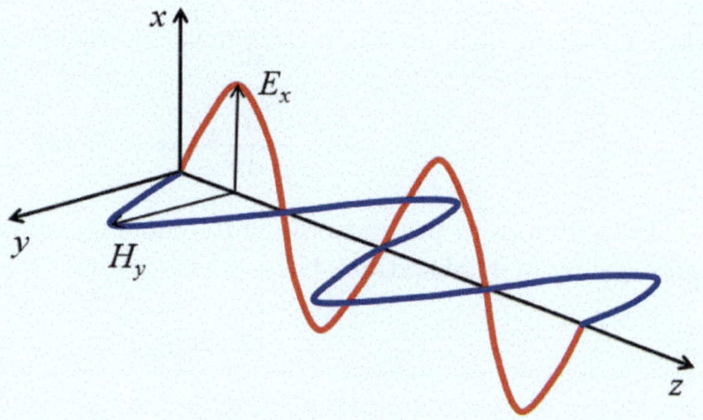

Fig. 3.1 Electric field, magnetic field, and the propagation of plane wave in the rectangular coordinate

$$\nabla \times E = \begin{vmatrix} a_x & a_y & a_z \\ \frac{\partial}{\partial x} & \frac{\partial}{\partial y} & \frac{\partial}{\partial z} \\ E_x & 0 & 0 \end{vmatrix} = \frac{\partial E_x}{\partial z} a_y = -j\omega\mu H_y a_y \tag{3.15}$$

$$\nabla \times H = \begin{vmatrix} a_x & a_y & a_z \\ \frac{\partial}{\partial x} & \frac{\partial}{\partial y} & \frac{\partial}{\partial z} \\ 0 & H_y & 0 \end{vmatrix} = -\frac{\partial H_y}{\partial z} a_x = (\sigma + j\omega\varepsilon) E_x a_x \tag{3.16}$$

Then we obtain

$$\frac{\mathrm{d}E_x(z)}{\mathrm{d}z} = -j\omega\mu H_y(z) \tag{3.17}$$

$$\frac{\mathrm{d}H_y(z)}{\mathrm{d}z} = -(\sigma + j\omega\varepsilon) E_x(z) \tag{3.18}$$

Differentiating the equations again, we obtain

$$\frac{\mathrm{d}^2 E_x(z)}{\mathrm{d}z^2} = -j\omega\mu \frac{\mathrm{d}H_y(z)}{\mathrm{d}z} = j\omega\mu(\sigma + j\omega\varepsilon) E_x(z) \tag{3.19}$$

$$\frac{\mathrm{d}^2 H_y(z)}{\mathrm{d}z^2} = -(\sigma + j\omega\varepsilon) \frac{\mathrm{d}E_x(z)}{\mathrm{d}z} = j\omega\mu(\sigma + j\omega\varepsilon) H_y(z) \tag{3.20}$$

Here we introduce the propagation constant γ, which is expressed as

$$\gamma = \sqrt{j\omega\mu(\sigma + j\omega\varepsilon)} = \alpha + j\beta \tag{3.21}$$

It is to be noted that γ is complex. The real part α is called the attenuation constant in unit (Np/m) and the imaginary part β is called the phase constant in unit (rad/m). Then we can write

$$\frac{\mathrm{d}^2 E_x(z)}{\mathrm{d}z^2} = \gamma^2 E_x(z) \tag{3.22}$$

$$\frac{\mathrm{d}^2 H_y(z)}{\mathrm{d}z^2} = \gamma^2 H_y(z) \tag{3.23}$$

The solutions of the 2nd-order differential equations take the form for the wave propagating in $+z$ direction:

$$E_x = E_m e^{-\gamma z} \tag{3.24}$$

$$H_y = H_m e^{-\gamma z} = \frac{E_m}{Z} e^{-\gamma z} \tag{3.25}$$

where E_m and H_m are the amplitudes of electric field and magnetic field, respectively. Z is called the intrinsic impedance or wave impedance, and refers to the ratio of electric field to magnetic field of the electromagnetic waves passing through a medium. Substituting the electric field of Eq. (3.24) into Eq. (3.17), we obtain

$$H_y = \frac{\sqrt{(\sigma + j\omega\varepsilon)j\omega\mu}}{j\omega\mu} E_m e^{-\gamma z} \tag{3.26}$$

The impedance can be obtained from Eq. (3.24) and Eq. (3.26):

$$Z = \frac{E_x}{H_y} = \sqrt{\frac{j\omega\mu}{\sigma + j\omega\varepsilon}} = |Z|\angle\theta_z = |Z|e^{j\theta_z} \tag{3.27}$$

where $|Z|$ is the magnitude of wave impedance and θ_z is the phase difference between the electric field and the magnetic field. It is to be noted that Z is complex, and E_x and H_y are not in-phase ($\theta_z > 0$) if $\sigma > 0$. If we rewrite Eq. (3.24) and Eq. (3.25) using α and β, those are

$$E_x = E_m e^{-(\alpha + j\beta)z} \tag{3.28}$$

$$H_y = \frac{E_m}{|Z|} e^{-(\alpha + j\beta)z} e^{-j\theta_z} \tag{3.29}$$

Reinserting $e^{j\omega t}$ factor and taking the real part, we obtain the following trigonometric forms of time-varying electric and magnetic fields:

$$E_x(z, t) = \mathrm{Re}\left(E_x e^{j\omega t}\right) = E_m e^{-\alpha z} \cos(\omega t - \beta z) \tag{3.30}$$

$$H_y(z, t) = \mathrm{Re}\left(H_y e^{j\omega t}\right) = \frac{E_m}{|Z|} e^{-\alpha z} \cos(\omega t - \beta z - \theta_z) \tag{3.31}$$

3.2.2 Propagation in Lossless Media

For free space, the values of electrical conductivity, permittivity, and permeability are $\sigma = 0$ (S/m), $\varepsilon_0 = 8.854 \times 10^{-12}$ (F/m), and $\mu_0 = 4\pi \times 10^{-7}$ (H/m), respectively. Since $\sigma = 0$, the attenuation constant and the phase constant are given by from Eq. (3.21)

$$\alpha = 0 \ (\mathrm{Np/m}) \tag{3.32}$$

$$\beta_0 = \omega\sqrt{\mu_0\varepsilon_0} \text{ (rad/m)} \tag{3.33}$$

Because $\alpha = 0$, the attenuation of wave does not occur. The intrinsic impedance is given by from Eq. (3.27)

$$Z_0 = \sqrt{\frac{\mu_0}{\varepsilon_0}} \simeq 377\ (\Omega) \tag{3.34}$$

and there is no phase difference between the electric field and the magnetic field ($\theta_z = 0$). Then the wave equations are expressed as

$$E_x(z, t) = E_m\cos(\omega t - \beta_0 z) \tag{3.35}$$

$$H_y(z, t) = \frac{E_m}{Z_0}\cos(\omega t - \beta_0 z) \tag{3.36}$$

At any point, the sinusoidal variation with time is period, given by $T = 2\pi/\omega = 1/f$, where f is the frequency. At any time, the variation with distance is wavelength, given by $\lambda_0 = 2\pi/\beta_0$. In response to the variation in both time and location, the speed of the travelling wave is determined by the movement of a point of a constant phase, which is called the phase velocity v_0. From the condition, $\omega t - \beta_0 z = constant$, we obtain

$$v_0 = \frac{dz}{dt} = \frac{\omega}{\beta_0} = \frac{1}{\sqrt{\mu_0\varepsilon_0}} \simeq 3 \times 10^8 \text{ (m/s)} \tag{3.37}$$

and

$$\lambda_0 = \frac{2\pi}{\beta_0} = \frac{1}{f\sqrt{\mu_0\varepsilon_0}} = \frac{v_0}{f} \simeq \frac{3 \times 10^8}{f} \text{ (m)} \tag{3.38}$$

For dielectrics, the permittivity and permeability are usually expressed by relative values with respect to those of free space, $\varepsilon_r = \varepsilon/\varepsilon_0$ and $\mu_r = \mu/\mu_0$, respectively. For perfect dielectrics with $\sigma = 0$, the following equations can be obtained:

$$\alpha = 0 \text{ (Np/m)} \tag{3.39}$$

$$\beta = \omega\sqrt{\mu\varepsilon} = \beta_0\sqrt{\varepsilon_r\mu_r} \text{ (rad/m)} \tag{3.40}$$

$$v = \frac{\omega}{\beta} = \frac{1}{\sqrt{\mu\varepsilon}} = \frac{v_0}{\sqrt{\varepsilon_r\mu_r}} \simeq \frac{3 \times 10^8}{\sqrt{\varepsilon_r\mu_r}} \text{ (m/s)} \tag{3.41}$$

$$\lambda = \frac{2\pi}{\beta} = \frac{1}{f\sqrt{\mu\varepsilon}} = \frac{\lambda_0}{\sqrt{\varepsilon_r\mu_r}} \text{ (m)} \tag{3.42}$$

Since $\varepsilon_r > 1$ for dielectrics, the wavelength is shorter and the phase velocity is lower than those in free space. The intrinsic impedance is given by

$$Z = \sqrt{\frac{\mu}{\varepsilon}} = Z_0\sqrt{\frac{\mu_r}{\varepsilon_r}} \simeq 377\sqrt{\frac{\mu_r}{\varepsilon_r}}\ (\Omega) \tag{3.43}$$

High-permeability materials (such as ferrites and magnetic materials with $\mu_r > 1$) have the higher intrinsic impedance.

3.2.3 *Propagation in Lossy Dielectrics*

In the medium where there is a loss ($\sigma > 0$), the Maxwell's equation based on Ampere's circuital law can be expressed as

$$\nabla \times H = (\sigma + j\omega\varepsilon)E = j\omega\varepsilon\left(1 - j\frac{\sigma}{\omega\varepsilon}\right)E \tag{3.44}$$

where $\sigma/\omega\varepsilon$ has the meaning of the proportion of material loss. Maxwell's equation includes the conduction current $J_c = \sigma E$ and the displacement current $J_d = j\omega\varepsilon E$, which is represented in Fig. 3.2. Conduction current means the loss of energy (ohmic loss), while the displacement current means the accumulation of energy. The ratio of conduction current to displacement current is defined as the loss tangent, expressed as

$$\tan\theta = \frac{\sigma}{\omega\varepsilon} \tag{3.45}$$

where θ is the loss angle of medium indicated in Fig. 3.2. The $\tan\theta$ is used to distinguish the conductors and dielectrics. If $\sigma/\omega\varepsilon < 0.01$, the material is called a dielectric, and if $\sigma/\omega\varepsilon > 100$, it is treated as a conductor.

In the case of dielectrics ($\sigma/\omega\varepsilon < 1$), the propagation constant γ can be approximated to

$$\gamma = j\omega\sqrt{\mu\varepsilon}\sqrt{1 - j\frac{\sigma}{\omega\varepsilon}}$$

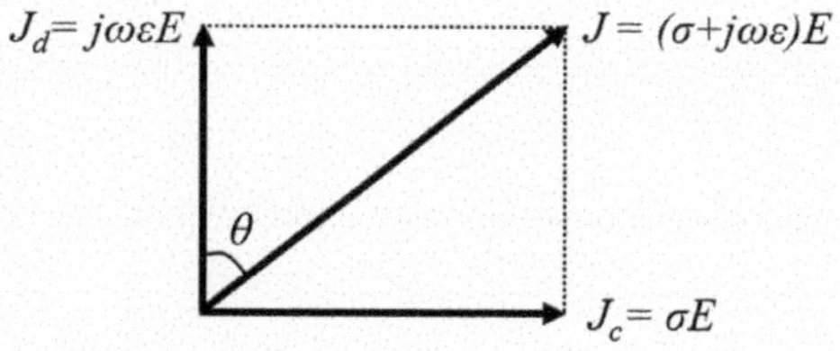

Fig. 3.2 Vector representation of conduction current J_c and displacement current J_d in complex plane

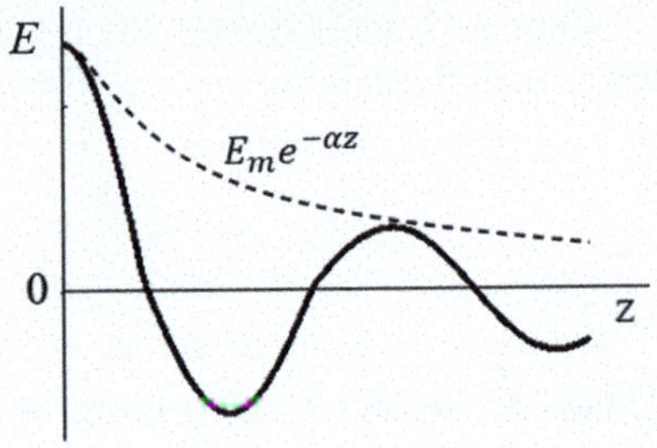

Fig. 3.3 Attenuation of propagation wave in a lossy medium

$$\simeq j\omega\sqrt{\mu\varepsilon}\left[1 - j\frac{\sigma}{2\omega\varepsilon} + \frac{1}{8}\left(\frac{\sigma}{\omega\varepsilon}\right)^2 - \cdots\right] \simeq j\omega\sqrt{\mu\varepsilon}\left[1 - j\frac{\sigma}{2\omega\varepsilon}\right] \quad (3.46)$$

using the binomial theorem for $x < 1$,

$$(1 + x)^n = 1 + nx + \frac{n(n-1)}{2!}x^2 + \cdots \quad (3.47)$$

From Eq. (3.46), we obtain

$$\alpha \simeq j\omega\sqrt{\mu\varepsilon}\left[-j\frac{\sigma}{2\omega\varepsilon}\right] = \frac{\sigma}{2}\sqrt{\frac{\mu}{\varepsilon}} \quad (3.48)$$

$$\beta \simeq \omega\sqrt{\mu\varepsilon} \quad (3.49)$$

It is to be noted that α (wave attenuation) is dominantly controlled by loss parameter σ, while β (which is related with wavelength) is insensitive to σ. In a similar way, the intrinsic impedance can be approximated to

$$Z = \sqrt{\frac{\mu}{\varepsilon}}\sqrt{\frac{1}{1 - j\left(\frac{\sigma}{\omega\varepsilon}\right)}} \simeq \sqrt{\frac{\mu}{\varepsilon}}\left[1 + j\frac{\sigma}{2\omega\varepsilon} - \frac{3}{8}\left(\frac{\sigma}{\omega\varepsilon}\right)^2 + \cdots\right]$$
$$\simeq \sqrt{\frac{\mu}{\varepsilon}}\left[1 + j\frac{\sigma}{2\omega\varepsilon}\right] \quad (3.50)$$

In the case of perfect dielectric ($\sigma = 0$), it becomes a lossless medium:

$$\gamma = j\omega\sqrt{\mu\varepsilon} \quad (3.51)$$

$$Z = \sqrt{\frac{\mu}{\varepsilon}} \quad (3.52)$$

Even in the dielectric medium of $\sigma/\omega\varepsilon \ll 1$, if there is a loss from dielectric resonance or relaxation, a complex permittivity ($\varepsilon = \varepsilon' - j\varepsilon''$) must be applied. In this case, if $\varepsilon''/\varepsilon' < 1$, the propagation constant is given by

$$\gamma = j\omega\sqrt{\mu(\varepsilon' - j\varepsilon'')} = j\omega\sqrt{\mu\varepsilon'}\sqrt{1 - j\frac{\varepsilon''}{\varepsilon'}} \simeq j\omega\sqrt{\mu\varepsilon'}\left[1 - j\frac{\varepsilon''}{2\varepsilon'}\right] \tag{3.53}$$

Then we obtain

$$\alpha \simeq \frac{\omega\sqrt{\mu\varepsilon'}}{2}\left(\frac{\varepsilon''}{\varepsilon'}\right) \tag{3.54}$$

$$\beta \simeq \omega\sqrt{\mu\varepsilon'} \tag{3.55}$$

which indicate that α is proportional to ε'', while β is insensitive to ε''. The intrinsic impedance can be expressed as

$$Z = \sqrt{\frac{\mu}{\varepsilon' - j\varepsilon''}} = \sqrt{\frac{\mu}{\varepsilon'}}\sqrt{\frac{1}{1 - j(\frac{\varepsilon''}{\varepsilon'})}} \simeq \sqrt{\frac{\mu}{\varepsilon'}}\left[1 + j\frac{\varepsilon''}{2\varepsilon'}\right] \tag{3.56}$$

If there is a loss in the medium ($\sigma > 0$ or $\varepsilon'' > 0$), the attenuation constant is $\alpha > 0$. Therefore, the exponential term $e^{-\alpha z}$ in Eq. (3.30) and Eq. (3.31) must be taken into account. As shown in Fig. 3.3, the magnitude $E_m e^{-\alpha z}$ of the propagation wave decreases exponentially with distance z. The unit of attenuation constant is neper per meter (Np/m). 1 (Np) is defined as an attenuation by a factor of $1/e$ ($\approx 37\%$) of the initial value. This is expressed in unit (dB) as

$$1\ (\text{Np}) = 20\log\left(\frac{1}{e^{-1}}\right) = 8.686\ (\text{dB}) \tag{3.57}$$

Also, in the lossy medium, the phase angle of intrinsic impedance is $\theta_z \neq 0$. That means that there is a phase difference between the electric field and the magnetic field.

3.2.4 *Propagation in Good Conductors*

In the case of conductors ($\sigma/\omega\varepsilon \gg 1$), the propagation constant is simplified to

$$\gamma = \sqrt{j\omega\mu(\sigma + j\omega\varepsilon)} \simeq \sqrt{j\omega\mu\sigma} = \sqrt{\omega\mu\sigma}\angle 45^\circ = \sqrt{\omega\mu\sigma}\sqrt{\frac{1}{2}}(1 + j1) \tag{3.58}$$

Therefore, the attenuation constant and the phase constant are given by

$$\alpha = \beta = \sqrt{\frac{\omega\mu\sigma}{2}} = \sqrt{\pi f \mu\sigma} \tag{3.59}$$

and the intrinsic impedance is simplified to

$$Z = \sqrt{\frac{j\omega\mu}{\sigma + j\omega\varepsilon}} \simeq \sqrt{\frac{j\omega\mu}{\sigma}} = \sqrt{\frac{2\pi f \mu}{\sigma}}\angle 45^\circ = \sqrt{\frac{2\pi f \mu}{\sigma}}\sqrt{\frac{1}{2}}(1 + j1) \tag{3.60}$$

The impedance is expressed by complex number of which argument is 45°, which means that the electric field of the wave passing through the conductor leads the magnetic field by 45°. Then the wave equations are expressed as

$$E_x(z, t) = E_m e^{-z\sqrt{\pi f \mu\sigma}} \cos\left(\omega t - z\sqrt{\pi f \mu\sigma}\right) \tag{3.61}$$

$$H_y(z, t) = \frac{E_m}{\sqrt{\frac{2\pi f \mu}{\sigma}}} e^{-z\sqrt{\pi f \mu\sigma}} \cos\left(\omega t - z\sqrt{\pi f \mu\sigma} - \frac{\pi}{4}\right) \tag{3.62}$$

In the conductive medium, the amplitude of propagation wave $E_m e^{-\alpha z}$ decreases exponentially with distance z. At a distance of $\delta = 1/\alpha$, the amplitude decreases to $1/e$ ($\approx$ 37%) of the initial value. Then δ is defined as the skin depth [4], which is given by for the conductor

$$\delta = \frac{1}{\alpha} = \frac{1}{\sqrt{\pi f \mu\sigma}} \tag{3.63}$$

The higher the frequency, conductivity, and permeability, the smaller the skin depth. In the case of pure cupper ($\sigma = 5.8 \times 10^7$ (S/m), $\mu = 4\pi \times 10^{-7}$ (H/m)), $\delta = 8.53$ mm at 60 Hz and $\delta = 2.1$ μm at 1 GHz. As frequency increases, current flow moves to the surface, resulting in the small skin depth. Skin effect reduces the effective cross section of the conductor and thus increases its effective resistance.

3.3 Reflection and Transmission of Plane Waves

3.3.1 Normal Incidence

As shown in Fig. 3.4, when a plane wave enters normally from medium 1 to medium 2 in a quadrature coordinate space, the difference in impedance causes some to reflect at the boundary and the rest to transmit. The mathematical treatment for this reflection

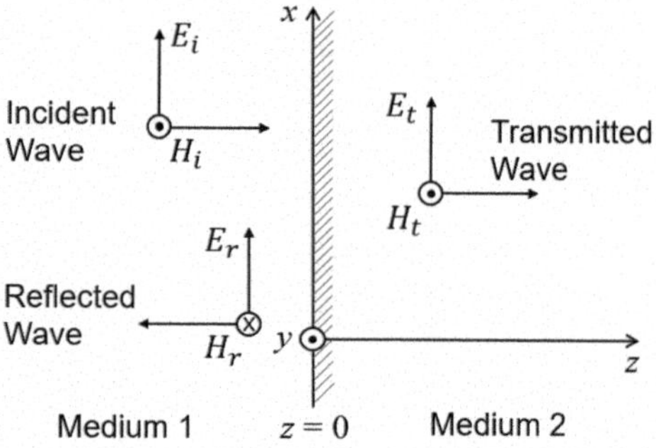

Fig. 3.4 Normal incidence, reflection, and transmission of plane waves

and transmission of plane waves for normal and oblique incidence is described in many text books on electromagnetics [5–7].

The electric and magnetic fields of incident wave (E_i, H_i), reflected wave (E_r, H_r), and transmitted wave (E_t, H_t) are expressed as

$$E_i = E_{i0}e^{-\gamma_1 z}a_x \tag{3.64}$$

$$H_i = H_{i0}e^{-\gamma_1 z}a_y = \frac{E_{i0}}{Z_1}e^{-\gamma_1 z}a_y \tag{3.65}$$

$$E_r = E_{r0}e^{\gamma_1 z}a_x \tag{3.66}$$

$$H_r = -H_{r0}e^{\gamma_1 z}a_y = -\frac{E_{r0}}{Z_1}e^{\gamma_1 z}a_y \tag{3.67}$$

$$E_t = E_{t0}e^{-\gamma_2 z}a_x \tag{3.68}$$

$$H_t = H_{t0}e^{-\gamma_2 z}a_y = \frac{E_{t0}}{Z_2}e^{-\gamma_2 z}a_y \tag{3.69}$$

where γ_1 and γ_2 are the propagation constants, Z_1 and Z_2 are the intrinsic impedances of medium 1 and medium 2, respectively. a_x and a_y are unit vectors to the x-axis and y-axis, respectively.

In medium 1, the incident wave and the reflected wave are combined to form a standing wave ($E_1 = E_i + E_r$), and the transmitted wave in medium 2 becomes a pure travelling wave ($E_2 = E_t$). Applying the boundary condition that the tangential components of electric and magnetic fields must be continuous at the interface ($z = 0$), we obtain

$$E_{i0} + E_{r0} = E_{t0} \tag{3.70}$$

$$H_{i0} - H_{r0} = H_{t0} \tag{3.71}$$

Using the intrinsic impedance of medium 1 and medium 2, Eq. (3.71) becomes

$$\frac{E_{i0}}{Z_1} - \frac{E_{r0}}{Z_1} = \frac{E_{t0}}{Z_2} \tag{3.72}$$

From these equations, the reflection coefficient Γ and the transmission coefficient T can be obtained as

$$\Gamma = \frac{E_{r0}}{E_{i0}} = \frac{Z_2 - Z_1}{Z_2 + Z_1} \tag{3.73}$$

$$T = \frac{E_{t0}}{E_{i0}} = \frac{2Z_2}{Z_2 + Z_1} \tag{3.74}$$

Therefore, the following relation is established between the reflection coefficient and the transmission coefficient:

$$1 + \Gamma = T \tag{3.75}$$

It can be seen that the reflection coefficient is proportional to the impedance difference between the two media, and has a value $0 \le |\Gamma| \le 1$.

For example, if medium 1 is a dielectric ($\sigma_1 = 0, \varepsilon_1, \mu_1$) and medium 2 is a metal ($\sigma_2 = \infty$), it will exhibit complete reflection ($\Gamma = -1$), because $Z_2 = 0$. In medium 1, the incident wave and the reflected wave are combined to form a standing wave, which is expressed as

$$E_1 = E_i + E_r = \left(E_{i0}e^{-\gamma_1 z} + E_{r0}e^{\gamma_1 z}\right)a_x \tag{3.76}$$

$$H_1 = H_i + H_r = \left(\frac{E_{i0}}{Z_1}e^{-\gamma_1 z} - \frac{E_{r0}}{Z_1}e^{\gamma_1 z}\right)a_y \tag{3.77}$$

Substituting $\Gamma = E_{r0}/E_{i0} = -1$, $\sigma_1 = 0$, $\alpha_1 = 0$, and $\gamma_1 = j\beta_1$, we get

$$\begin{aligned} E_1 &= -E_{i0}\left(e^{j\beta_1 z} - e^{-j\beta_1 z}\right)a_x \\ &= -E_{i0}\left[(\cos(\beta_1 z) + j\sin(\beta_1 z)) - (\cos(-\beta_1 z) + j\ \sin(-\beta_1 z))\right]a_x \\ &= -2jE_{i0}\sin(\beta_1 z)a_x \end{aligned} \tag{3.78}$$

$$\begin{aligned} H_1 &= \frac{E_{i0}}{Z_1}\left(e^{-j\beta_1 z} + e^{j\beta_1 z}\right)a_y \\ &= \frac{E_{i0}}{Z_1}\left[(\cos(-\beta_1 z) + j\sin(-\beta_1 z)) + (\cos(\beta_1 z) + j\ \sin(\beta_1 z))\right]a_y \\ &= 2\frac{E_{i0}}{Z_1}\cos(\beta_1 z)a_y \end{aligned} \tag{3.79}$$

Reinserting $e^{j\omega t}$ factor and taking the real part, we obtain the time-varying electric and magnetic fields of the standing wave:

$$E_1(t) = \mathrm{Re}\left(E_1 e^{j\omega t}\right) = 2E_{i0}\sin(\beta_1 z)\sin(\omega t)a_x \tag{3.80}$$

$$H_1(t) = \mathrm{Re}\left(H_1 e^{j\omega t}\right) = \frac{2E_{i0}}{Z_1}\cos(\beta_1 z)\cos(\omega t)a_y \tag{3.81}$$

Figure 3.5a shows the $E_1(t)$ of standing waveform. We see that a point of a particular phase (e.g., point P) oscillates up and down with time. At time $t = 0$, $E_1 = 0$ (waveform ①). At $t = T/4$ (where T denotes the period), $E_1 = 2E_{io}\sin\beta_1 z$ (waveform ②). At $t = T(3/4)$, $E_1 = -2E_{io}\sin\beta_1 z$ (waveform ③). At a distance of $-\beta_1 z = n\pi$ (i.e., $z = -n\lambda/2$) where $n = 0, 1, 2, \cdot\cdot\cdot$, $E_1 = 0$. At $-\beta_1 z = (2n+1)\pi/2$(i.e., $z = -(2n+1)\lambda/4$), E_1 has the maximum amplitude ($\pm 2E_{i0}$). Figure 3.5b shows the $H_1(t)$ waveform with time. ① denotes the waveform at $t = 0$, ② denotes the waveform ($H_1 = 0$) at $t = T/4$, and ③ denotes the waveform at $t = T/2$. As opposed to the electric field, it shows the maximum ($\pm 2E_{i0}/Z_1$) at $-\beta_1 z = n\pi$ (i.e., $z = -n\lambda/2$) and $H_1 = 0$ at $-\beta_1 z = (2n+1)\pi/2$(i.e., $z = -(2n+1)\lambda/4$).

At the points of $z = -(2n+1)\lambda/4$, the electric field is maximum and the magnetic field is minimum, at which the impedance is at its maximum, corresponding to the high-impedance surface of the short-circuited transmission line (see Fig. 2.16).

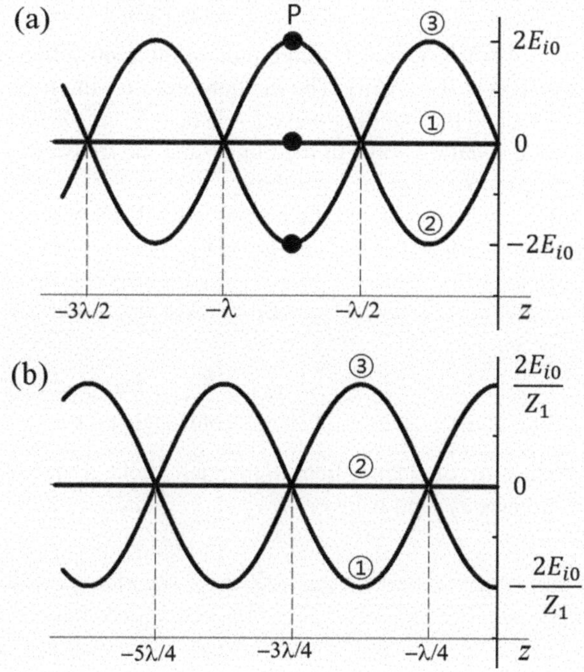

Fig. 3.5 Representation of standing wave in front of a perfect reflector: **a** electric field and **b** magnetic field

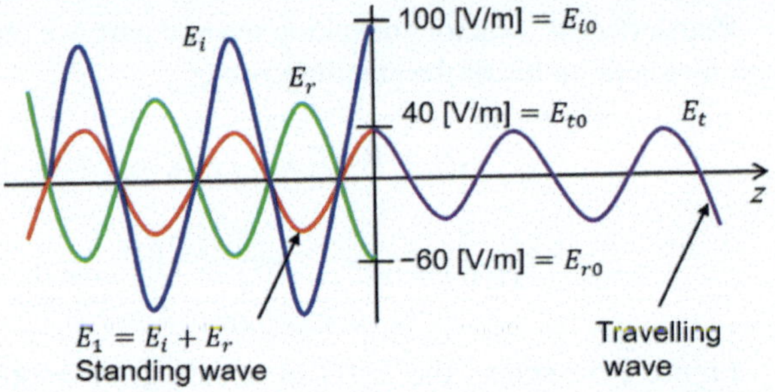

Fig. 3.6 Illustration of reflection and transmission of electromagnetic wave, showing the incident field E_i, reflected field E_r, transmitted field E_t, and standing wave $E_1 = E_i + E_r$

Providing resistance by film coating on this surface or utilizing the intrinsic material loss at this layer thickness can make the impedance-matched EM wave absorbers, which will be explained in Chap. 5. Conversely, at the points of $z = -n\lambda/2$, the electric field is zero and the magnetic field is maximum, producing the zero-impedance surface.

Another example is the incidence of electromagnetic waves with an electric field with magnitude $E_{i0} = 100$ (V/m) from the lossless medium 1 ($\alpha_1 = 0, Z_1 = 400\ \Omega$) to the lossless medium 2 ($\alpha_2 = 0, Z_2 = 100\ \Omega$). It can be calculated that (a) reflection coefficient: $\Gamma = (100 - 400)/(100 + 400) = -0.6$, (b) magnitude of the reflected electric field: $E_{r0} = \Gamma E_{i0} = -0.6 \times 100 = -60$ (V/m), (c) magnitude of the incident magnetic field: $H_{i0} = E_{i0}/Z_1 = 100/400 = 0.25$ (A/m), (d) magnitude of the reflected magnetic field: $H_{r0} = -E_{r0}/Z_1 = -(-60/400) = 0.15$ (A/m), (e) magnitude of the transmitted electric field: $E_{t0} = TE_{i0} = [(2 \times 100)/(100 + 400)] \times 100 = 40$ (V/m), and (f) magnitude of the transmitted magnetic field: $H_{t0} = E_{t0}/Z_2 = 40/100 = 0.4$ (A/m). The incident field E_i, the reflected field E_r, the transmitted field E_t, and the standing wave $E_1 = E_i + E_r$ are sketched in Fig. 3.6. It is to be noted that the standing wave in medium 1 and the transmitted travelling wave in medium 2 are continuous at the boundary.

The electric field and the magnetic field at a point $z = -l$ are the sum of incident wave and reflected wave and given by

$$E_1 = E_{i0}\left(e^{-j\beta_1 l} + \Gamma e^{j\beta_1 l}\right) \tag{3.82}$$

$$H_1 = \frac{E_{i0}}{Z_1}\left(e^{-j\beta_1 l} - \Gamma e^{j\beta_1 l}\right) \tag{3.83}$$

The ratio of electric field to magnetic field at the point $z = -l$ is defined as input impedance Z_{in} and is given by

$$Z_{in} = \frac{E_1}{H_1}\Big|_{z=-l} = Z_1 \frac{e^{-j\beta_1 l} + \Gamma e^{j\beta_1 l}}{e^{-j\beta_1 l} - \Gamma e^{j\beta_1 l}} \tag{3.84}$$

Substituting Eq. (3.73) into Eq. (3.84) and using Euler's formula ($e^{\pm j\beta_1 l} = \cos\beta_1 l \pm j\sin\beta_1 l$), we obtain

$$Z_{in} = Z_1 \frac{Z_2 + jZ_1 \tan \beta_1 l}{Z_1 + jZ_2 \tan \beta_1 l} \tag{3.85}$$

If medium 2 is a metal ($Z_2 = 0$),

$$Z_{in} = jZ_1 \tan \beta_1 l \tag{3.86}$$

This equation is identical to that for the short-circuited transmission line (Eq. (2.99)), and shows the same pure reactance variation with distance l (see Fig. 2.16). At $l = (2n+1)\lambda/4$ (where $n = 0, 1, 2, \cdots$), $Z_{in} = \pm\infty$, so it forms a high-impedance surface. At $l = n\lambda/2$, $Z_{in} = 0$, and a low-impedance surface is resulted.

In medium 1, the ratio of the maximum and the minimum values of the electric field, or magnetic field is called the standing wave ratio (SWR) s, and is given by

$$s = \frac{|E_1|_{max}}{|E_1|_{min}} = \frac{|H_1|_{max}}{|H_1|_{min}} = \frac{1+|\Gamma|}{1-|\Gamma|} = \frac{|E_{i0}+E_{r0}|}{|E_{i0}-E_{r0}|} \tag{3.87}$$

If $|\Gamma| = 0$ (non-reflection), $s = 1$. If $|\Gamma| = 1$ (complete reflection), $s = \infty$. Solving the Eq. (3.87) for the magnitude of reflection coefficient gives

$$|\Gamma| = \frac{s-1}{s+1} \tag{3.88}$$

$|\Gamma|$ can be obtained by measuring s.

3.3.2 *Oblique Incidence*

When dealing with wave reflection at an oblique incidence, it is divided into parallel polarization and perpendicular polarization according to the orientation of the electric field relative to the incidence plane. The plane of incidence is defined as a plane consisting of a unit normal vector with respect to the boundary and a propagation vector of the incident wave. When an electric field lies on the plane of incidence, we call it parallel polarization. In this case, it is also called the transverse magnetic (TM) polarization because the magnetic field is parallel to the interface. In the opposite case, where the electric field is perpendicular to the plane of incidence, we call it perpendicular polarization. It is also known as the transverse electrical (TE) polarization because the electric field is parallel to the interface.

Parallel (TM) polarization is illustrated in Fig. 3.7 where the electric field lies on x–z plane of incidence. The angle of incidence, the angle of reflection, and the angle

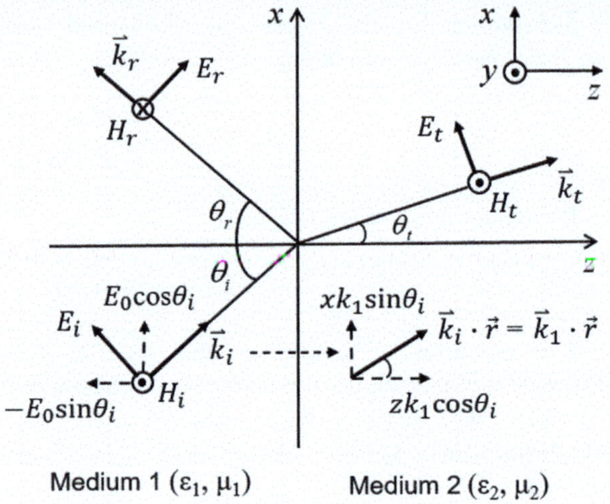

Fig. 3.7 Oblique incidence of plane wave for parallel (TM) polarization

of transmission are referred to θ_i, θ_r, and θ_t, respectively. The normal and tangential components of incident wave E_i are shown in Fig. 3.7 and expressed as

$$E_{i0} = E_0(\cos\theta_i a_x - \sin\theta_i a_z) \tag{3.89}$$

Propagation of incident wave is also shown in Fig. 3.7 and is expressed as

$$\begin{aligned}\overrightarrow{k_i} \cdot \vec{r} = \overrightarrow{k_1} \cdot \vec{r} &= (k_1\sin\theta_i a_x + 0 + k_1\cos\theta_i a_z) \cdot \left(xa_x + ya_y + za_z\right) \\ &= xk_1\sin\theta_i + zk_1\cos\theta_i\end{aligned} \tag{3.90}$$

where $\overrightarrow{k_i}$ is the propagation vector and $\vec{r}$ is the position vector. k_1 is the wave number in medium 1, given by $k_1 = \omega\sqrt{\mu_1\varepsilon_1}$. a_x, a_y, a_z are unit vectors to the axis x, y, z, respectively.

Then the electric field and the magnetic field of incident wave can be expressed as [5]

$$E_i = E_{i0}e^{-j\overrightarrow{k_i}\cdot\vec{r}} = E_0(\cos\theta_i a_x - \sin\theta_i a_z)e^{-jk_1(x\sin\theta_i + z\cos\theta_i)} \tag{3.91}$$

$$H_i = H_{i0}e^{-j\overrightarrow{k_i}\cdot\vec{r}} = \frac{E_0}{Z_1}e^{-jk_1(x\sin\theta_i + z\cos\theta_i)}a_y \tag{3.92}$$

where Z_1 is the wave impedance in medium 1, given by $Z_1 = \sqrt{\mu_1/\varepsilon_1}$. In a same way, the reflected field and the transmitted field can be expressed as

$$E_r = E_0\Gamma(\cos\theta_r a_x + \sin\theta_r a_z)e^{-jk_1(x\sin\theta_r - z\cos\theta_r)} \tag{3.93}$$

$$H_r = -\frac{E_0\Gamma}{Z_1} e^{-jk_1(x\sin\theta_r - z\cos\theta_r)} a_y \quad (3.94)$$

$$E_t = E_0 T(\cos\theta_t a_x - \sin\theta_t a_z) e^{-jk_2(x\sin\theta_t + z\cos\theta_t)} \quad (3.95)$$

$$H_t = \frac{E_0 T}{Z_2} e^{-jk_2(x\sin\theta_t + z\cos\theta_t)} a_y \quad (3.96)$$

where k_2 is a wave number and Z_2 is a wave impedance of medium 2, given by $k_2 = \omega\sqrt{\mu_2\varepsilon_2}$ and $Z_2 = \sqrt{\mu_2/\varepsilon_2}$, respectively.

Since the tangential components of E and H must be continuous at the boundary $z = 0$, we obtain

$$E_i(z = 0) + E_r(z = 0) = E_t(z = 0) \quad (3.97)$$

$$H_i(z = 0) + H_r(z = 0) = H_t(z = 0) \quad (3.98)$$

Applying the boundary conditions leads to

$$\begin{aligned} E_0(\cos\theta_i a_x) e^{-jk_1(x\sin\theta_i)} &+ E_0\Gamma(\cos\theta_r a_x) e^{-jk_1(x\sin\theta_r)} \\ &= E_0 T(\cos\theta_t a_x) e^{-jk_2(x\sin\theta_t)} \end{aligned} \quad (3.99)$$

$$\frac{1}{Z_1} e^{-jk_1(x\sin\theta_i)} - \frac{\Gamma}{Z_1} e^{-jk_1(x\sin\theta_r)} = \frac{T}{Z_2} e^{-jk_2(x\sin\theta_t)} \quad (3.100)$$

For all x values, E_x and H_y must be continuous at $z = 0$, so that the following phase matching conditions are established:

$$k_1 \sin\theta_i = k_1 \sin\theta_r = k_2 \sin\theta_t \quad (3.101)$$

This is called the Snell's law of reflection and refraction. It is clear that the angle of reflection θ_r equals the angle of incidence θ_i:

$$\theta_r = \theta_i \quad (3.102)$$

Applying the Snell's law to Eq. (3.99) and Eq. (3.100) yields

$$\Gamma = \frac{Z_2\cos\theta_t - Z_1\cos\theta_i}{Z_2\cos\theta_t + Z_1\cos\theta_i} \quad (3.103)$$

$$T = \frac{2Z_2\cos\theta_i}{Z_2\cos\theta_t + Z_1\cos\theta_i} \quad (3.104)$$

These are called the Fresnel's equations. Using the relation between the angle of refraction θ_t and the angle of incidence θ_i,

$$\sin\theta_t = \frac{k_1}{k_2}\sin\theta_i = \frac{\sqrt{\mu_1\varepsilon_1}}{\sqrt{\mu_2\varepsilon_2}}\sin\theta_i \tag{3.105}$$

$$\cos\theta_t = \sqrt{1-\sin^2\theta_t} = \sqrt{1-\frac{\mu_1\varepsilon_1}{\mu_2\varepsilon_2}\sin^2\theta_i} \tag{3.106}$$

we obtain the reflection coefficient and the transmission coefficient expressed as a function of angle of incidence θ_i:

$$\Gamma = \frac{\sqrt{\frac{\mu_2}{\varepsilon_2}}\sqrt{1-\frac{\mu_1\varepsilon_1}{\mu_2\varepsilon_2}\sin^2\theta_i} - \sqrt{\frac{\mu_1}{\varepsilon_1}}\cos\theta_i}{\sqrt{\frac{\mu_2}{\varepsilon_2}}\sqrt{1-\frac{\mu_1\varepsilon_1}{\mu_2\varepsilon_2}\sin^2\theta_i} + \sqrt{\frac{\mu_1}{\varepsilon_1}}\cos\theta_i} \tag{3.107}$$

$$T = \frac{2\sqrt{\frac{\mu_2}{\varepsilon_2}}\cos\theta_i}{\sqrt{\frac{\mu_2}{\varepsilon_2}}\sqrt{1-\frac{\mu_1\varepsilon_1}{\mu_2\varepsilon_2}\sin^2\theta_i} + \sqrt{\frac{\mu_1}{\varepsilon_1}}\cos\theta_i} \tag{3.108}$$

From Eq. (3.103), it is possible $\Gamma = 0$ if the numerator is zero. The incident angle at which this takes place is called the Brewster angle θ_b, which can be obtained using

$$Z_2\cos\theta_t = Z_1\cos\theta_b \tag{3.109}$$

Applying Snell's law ($k_1\sin\theta_i = k_2\sin\theta_t$), we get

$$\sin\theta_b = \sqrt{\frac{1-(\frac{Z_2}{Z_1})^2}{1-(\frac{Z_2}{Z_1})^2(\frac{k_1}{k_2})^2}} = \sqrt{\frac{1-\frac{\mu_2\varepsilon_1}{\mu_1\varepsilon_2}}{1-(\frac{\varepsilon_1}{\varepsilon_2})^2}} \tag{3.110}$$

For the non-magnetic medium ($\mu_1 = \mu_2 = \mu_0$), it is expressed as

$$\sin\theta_b = \sqrt{\frac{\varepsilon_2}{\varepsilon_1+\varepsilon_2}} \tag{3.111}$$

Perpendicular (TE) polarization is the case when the electric field is perpendicular to the plane of incidence (x–z plane), as illustrated in Fig. 3.8. The electric field is then parallel to the interface. The incident and reflected fields in medium 1 are given by [5]

$$E_i = E_0 e^{-jk_1(x\sin\theta_i + z\cos\theta_i)} a_y \tag{3.112}$$

$$H_i = \frac{E_0}{Z_1}(-\cos\theta_i a_x + \sin\theta_i a_z)e^{-jk_1(x\sin\theta_i + z\cos\theta_i)} \tag{3.113}$$

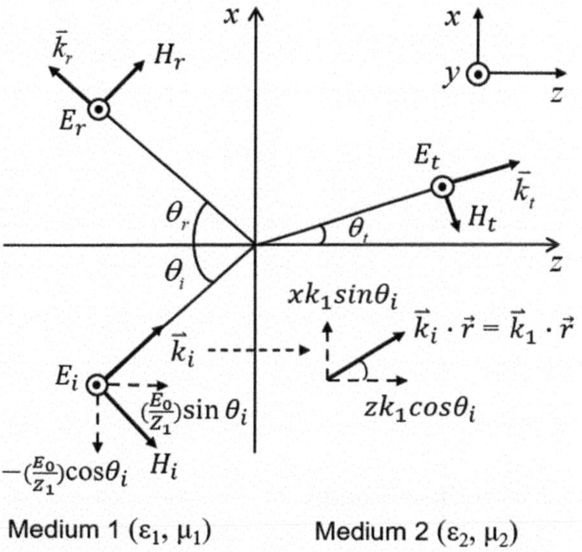

Fig. 3.8 Oblique incidence of plane wave for perpendicular (TE) polarization

$$E_r = E_0\Gamma\, e^{-jk_1(x\sin\theta_r - z\cos\theta_r)}a_y \tag{3.114}$$

$$H_r = \frac{E_0\Gamma}{Z_1}(\cos\theta_r a_x + \sin\theta_r a_z)e^{-jk_1(x\sin\theta_r - z\cos\theta_r)} \tag{3.115}$$

The transmitted fields in medium 2 are expressed as

$$E_t = E_0 T e^{-jk_2(x\sin\theta_t + z\cos\theta_t)}a_y \tag{3.116}$$

$$H_t = \frac{E_0 T}{Z_2}(-\cos\theta_t a_x + \sin\theta_t a_z)e^{-jk_2(x\sin\theta_t + z\cos\theta_t)} \tag{3.117}$$

Applying the boundary condition that the tangential components of E and H are continuous at $z = 0$ and Snell's law ($\theta_r = \theta_i$), we obtain the reflection coefficient Γ and the transmission coefficient T:

$$\Gamma = \frac{Z_2\cos\theta_i - Z_1\cos\theta_t}{Z_2\cos\theta_i + Z_1\cos\theta_t} = \frac{\sqrt{\frac{\mu_2}{\varepsilon_2}}\cos\theta_i - \sqrt{\frac{\mu_1}{\varepsilon_1}}\cos\theta_t}{\sqrt{\frac{\mu_2}{\varepsilon_2}}\cos\theta_i + \sqrt{\frac{\mu_1}{\varepsilon_1}}\cos\theta_t} \tag{3.118}$$

$$T = \frac{2Z_2\cos\theta_i}{Z_2\cos\theta_i + Z_1\cos\theta_t} = \frac{2\sqrt{\frac{\mu_2}{\varepsilon_2}}\cos\theta_i}{\sqrt{\frac{\mu_2}{\varepsilon_2}}\cos\theta_i + \sqrt{\frac{\mu_1}{\varepsilon_1}}\cos\theta_t} \tag{3.119}$$

These are the Fresnel's equations for perpendicular (TE) polarization. Using the relation of Eq. (3.106), we can express Γ and T as a function of angle of incidence θ_i:

$$\Gamma = \frac{\sqrt{\frac{\mu_2}{\varepsilon_2}}\cos\theta_i - \sqrt{\frac{\mu_1}{\varepsilon_1}}\sqrt{1 - \frac{\mu_1\varepsilon_1}{\mu_2\varepsilon_2}\sin^2\theta_i}}{\sqrt{\frac{\mu_2}{\varepsilon_2}}\cos\theta_i + \sqrt{\frac{\mu_1}{\varepsilon_1}}\sqrt{1 - \frac{\mu_1\varepsilon_1}{\mu_2\varepsilon_2}\sin^2\theta_i}} \tag{3.120}$$

$$T = \frac{2\sqrt{\frac{\mu_2}{\varepsilon_2}}\cos\theta_i}{\sqrt{\frac{\mu_2}{\varepsilon_2}}\cos\theta_i + \sqrt{\frac{\mu_1}{\varepsilon_1}}\sqrt{1 - \frac{\mu_1\varepsilon_1}{\mu_2\varepsilon_2}\sin^2\theta_i}} \tag{3.121}$$

3.4 Further Reading

The topics in this chapter, including sinusoidal plane wave and its reflection and transmission for normal and oblique incidence, are presented in greater detail in advanced textbooks on electromagnetics [5–8] and in the public domain [9, 10].

References

1. Electromagnetic wave equation on Wikipedia. https://en.wikipedia.org/wiki/Electromagnetic_wave_equation
2. Maxwell's equations on Wikipedia. https://en.wikipedia.org/wiki/Maxwell%27s_equations
3. Ponting vector on Wikipedia. https://en.wikipedia.org/wiki/Poynting_vector
4. Skin effect on Wikipedia. https://en.wikipedia.org/wiki/Skin_effect
5. M.N.O. Sadiku, *Elements of Electromagnetics*, 2nd edn. (Oxford University Press, New York, 1995), pp.453–517
6. W.H. Hayt Jr., *Engineering Electromagnetics*, 5th edn. (McGraw Hill, New York, 1989), pp.336–380
7. J.D. Kraus, *Electromagnetics*, 4th edn. (McGraw–Hill, New York, 1992), pp.609–617
8. D.M. Pozar, *Microwave Engineering*, 4th edn. (Wiley, USA, 2011), pp.6–63
9. Sinusoidal plane wave on Wikipedia. https://en.wikipedia.org/wiki/Sinusoidal_plane_wave
10. Fresnel equations on Wikipedia. https://en.wikipedia.org/wiki/Fresnel_equations

Chapter 4
Complex Permittivity and Permeability

The most basic material parameters for the design of EM wave absorbers are dielectric constant and dielectric loss, permeability and magnetic loss. These material parameters are expressed in complex numbers and are referred to as complex permittivity and complex permeability. This chapter is about the dielectric polarization and magnetization of materials at the atomic and molecular levels, and the associated high-frequency behavior is explained, including material loss. Dielectric polarization mechanisms (electronic polarization, ionic polarization, orientation polarization, space charge polarization) are described, with the addition of domain polarization in ferroelectrics and defect-dipole polarization in ionic solids. Debye relaxation and dielectric resonance are described, which are the two main mechanisms for the electromagnetic absorption in dielectrics. Magnetization is the process of domain wall movement and magnetic dipole rotation. The magnetic loss mechanisms generated in this process are divided into domain wall resonance, gyromagnetic resonance, and magnetic relaxation. In each case, the theoretical derivation of complex permeability is described, and the dominant loss mechanisms depending on the materials and frequency are discussed.

4.1 Dielectric Polarization

Dielectric polarization refers to the phenomenon in which when an electric field is applied to a dielectric, the displacement or movement of charge or the orientation (rotation) of dipole moment occurs in the medium [1, 2]. Due to the phenomenon of dielectric polarization, the bound charge is created on both surfaces of dielectrics. Four major types of polarization occur in dielectrics: (a) electronic polarization due to a displacement of the negatively charged electron cloud relative to the positive nucleus of an atom, (b) ionic polarization by displacement of cations and anions in opposite directions in ionic materials, (c) orientation polarization by rotation or

S. Kim, *Fundamentals and Principles of Electromagnetic Wave Absorbers*, Springer Series in Advanced Microelectronics 66, https://doi.org/10.1007/978-981-97-6477-8_4

alignment of permanent electric dipoles, and (d) interfacial or space change polarization due to the accumulation or depletion of free charges at the interphases in a multi-phase dielectric material.

(1) **Electronic Polarization**

Electronic polarization is caused by the displacement of electron cloud surrounding an atomic nucleus. In the absence of an external field ($E = 0$), the centers of positive charge and negative charge coincide, as shown in Fig. 4.1a, so there is no dipole moment. When placed in an external electric field, as shown in Fig. 4.1b, the nucleus and electron cloud move in opposite directions to create a dipole moment.

This type of polarization is common in all atoms. The magnitude of this moment is very small and thus the strength of polarization is weak. However, the electronic polarization has a very high response rate to the external field, which can occur up to an AC field of 10^{16} Hz (visible light band).

(2) **Ionic Polarization**

When an ionic material is placed in electric field, an electric dipole is induced by the relative shift of positive (+) ions and negative (−) ions. Using NaCl ionic crystals as an example, as shown in Fig. 4.2a, Coulomb attractions are balanced between Na^+ ions and Cl^- ions in the absence of an external electric field ($E = 0$). Since the ions are at equilibrium states, there is no dipole moment as a whole.

When the field E is applied (Fig. 4.2b), the Cl^- ion moves from the equilibrium position in the opposite direction to the field. At this time, the restoring force to its

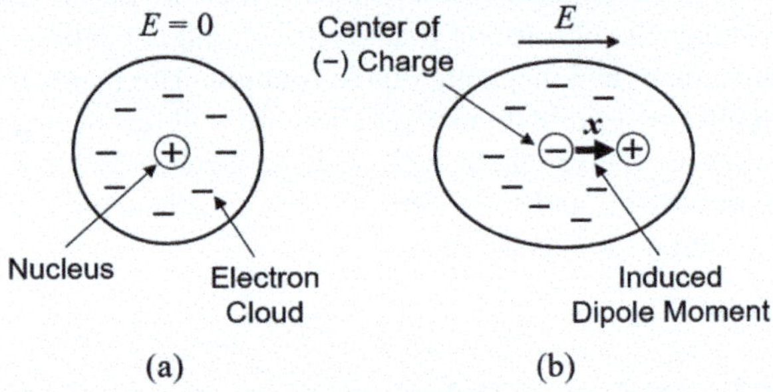

Fig. 4.1 Electronic polarization of atoms

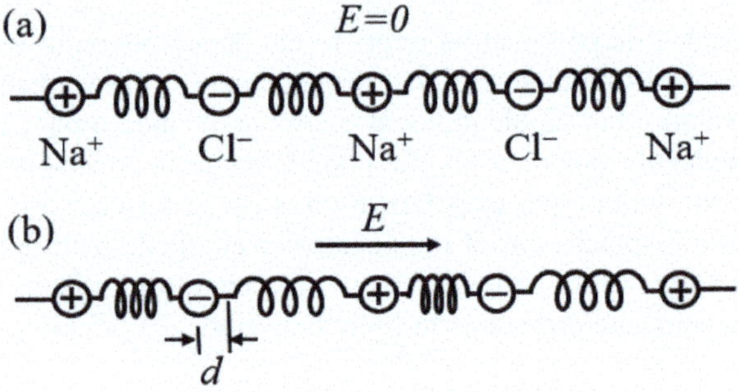

Fig. 4.2 Ionic polarization in NaCl crystals

original state is in equilibrium with electrostatic force acting by the external field. Since the displacement is d and the charge is e, the dipole moment caused by the ionic polarization is $\mu_i = ed$. It appears in ceramic compounds, which are ion-bonding materials, and may have a high dielectric constant due to the large displacement, depending on the structure. This type of polarization occurs up to an AC field of 10^{12} Hz (infrared region).

(3) Dipolar and Orientation Polarization

The polarization caused by orientation of polar molecules in an applied electric field is called the dipole orientation polarization. Orientation polarization appears in materials that possess a permanent dipole, such as H_2O, HCl. In the collection of permanent dipoles, the direction of dipoles is random in the absence of an external electric field ($E = 0$) and the polarization will be zero as a whole (Fig. 4.3a). When a peripheral electric field is applied, the dipoles rotate in the direction of the field, resulting in polarization (Fig. 4.3b). The magnitude of the polarization increases in proportion to the applied electric field, but decreases with the increase in temperature due to the thermal equilibrium of molecules.

The rotation of permanent dipoles is easy in liquid or gaseous states, but difficult to occur in solids. In some ionic crystals or glasses, however, a pair of charged defects may behave like a permanent dipole and exhibits a similar result to the orientation polarization [3, 4]. This is called the ion jump polarization because it involves the movement of ions, as illustrated in Fig. 4.4. In the liquid or gaseous state, the dipole orientation reacts to alternating electric fields up to 10^{10} Hz (microwave region), but the ion-jumping polarization in solids is a slow process and is observed at frequencies as low as 10^3 Hz. Due to the thermally activating and diffusional process, the higher the temperature and the smaller the moving ions, the faster the ion jump polarization.

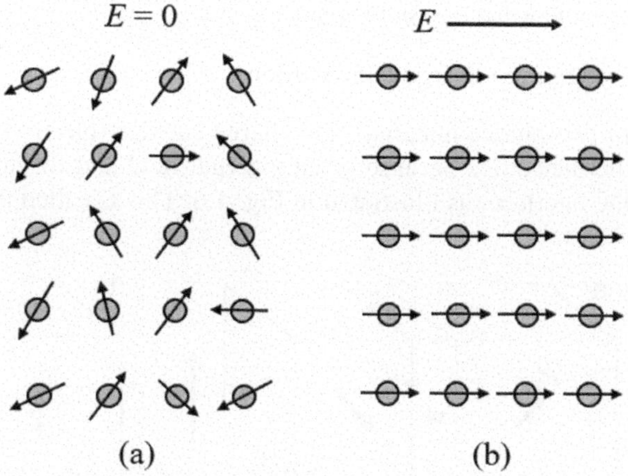

Fig. 4.3 Orientation polarization of the permanent dipoles

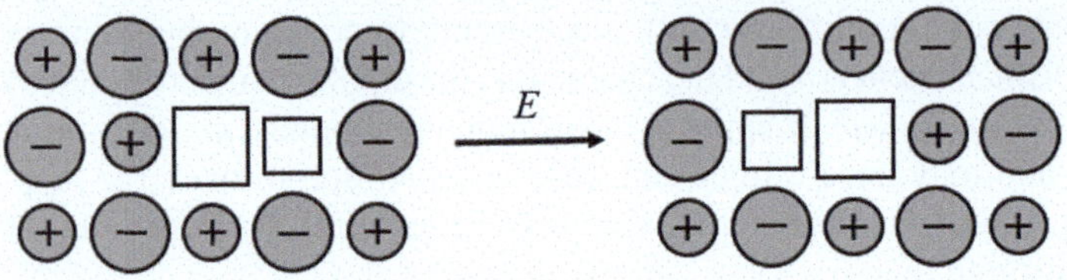

Fig. 4.4 Ion jump polarization in solid crystals

(4) **Ferroelectric Polarization**

Ferroelectrics are dielectric materials that possess regions with spontaneous polarization. Barium titanate or lead titanate is the typical ferroelectric ceramics with its origin of spontaneous polarization concerned with the relative displacement of the titanium ion to the oxygen octahedral cage. The regions with uniform-oriented polarization are called the ferroelectric domains, and the interfaces between them are called the domain walls. The dielectric response of ferroelectrics is strongly influenced by the configuration of domains and the mobility of domain walls [5, 6].

Domains in ferroelectrics can be categorized into two types, i.e., 90 and 180° ferroelectric domains [7, 8], as shown in Fig. 4.5. In the absence of an external electric field, all the domains have the same spontaneous polarization P_S, but oriented in opposite direction such that no net polarization is present as a whole. Under the action of an electric field E, the domain wall is displaced a distance l from its equilibrium position and produces the polarization. This is called the domain wall movement polarization, and it is the dominant polarization mechanism in ferroelectric materials. In a weak alternating electric field, the domain walls perform small oscillations around their equilibrium positions. When the domain wall oscillates with the same frequency of the external field, a resonance or relaxation occurs and its frequency is in microwave region [9, 10].

(5) **Space Charge Polarization**

In two-phases materials, the charge carriers (electrons, holes, ions) migrate inside the dielectrics because of an external field and forming a spatial charge region at the interface, as illustrated in Fig. 4.6. This is called the interfacial or space charge

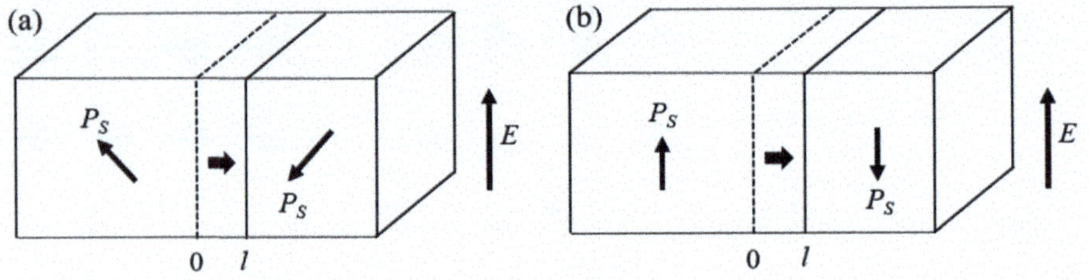

Fig. 4.5 The domain wall is displaced a distance l from its equilibrium position due to the action of external field E: **a** 90° domains and **b** 180° domains

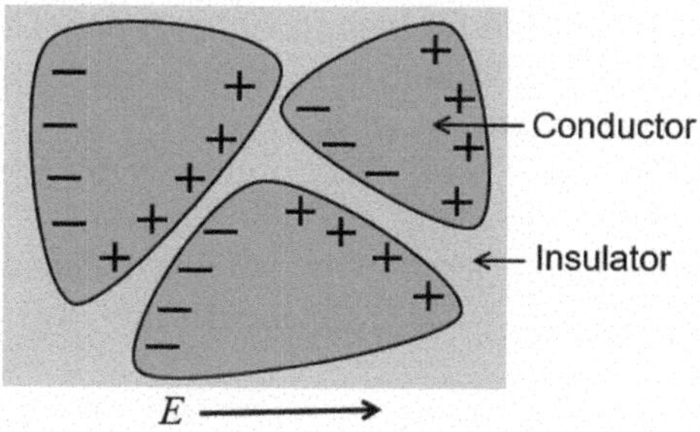

Fig. 4.6 Space charge polarization in two-phases materials

polarization. This type of polarization appears in multi-phase materials with greatly different dielectric constant or electrical conductivity, and also occurs when charge carriers accumulate at grain boundaries, interfaces, or specimen surfaces of electrodes. Compared to the other polarization mechanisms, this polarization is much bigger and reacts to AC electric fields at low frequencies up to 10^4 Hz.

It has been reported that the interfacial polarization can occur in the heterostructure of microscopic scale, which was reviewed in Ref. [11]. Defect sites in dielectrics can trap charge carriers and break the balance of charge distribution, leading to the dielectric polarization. The common defect sites are point defects including atomic vacancies and heteroatoms doping [12–15]. Charge carriers generated due to the absence of atoms in lattice points will be trapped at vacancies and produce the electric dipoles. Similarly, doping of atoms with different valance state may change the charge distribution and lead to the formation of electric dipoles. Microscopic heterojunction at the interface of metal–semiconductor in multi-shell materials also provides the sites of spatial charge region, which promotes the interfacial polarization [16–18]. A representative example can be found in the yolk–shell material of Ni@C@ZnO, where Schottky barrier is formed at the interface of conductive Ni@C core and ZnO semiconductor shell [16].

4.2 Complex Permittivity

When an electric field is applied to a substance, the bound charges in the material cause electric polarization, and the amounts of charge that the medium can store is proportional to this polarization. The dielectric permittivity of a substance is a physical unit representing an electrical polarization, and more charges can be stored in proportion to this value. In electromagnetism, an electric displacement field *D* is defined as the number of electric flux lines passing through a unit area. *D* is proportional to the applied electric field *E*, and the proportional constant is the dielectric permittivity ε, which is given by

$$D = \varepsilon E \tag{4.1}$$

The SI unit for permittivity is (F/m). The unit of the electric displacement D is (C/m^2), and the unit of the electric field E is (V/m).

The vacuum permittivity ε_0 is the ratio D/E in free space and has the value $\varepsilon_0 = 8.854 \times 10^{-12}$ F/m. The permittivity of a homogeneous material is usually given in a relative value to that of free space, which is often referred to as a dielectric constant. The absolute permittivity ε can be obtained by multiplying the relative permittivity (dielectric constant) ε_r by the permittivity of vacuum ε_0,

$$\varepsilon = \varepsilon_0 \varepsilon_r \tag{4.2}$$

In general, the response of dielectric materials to external fields (material's polarization) does not change instantaneously when an electric field is applied. There can be a phase delay δ between polarization and external field. Thus, the dielectric permittivity is not just a constant, but shows a dispersion with respect to the angular frequency ω of the external field and is expressed as a complex number:

$$\varepsilon(\omega) = \frac{D_0}{E_0} e^{-j\delta} = |\varepsilon(\omega)| e^{-j\delta} = |\varepsilon(\omega)|(\cos\delta - j\sin\delta) \tag{4.3}$$

where D_0 and E_0 represent the amplitude of displacement and electric field, respectively, and $j = \sqrt{-1}$ is an imaginary unit. We call this complex permittivity as a result of the response of materials to alternating applied fields and expressed as

$$\varepsilon(\omega) = \varepsilon'(\omega) - j\varepsilon''(\omega) \tag{4.4}$$

In the above equation, ε' is the real part and ε'' is the imaginary part of the complex permittivity. The real part is associated with energy storage and the imaginary part is associated with energy loss by the medium.

Substances can be classified into dielectrics and metals according to the magnitude of dielectric constant ε and conductivity σ of the material. In the case of metals, $\sigma/\omega\varepsilon \gg 1$, and, in the case of dielectrics, $\sigma/\omega\varepsilon \ll 1$. A common material does not belong to the extremes of these quantities. A perfect dielectric is a material in which $\sigma = 0$ and only displacement currents flow. Thus, the material only stores and discharges the electrical energy like a capacitor. If the conduction current in a medium is not negligible ($\sigma > 0$), the total current density is given by

$$J_{tot} = J_c + J_d = \sigma E + j\omega\varepsilon'(\omega)E = j\omega\varepsilon(\omega)E \tag{4.5}$$

where, J_c and J_d are conduction current density and displacement current density, respectively. If there is no change in the applied field, that is, if the frequency is zero, the displacement current does not flow. In addition, $\varepsilon(\omega)$ is complex permittivity and expressed as

$$\varepsilon(\omega) = \varepsilon' - j\frac{\sigma}{\omega} \tag{4.6}$$

4.3 Dielectric Absorption

In general, the process of electromagnetic wave absorption by dielectric materials takes place in various mechanisms, such as Debye relaxation and dielectric resonance, depending on the frequency.

4.3.1 Debye Relaxation

In a material in which a permanent dipole is present, dielectric polarization occurs by the rotation of dipoles. If the frequency of applied field is low enough, the dipoles can always reach equilibrium with the oscillating field. However, if the viscosity of medium is large enough that the dipole cannot keep up with the applied field, the energy of the field is absorbed and lost. This mechanism by which the dipole is relaxed is called the dielectric relaxation, and in the case of ideal (non-interacting) dipoles, it can be described by the classical Debye relaxation process [1, 19].

The easiest way to explain the relaxation phenomenon is to interpret the change in dipole moment or polarization when the driving force (electric field) is suddenly applied to the state of random distribution of dipoles. As illustrated in Fig. 4.7, instantaneous application of the external electric field E_0 will induce the polarization P_∞ from 0 by dipole orientation. However, it does not occur immediately, but it takes a finite time interval before the polarization attains its maximum value. This can be expressed as the following exponential law [20]:

$$P(t) = P_\infty\left(1 - e^{-t/\tau}\right) \tag{4.7}$$

where τ is called the relaxation time which is required to take the polarization to reach $1 - e^{-1} = 0.632$ of the equilibrium value P_∞.

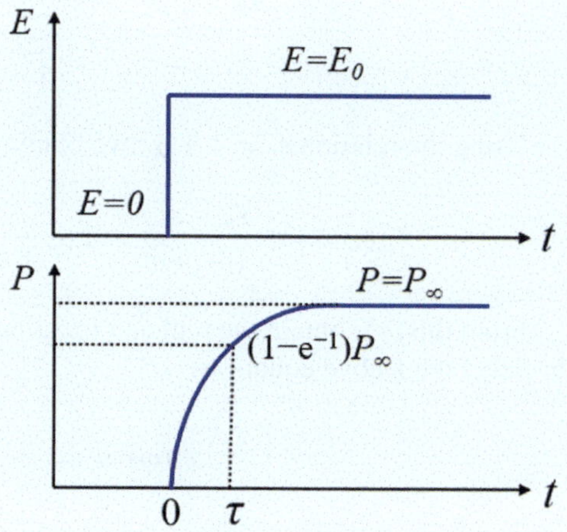

Fig. 4.7 Relaxation of dipole moments (polarization) with field application

Differentiating Eq. (4.7), we obtain the rate of polarization buildup:

$$\frac{\mathrm{d}P(t)}{\mathrm{d}t} = \frac{P_\infty \mathrm{e}^{-t/\tau}}{\tau} \tag{4.8}$$

Substituting Eq. (4.7) into Eq. (4.8) yields

$$\frac{\mathrm{d}P(t)}{\mathrm{d}t} = \frac{P_\infty - P(t)}{\tau} \tag{4.9}$$

When applying an alternating electric field $E = E_0\mathrm{e}^{j\omega t}$, the polarization is expressed as

$$P(t) = \varepsilon_0 \chi E = \varepsilon_0 \chi E_0 \mathrm{e}^{j\omega t} \tag{4.10}$$

where χ is the electric susceptibility. The maximum polarization after sufficient time to reach equilibrium is given by [20]

$$P_\infty = \varepsilon_0(\varepsilon_\mathrm{s} - \varepsilon_\infty)E \tag{4.11}$$

where ε_s and ε_∞ are the dielectric constants at DC ($\omega = 0$) and at infinite frequency ($\omega = \infty$), respectively. Substituting Eq. (4.11) into Eq. (4.9), we obtain

$$\frac{\mathrm{d}P(t)}{\mathrm{d}t} = \frac{1}{\tau}\left[\varepsilon_0(\varepsilon_\mathrm{s} - \varepsilon_\infty)E_0\mathrm{e}^{j\omega t} - \varepsilon_0 \chi E_0 \mathrm{e}^{j\omega t}\right] \tag{4.12}$$

This is the same as the differential expression of Eq. (4.10), so we get

$$\chi j\omega\tau = (\varepsilon_\mathrm{s} - \varepsilon_\infty) - \chi \tag{4.13}$$

Therefore, the susceptibility is given by

$$\chi = \frac{\varepsilon_\mathrm{s} - \varepsilon_\infty}{1 + j\omega\tau} \tag{4.14}$$

Using the relation $\varepsilon_\mathrm{r} = 1 + \chi$, we obtain the relative permittivity:

$$\varepsilon_\mathrm{r} = 1 + \frac{\varepsilon_\mathrm{s} - \varepsilon_\infty}{1 + j\omega\tau} = 1 + \frac{\varepsilon_\mathrm{s} - \varepsilon_\infty}{1 + \omega^2\tau^2} + j\frac{(\varepsilon_\mathrm{s} - \varepsilon_\infty)\omega\tau}{1 + \omega^2\tau^2} \tag{4.15}$$

From this, we obtain the real and imaginary parts of complex permittivity, which is called the Debye equations:

$$\varepsilon_\mathrm{r}'(\omega) = \varepsilon_\infty + \frac{\varepsilon_\mathrm{s} - \varepsilon_\infty}{1 + (\omega\tau)^2} \tag{4.16}$$

$$\varepsilon_r''(\omega) = \frac{\omega\tau(\varepsilon_s - \varepsilon_\infty)}{1 + (\omega\tau)^2} \tag{4.17}$$

This is shown in Fig. 4.8. At low frequencies ($\omega\tau \ll 1$), the relaxation rate (rotational speed of dipole), which is equal to $1/\tau$, is much faster than the frequency of external field. Therefore, the polarization is in phase with the field, and $\varepsilon_r' = \varepsilon_s$ and $\varepsilon_r'' = 0$. At high frequencies ($\omega\tau \gg 1$), the relaxation rate $1/\tau$ is much lower than the field frequency. Therefore, the polarization does not keep up with the changes in the oscillating field, and $\varepsilon_r' = \varepsilon_\infty = 1$ and $\varepsilon_r'' = 0$. When $\omega\tau = 1$, ε_r'' shows a peak, which is called the Debye relaxation peak. When the relaxation rate is equal to the angular frequency, the dielectric loss is maximized and the polarization lags the electric field by 90° phase difference. In this frequency interval, ε_r' gradually decreases. At $\omega\tau = 1$, ε_r' and ε_r'' are expressed as

$$\varepsilon_r' = \frac{\varepsilon_s + \varepsilon_\infty}{2} \tag{4.18}$$

$$\varepsilon_r'' = \frac{\varepsilon_s - \varepsilon_\infty}{2} \tag{4.19}$$

The typical relaxation time is about 10^{-11} s which corresponds to the frequency in GHz range. In the case of water, of which polarization mechanism is dipole orientation, it shows the maximum dielectric loss in microwave frequency range. In solid materials with multiple dipoles and different relaxation times, the dispersion peak widens. The relaxation time τ can vary considerably depending on the materials and temperature. When heavy atoms are included, τ tends to get larger, and τ becomes smaller as the temperature increases.

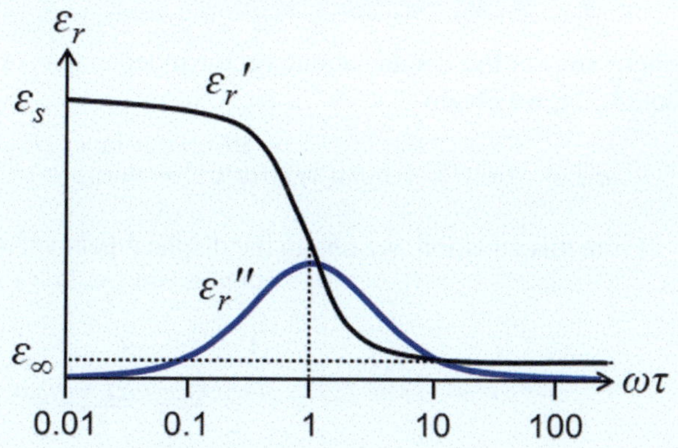

Fig. 4.8 The frequency dispersion of real and imaginary parts of complex permittivity of Debye relaxation

4.3.2 Dielectric Resonance

When atoms, ions, and electrons perform rotational or oscillating motions, a resonance occurs near the intrinsic vibrational frequency of each movement. The electronic polarization caused by the displacement of electron cloud surrounding the nucleus is a typical example of dielectric resonance. When an electric field is applied to a material, the electron cloud of the atom is shifted by the field and produces a dipole moment, as illustrated earlier in Fig. 4.1. The movement of electrons can then be treated as a harmonic oscillator under the action of a restoring force. If the displacement vector from the nucleus to the center of electron cloud with charge e is x, the polarization vector P is given by the dipole moment expressed as

$$P = -ex \tag{4.20}$$

The equation of motion of the driven harmonic oscillator can be expressed as

$$m\frac{\mathrm{d}^2x(\omega,t)}{\mathrm{d}t^2} + m\eta_0\frac{\mathrm{d}x(\omega,t)}{\mathrm{d}t} + m\omega_0^2x(\omega,t) = -eE(\omega,t) \tag{4.21}$$

where m: mass of electron, ω: angular frequency of applied electric field, η_0: damping factor, and ω_0: resonance frequency of the oscillator. The first term $\left(m\left(\mathrm{d}^2x/\mathrm{d}t^2\right)\right)$ of the left-hand side of Eq. (4.21) is the acceleration force, the second term $(m\eta_0(\mathrm{d}x/\mathrm{d}t))$ is the viscous force, the third term $(m\omega_0^2x)$ is the restoring force (Hooke's force), and right-hand side term is the external force by the applied alternating electric field.

If the motion is periodic, repeating itself in a sinusoidal fashion, the displacement is also changed with a function of sinusoidal wave, that is

$$x(\omega,t) = x(\omega)\cos\omega t = \mathrm{Re}\left[x(\omega)\mathrm{e}^{j\omega t}\right] \tag{4.22}$$

where $x(\omega)$ is the displacement vector independent of time. Substituting this into Eq. (4.21), we obtain

$$m(j\omega)^2x(\omega)\mathrm{e}^{j\omega t} + m\eta_0(j\omega)x(\omega)\mathrm{e}^{j\omega t} + m\omega_0^2x(\omega)\mathrm{e}^{j\omega t} = -eE(\omega)\mathrm{e}^{j\omega t} \tag{4.23}$$

From this equation, we obtain the displacement vector of a single atom expressed as

$$x(\omega) = -\frac{1}{m}\frac{e}{\left(\omega_0^2-\omega^2\right)+j\eta_0\omega}E(\omega) \tag{4.24}$$

Then we get the polarization vector of a single atom:

$$P(\omega) = -ex(\omega) = \frac{1}{m}\frac{e^2}{\left(\omega_0^2-\omega^2\right)+j\eta_0\omega}E(\omega) \tag{4.25}$$

For all N atoms per unit volume, the total polarization vector is given by

$$NP(\omega) = \varepsilon_0 \chi E(\omega) \tag{4.26}$$

where χ is the electric susceptibility. Then we get

$$\chi(\omega) = \frac{Ne^2}{\varepsilon_0 m} \frac{1}{\left(\omega_0^2 - \omega^2\right) + j\eta_0\omega} \tag{4.27}$$

Using the relation $\varepsilon_r = 1 + \chi$, we obtain

$$\varepsilon_r = 1 + \frac{\omega_p^2}{\omega_0^2 - \omega^2 + j\eta_0\omega} \tag{4.28}$$

where ω_p is the plasma frequency ($= \sqrt{Ne^2/\varepsilon_0 m}$) which is known as the characteristic frequency of oscillation around their equilibrium positions of electrons in a plasma (i.e., neutral gas of charged particles) [21].

The real and imaginary parts of complex permittivity $\varepsilon_r = \varepsilon_r' - j\varepsilon_r''$ are given by the following equations, and their frequency dispersion is shown in Fig. 4.9:

$$\varepsilon_r' = 1 + \frac{\omega_p^2\left(\omega_0^2 - \omega^2\right)}{\left(\omega_0^2 - \omega^2\right)^2 + \eta_0^2\omega^2} \tag{4.29}$$

$$\varepsilon_r'' = \frac{\omega_p^2\eta_0\omega}{\left(\omega_0^2 - \omega^2\right)^2 + \eta_0^2\omega^2} \tag{4.30}$$

At low frequencies ($\omega \to 0$), $\varepsilon_r' = \varepsilon_s = 1 + \left(\omega_p^2/\omega_0^2\right)$ and $\varepsilon_r'' = 0$, and there is no phase difference between the polarization and the electric field. At the resonance frequency ($\omega = \omega_0$), $\varepsilon_r' = 1$ and ε_r'' shows the maximum peak, and the phase

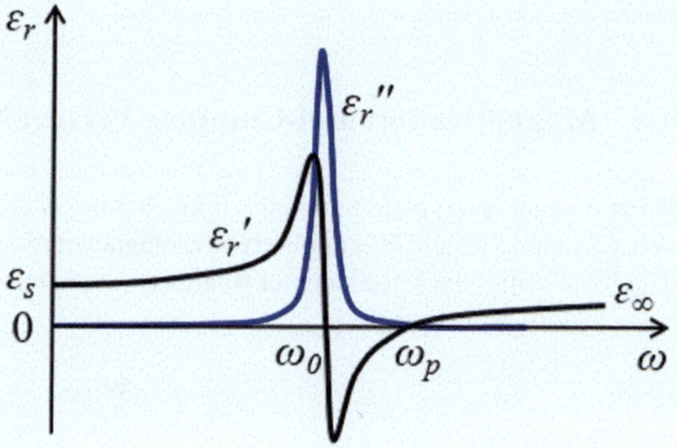

Fig. 4.9 Frequency dispersion of complex permittivity by resonance effect

difference between polarization and electric field is 90°. At high frequencies ($\omega \rightarrow \infty$), $\varepsilon_r' = \varepsilon_\infty = 1$ and $\varepsilon_r'' = 0$, and the phase difference between the polarization and the electric field is 180°. Using ε_s and ε_∞, complex permittivity can be expressed as

$$\varepsilon_r = \varepsilon_\infty + \frac{(\varepsilon_s - \varepsilon_\infty)\omega_0^2}{\omega_0^2 - \omega^2 + j\eta_0\omega} \tag{4.31}$$

where it is noted that $\omega_p^2 = (\varepsilon_s - 1)\omega_0^2 = (\varepsilon_s - \varepsilon_\infty)\omega_0^2$ and $\varepsilon_s - \varepsilon_\infty$ represents the strength of harmonic oscillator.

4.3.3 *Frequency Dependence of Dielectric Loss*

The dielectric constant decreases with increasing frequency. At frequencies where dielectric relaxation or dielectric resonance occurs, dielectric loss spectrum shows peaks. Dielectric loss due to space charge polarization in heterogeneous materials, ion jump or ion movement in ceramics, and electron hopping in semiconductors show the relaxation-type spectra of Drude dispersion model shown in Fig. 4.8. Polarization by ion vibration and electronic polarization show the resonance-type spectra of Lorenz dispersion model shown in Fig. 4.9. Ion jump polarization is the movement of ions in an ionic solid from their equilibrium position on application of electric field and predominates at low frequencies below 10^3 Hz. Space charge or interfacial polarization occurs due to the impediment of charge carriers at the interface of heterogeneous materials below 10^6 Hz. Alignment of spontaneous dipoles by domain wall movement in ferroelectric materials is observed at frequencies less than 10^9 Hz (microwave region). In the case of polar molecules such as water, dipolar or oriented polarization is dominant at microwave region. The ion vibration and deformation losses become important in the infrared band. The electronic polarization loss gives rise to absorption in the visible spectrum. When the frequency is below about 10^{10} Hz, ion jump and dipole relaxation, defect-induced polarization, and space charge polarization are the main contribution sources to dielectric losses [11].

4.4 Magnetization and Complex Permeability

A bar magnet is a typical magnetic dipole having N and S poles that always coexist. When a magnetic field H is applied to the magnet, it rotates, as illustrated in Fig. 4.10a. The force with which the magnet rotates is defined as the magnetic dipole moment μ_m which is expressed as

$$\mu_m = qd \text{ (Wb m)} \tag{4.32}$$

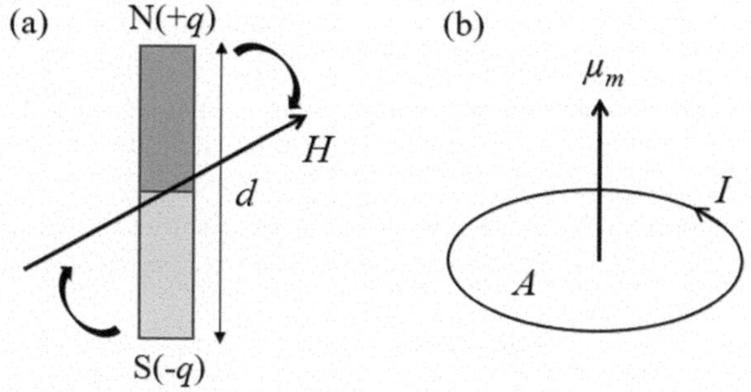

Fig. 4.10 Magnetic dipole moment: **a** bar magnet and **b** loop current

where q is the pole strength in unit (Wb) and d is the magnet length.

From Ampere's circuital law, electric currents produce a magnetic field. Magnetic dipole moment in the current loop, as shown in Fig. 4.10b, is given by

$$\mu_{\mathrm{m}} = IA \left(\mathrm{Am}^2\right) \tag{4.33}$$

where I is the current and A is the area of the current loop. The loop of currents and the bar magnet are equivalent. Placing a substance inside a solenoid, the magnetic moment of the constituent atoms is produced in response to magnetic field H. Magnetization intensity (or simply, magnetization) M is defined by the magnetic dipole moment per unit volume and expressed as

$$M = \chi_{\mathrm{m}} H \ (\mathrm{A/m}) \tag{4.34}$$

where we define χ_{m} the magnetic susceptibility, in analogy to the electric susceptibility of Eq. (4.10).

In electromagnetism, permeability is a value that indicates the measure of magnetization in response to an applied magnetic field. The typical symbol is the Greek letter μ, and the unit is (H/m). When the magnetic body is magnetized, the magnetic flux density B is proportional to the applied magnetic field H, and is expressed as

$$B = \mu H = \mu_0 (H + M) \tag{4.35}$$

The permeability of the vacuum, usually denoted by μ_0, has the value:

$$\mu_0 = 4\pi \times 10^{-7} \ (\mathrm{H/m}) \tag{4.36}$$

The ratio of the permeability of medium to that of vacuum is defined as the relative permeability μ_{r}, expressed as

$$\mu_{\mathrm{r}} = \frac{\mu}{\mu_0} \tag{4.37}$$

From Eqs. (4.34) and (4.35),

$$\mu_r = 1 + \chi_m \tag{4.38}$$

If the frequency is low enough, magnetization reacts in phase to the changes in the external magnetic field. The permeability at this low frequency is called the static permeability. As the frequency increases, magnetization follows the external magnetic field with some lag time, and the phase difference δ between B the H begins to increase. Then the permeability is expressed as a complex number:

$$\mu(\omega) = \frac{B(\omega)}{H(\omega)} = \frac{B_0 e^{j(\omega t - \delta)}}{H_0 e^{j\omega t}} = \frac{B_0}{H_0} e^{-j\delta} = \frac{B_0}{H_0}(\cos\delta - j\sin\delta) \tag{4.39}$$

This is defined as the complex permeability with real part μ' and imaginary part μ'', written in rectangular form:

$$\mu(\omega) = \mu'(\omega) - j\mu''(\omega) \tag{4.40}$$

The real part is associated with the energy accumulation by medium, and the imaginary part is associated with the energy loss caused by medium. If there is no phase difference between magnetization and applied field ($\delta = 0°$), $\mu = \mu'$ and $\mu'' = 0$. If $\delta = 90°$, the loss is at maximum ($\mu = j\mu''$ and $\mu' = 0$). The ratio of the imaginary to the real part of the complex permeability is called the loss tangent:

$$\frac{\mu''}{\mu'} = \frac{\left(\frac{B_0}{H_0}\right)\sin\delta}{\left(\frac{B_0}{H_0}\right)\cos\delta} = \tan\delta \tag{4.41}$$

4.5 Magnetic Loss Mechanisms

Energy losses caused by an alternating magnetic field include eddy current loss, hysteresis loss, and residual loss. If each of the power losses is referred to P_e, P_h, and P_r, respectively, the total loss is given by

$$P = P_e + P_h + P_r \tag{4.42}$$

Changes in magnetic flux inside a magnetic body induce the electromotive force (*emf*) by Faraday's law of electromagnetic induction. If the magnetic material is electrically conductive, this *emf* circulates an electric current within the material. This circulating current is called the eddy current, and the resulting power loss is called the eddy current loss. To reduce this loss, ceramic magnetic materials (ferrites) are commonly used, which are nonconductors. Hysteresis loss is an energy loss proportional to the area within the closed hysteresis loop during the magnetization cycle.

The hysteresis loop equals the energy which is necessary to reverse the magnetization. The shape and area of loop depend on the internal structure and composition of the ferromagnetic materials. If the applied field is so weak that the magnetization is reversible, the hysteresis loss can be ignored. The residual loss, which has a meaning of remained loss excluding hysteresis loss and eddy current loss from total magnetic loss, includes the magnetic resonance loss or magnetic relaxation loss that accompanies the oscillating magnetization process.

The relative portion and frequency dependence of the three loss factors are strongly dependent on materials (metals or ferrites). Hysteresis loss is common in both materials. Metals have a very large eddy current loss as the frequency increases. In the case of ferrites, the eddy current loss is small because of their insulating properties, and the residual loss becomes great in the high-frequency region. The residual loss of ferrites consists of resonance by magnetic domain wall movement or by domain rotation, and magnetic relaxation loss, all of which are functions of frequency.

4.5.1 Eddy Current Loss

According to Faraday's law, the electromotive force is induced by the time change of magnetic flux, $e = -\delta\Phi/\delta t$. As shown in Fig. 4.11, the electromotive force e is represented by the line integral of the electric field E along the closed path C, and the magnetic flux Φ is equal to the surface integral of the magnetic flux density B over the area S enclosed by path C. When the magnetic flux changes with time, an *emf* is induced in the material, that is expressed as

$$e = \oint_C E\mathrm{d}l = -\oint_S \frac{\mathrm{d}B}{\mathrm{d}t}\mathrm{d}S \tag{4.43}$$

If the path C is a circle of radius r, the line and surface integrals yield

$$E(r) = -\frac{r}{2}\left(\frac{\mathrm{d}B}{\mathrm{d}t}\right) \tag{4.44}$$

Using the Ohm's law, $J = E/\rho$ (where J is the current density and ρ is the electrical resistivity), the produced loss per unit volume for the circumferential magnetic body with a radius r_0 is given by [22]

$$P_\mathrm{e} = \frac{1}{\pi r_0^2}\int_0^{r_0} J(r)e(r)\mathrm{d}r = \frac{1}{\pi r_0^2}\int_0^{r_0}\left[\frac{E(r)}{\rho}\right][2\pi rE(r)]\mathrm{d}r = \frac{r_0^2}{8\rho}\left(\frac{\mathrm{d}B}{\mathrm{d}t}\right)^2 \tag{4.45}$$

When the applied magnetic field is in the form of sine wave, the magnetic flux density is expressed as $B = B_0 \sin 2\pi ft$, and the average value of $(\mathrm{d}B/\mathrm{d}t)^2$ becomes

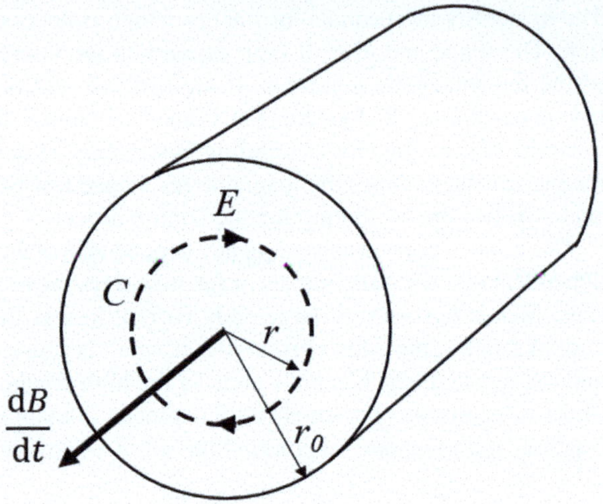

Fig. 4.11 Eddy current in a circumferential magnetic body

$2\pi f^2 B_0^2$. Then we get

$$P_e = \frac{\pi^2}{16} \frac{d^2 f^2 B_0^2}{\rho} \tag{4.46}$$

where d is the diameter $(= 2r_0)$ of cylinder. The eddy current loss is proportional to the square of size d, frequency f, magnitude of magnetic flux density B_0, and inversely proportional to the electrical resistivity ρ. To reduce the eddy current loss, the specimen diameter should be reduced (such as thin sheets or laminates) and a high-resistivity materials (such as ferrites) should be used.

4.5.2 *Domain Wall Resonance*

When a magnetic material is subject to an external magnetic field, magnetization proceeds by the movement of domain walls and the rotation of magnetization. Ferromagnets or ferrimagnets are divided into regions called domains where the magnetization is uniform in the direction of easy axis. If the magnetic field changes, the walls move, increasing the relative sizes of the domains with the magnetization orienting to the applied field. If the magnetic field is so large that the domain wall movement is completed, further magnetization is carried out by rotating the magnetization in the direction of external field. If it is a single domain (e.g., fine particulates) at the initial state, magnetization occurs only by the rotational process.

The domain wall has a characteristic of inertial motion. The wall in moving by the external field continues to move to a certain extent even if the external field is removed momentarily. The inertial motion of domain walls can be treated as a harmonic oscillator in vibrating systems. Applying an alternating magnetic field to

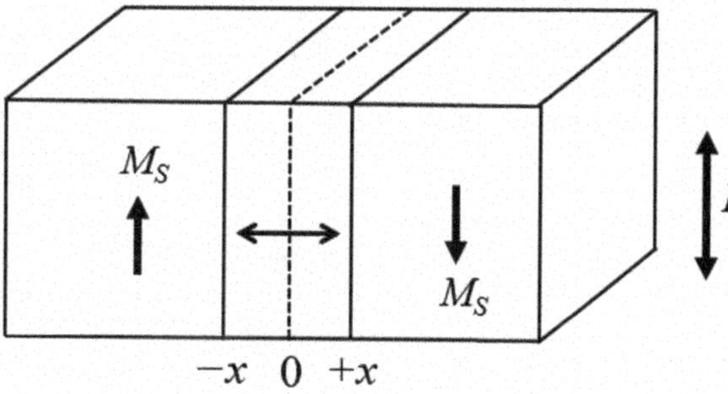

Fig. 4.12 Magnetization by domain wall movement

this domain wall in inertial motion may cause a resonance phenomenon, which is called the domain wall resonance, and results in energy loss.

As illustrated in Fig. 4.12, when the 180° domain wall moves x by a magnetic field H, the equation of motion can be expressed as that for driven damped harmonic oscillators [22]:

$$m_w \frac{d^2x}{dt^2} + \beta \frac{dx}{dt} + \alpha x = 2M_s H \tag{4.47}$$

The first term $(m_w(d^2x/dt^2))$ of the left-hand side of Eq. (4.47) corresponds to the inertial force of domain wall, where m_w is the wall mass per unit area. The second term $(\beta(dx/dt))$ is the viscous damping force, where β is the wall damping coefficient. The third term (αx) is the restoring force, where α is the elastic constant for the wall movement. The right-hand side term is the driving force for domain wall movement by the applied external field H.

The wall displacement x produces a change in net magnetization of $M = 2M_s x$ per unit volume, where M_s is the saturation magnetization. From this, we obtain $x = M/2M_s$. Substituting this into Eq. (4.47) leads to

$$m_w \frac{d^2M}{dt^2} + \beta \frac{dM}{dt} + \alpha M = 4M_s^2 H \tag{4.48}$$

Harmonic magnetic drive field, given by $H = H_0 e^{j\omega t}$ of angular frequency ω, vibrates the domain wall and changes the magnetization M. During this process, if the magnetization lags behind the drive field by phase angle δ, we can express the magnetization as $M = M_0 e^{j(\omega t - \delta)}$. Substituting this into Eq. (4.48) yields

$$M_0 e^{-j\delta}\left(-m_w\omega^2 + j\beta\omega + \alpha\right) = 4M_s^2 H_0 \tag{4.49}$$

Then the magnetic susceptibility χ_m is given by

$$\chi_m = \frac{M}{H} = \frac{M_0 e^{j(\omega t-\delta)}}{H_0 e^{j\omega t}} = \frac{M_0 e^{-j\delta}}{H_0} = \frac{4M_s^2}{-m_w\omega^2 + j\beta\omega + \alpha} \tag{4.50}$$

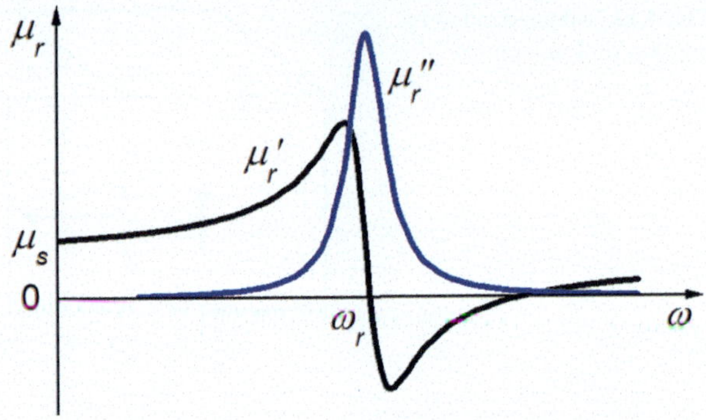

Fig. 4.13 Frequency dispersion of complex permeability caused by domain wall resonance

Since the susceptibility is complex ($\chi_m = \chi_m' - j\chi_m''$), the real and imaginary parts can be expressed as

$$\chi_m' = \frac{4M_s^2(\alpha - m_w\omega^2)}{(\alpha - m_w\omega^2)^2 + \beta^2\omega^2} \quad (4.51)$$

$$\chi_m'' = \frac{4M_s^2\beta\omega}{(\alpha - m_w\omega^2)^2 + \beta^2\omega^2} \quad (4.52)$$

When $\alpha - m_w\omega^2 = 0$, $\chi_m' = 0$ and χ_m'' becomes maximum, that is called the domain wall resonance.

Using the relation $\mu_r = 1 + \chi_m$, frequency dispersion of complex permeability $\mu_r' = 1 + \chi_m'$ and $\mu_r'' = \chi_m''$ can be sketched in Fig. 4.13, showing the dispersion pattern of Lorentzian harmonic oscillator. Real part μ_r' has a static permeability μ_s at low frequencies and then decreases sharply near the resonance frequency to a value of 0, and then it shows a negative value and gradually converges to a value of 1 at high frequencies. Imaginary part μ_r'' shows a peak at the resonance frequency at which the energy loss caused by the domain wall motion is maximum. In the case of spinel ferrites, the resonance frequency, $f_r = \omega_r/2\pi$, is in the range of 1 ~ 100 MHz [22].

4.5.3 Gyromagnetic Resonance

At higher frequencies beyond the domain wall resonance, the magnetization proceeds dominantly by the rotation of magnetic moments. For a magnetic moment placed in a magnetic field, it experiences a torque that aligns in the direction of the magnetic field. As illustrated in Fig. 4.14a, when the magnetization M_s is directed towards the easy axis by the anisotropy field H_a, it approaches by precession around the anisotropy field rather than instantaneously settle down in the direction of magnetic field. This is

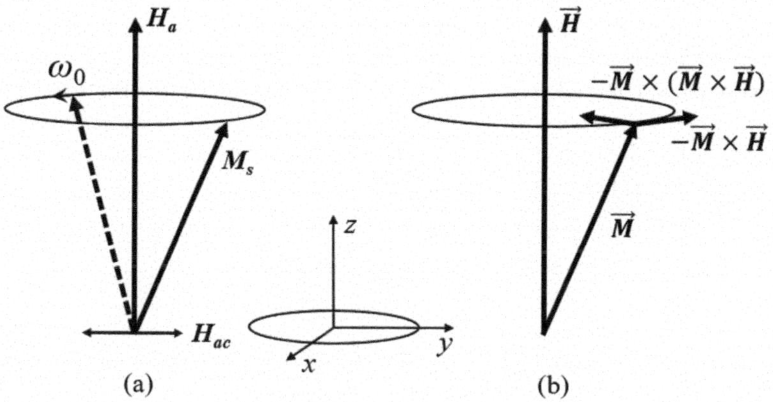

Fig. 4.14 **a** The precession of the magnetization vector M_s around the anisotropy field H_a and **b** the vector notation of the Landau–Lifshitz equation

called the Larmor precession and the precession angular velocity (Larmor frequency) is given by

$$\omega_0 = \gamma H_a \tag{4.53}$$

where γ is the gyromagnetic ratio given by $\gamma = 1.105 \times 10^5 g$ (m/A s) and g factor has a value $g = 2$ for the spin magnetic moment and $g = 1$ for the orbital magnetic moment.

When the magnetization converges in the direction of magnetic field in a precession motion, the presence of a damping factor speeds up the rate of convergence. As illustrated in Fig. 4.14b, the equation of precession motion including the damping action is given by the Landau–Lifshitz equation [23]:

$$\frac{d\vec{M}}{dt} = -\gamma\left[\vec{M}\times\vec{H}\right] - \frac{\alpha\gamma}{M}\left[\vec{M}\times\left[\vec{M}\times\vec{H}\right]\right] \tag{4.54}$$

or by the Landau–Lifshitz–Gilbert equation [23]:

$$\frac{d\vec{M}}{dt} = -\gamma\left[\vec{M}\times\vec{H}\right] + \frac{\alpha}{M}\left[\vec{M}\times\frac{d\vec{M}}{dt}\right] \tag{4.55}$$

The first term of the right-hand side of Eq. (4.54) corresponds to the precession motion of the magnetization vector $\vec{M}$, and the second term refers to the convergence motion in the direction of the magnetic field $\vec{H}$ by the damping term α. The greater the α, the faster $\vec{M}$ approaches to the direction $\vec{H}$.

If AC field $H_{ac} = h = h_0 e^{j\omega t}$ is applied perpendicular to the anisotropy field H_a which is directed along z-axis, each component of magnetic field is denoted as $H_x = h_x = h_0 e^{j\omega t}$, $H_y = h_y = h_0 e^{j\omega t}$, $H_z = H_a$, where h_x and h_y are x- and y-component of h, respectively. This field produces a magnetization of which components are $M_x = m_x = m_0 e^{j\omega t}$, $M_y = m_y = m_0 e^{j\omega t}$, $M_z \simeq M_s$, where M_s is the saturation magnetization, and m_x and m_y are the additional AC magnetization (in the x–y plane) generated by h. Substituting these into Eq. (4.55) yields the component equations of motion. Using the relations, $\omega_0 = \gamma H_a$, $H_a \gg h$, and $M_s \gg m$, we obtain the following equations [22, 24]:

$$\frac{dm_x}{dt} = j\omega m_x = -(\omega_0 + j\omega\alpha)m_y + \gamma M_s h_y \tag{4.56}$$

$$\frac{dm_y}{dt} = j\omega m_y = (\omega_0 + j\omega\alpha)m_x - \gamma M_s h_x \tag{4.57}$$

$$\frac{dm_z}{dt} = 0 \tag{4.58}$$

Solving (4.56) and (4.57) for m_x and m_y gives

$$\left[(\omega_0 + j\omega\alpha)^2 - \omega^2\right]m_x = (\omega_0 + j\omega\alpha)\gamma M_s h_x + j\omega\gamma M_s h_y \tag{4.59}$$

$$\left[(\omega_0 + j\omega\alpha)^2 - \omega^2\right]m_y = -j\omega\gamma M_s h_x + (\omega_0 + j\omega\alpha)\gamma M_s h_y \tag{4.60}$$

which shows the linear relationship between h and m. Equations (4.59) and (4.60) can be written with a tensor susceptibility $[\chi]$:

$$\begin{bmatrix} m_x \\ m_y \\ m_z \end{bmatrix} = [\chi]\begin{bmatrix} h_x \\ h_y \\ h_z \end{bmatrix} = \begin{bmatrix} \chi_{xx} & \chi_{xy} & 0 \\ \chi_{yx} & \chi_{yy} & 0 \\ 0 & 0 & 0 \end{bmatrix}\begin{bmatrix} h_x \\ h_y \\ h_z \end{bmatrix} \tag{4.61}$$

where the elements of $[\chi]$ are given by

$$\chi_{xx} = \chi_{yy} = \frac{\gamma M_s(\omega_0 + j\omega\alpha)}{(\omega_0 + j\omega\alpha)^2 - \omega^2} \tag{4.62}$$

$$\chi_{xy} = -\chi_{yx} = \frac{j\gamma M_s \omega}{(\omega_0 + j\omega\alpha)^2 - \omega^2} \tag{4.63}$$

which can be expressed as a function of complex number, $\chi_{xx} = \chi'_{xx} - j\chi''_{xx}$ and $\chi_{xy} = \chi'_{xy} - j\chi''_{xy}$, respectively. Then the complex magnetic susceptibility, $\chi_m = \chi'_m - j\chi''_m$, is expressed as

$$\chi'_m = \chi'_{xx} = \frac{\gamma M_s \omega_0\left[\omega_0^2 - \omega^2(1-\alpha^2)\right]}{\left[\omega_0^2 - \omega^2(1+\alpha^2)\right]^2 + 4\omega^2\omega_0^2\alpha^2} \tag{4.64}$$

$$\chi''_m = \chi''_{xx} = \frac{\gamma M_s \omega \alpha \left[\omega_0^2 + \omega^2(1+\alpha^2)\right]}{\left[\omega_0^2 - \omega^2(1+\alpha^2)\right]^2 + 4\omega^2\omega_0^2\alpha^2} \tag{4.65}$$

The complex permeability, $\mu_r = 1 + \chi_m = \mu'_r - j\mu''_r$, is then given by

$$\mu'_r = 1 + \chi'_m = 1 + \frac{\gamma M_s \omega_0 \left[\omega_0^2 - \omega^2(1-\alpha^2)\right]}{\left[\omega_0^2 - \omega^2(1+\alpha^2)\right]^2 + 4\omega^2\omega_0^2\alpha^2} \tag{4.66}$$

$$\mu''_r = \chi''_m = \frac{\gamma M_s \omega \alpha \left[\omega_0^2 + \omega^2(1+\alpha^2)\right]}{\left[\omega_0^2 - \omega^2(1+\alpha^2)\right]^2 + 4\omega^2\omega_0^2\alpha^2} \tag{4.67}$$

The equation for μ_r has the same form of Lorentzian frequency dispersion and can be sketched in the similar pattern of Fig. 4.13. At the resonance frequency corresponding to Larmor frequency, μ'_r decreases sharply and μ''_r shows a peak. The energy loss caused by Larmor precession is maximum. This effect is known as the gyromagnetic resonance. The magnetic resonance caused by the anisotropy field inherent in a material is called the natural resonance. As the anisotropy constant increases, the resonance frequency is raised to a higher frequency.

Figure 4.15 shows the complex permeability spectra of $Ni_xZn_{1-x}Fe_2O_4$ ferrites, indicating the typical pattern of natural resonance [25]. For an example of $x = 0.64$, the μ'_r shows a constant value of about 85 at low frequencies, which corresponds to static permeability μ_s, and drops rapidly around 60 MHz where μ''_r has a peak value of 50. It is noted that the higher the static permeability, the lower the resonance frequency. In the case of positive cubic anisotropy constant ($K_1 > 0$), the anisotropy field is defined as $H_a = 2K_1/M_s$, and the static permeability of polycrystals obtained by rotational magnetization is given by $\mu_s = M_s^2/3K_1\mu_0$ [22], then we get

$$f_r\mu_s = \left(\frac{\gamma}{2\pi}\frac{2K_1}{M_s}\right)\left(\frac{M_s^2}{3K_1\mu_0}\right) = \frac{\gamma}{3\pi\mu_0}M_s \tag{4.68}$$

In this equation, the right-hand side is a constant value containing the saturation magnetization which is determined mostly by the composition of material. It can be seen that the resonance frequency and the static permeability on the left-hand side are in a reciprocal relationship to each other. As the permeability increases, the resonance frequency decreases, and conversely, when the permeability is lowered, the resonance frequency increases. For soft ferrites with $M_s \approx 0.3$ T, approximately the following relationship is established [22, 25]:

$$f_r\mu_s \simeq 5.6 \times 10^9\,\text{Hz} = 5600\,\text{MHz} \tag{4.69}$$

In Fig. 4.15, we see the dotted line which connects the half-value points of static (low-frequency) permeability and it satisfies Eq. (4.69). This is called the Snoek's limit. Ferrites with cubic crystal anisotropy cannot in principle have a permeability greater than the limits of Snoek.

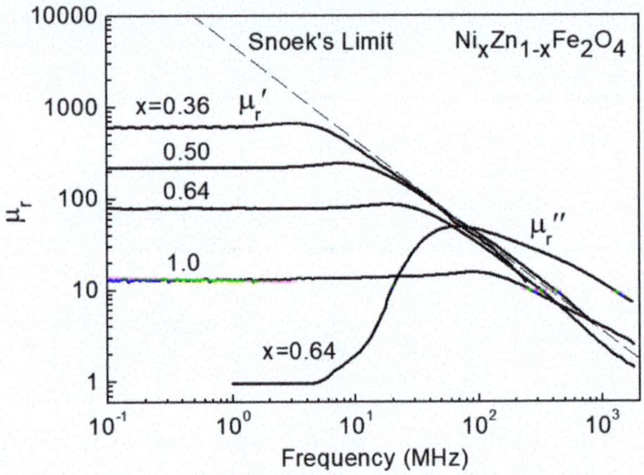

Fig. 4.15 Frequency dependence of complex permeability for polycrystalline $Ni_xZn_{1-x}Fe_2O_4$ ferrites (data from Ref. [25])

However, when the anisotropy constant becomes large, such as hexagonal ferrites, it is not bound by the limits of Snoek. The most typical case is the M-type hexagonal ferrite ($(BaSr)Fe_{12}O_{19}$), which has a very large anisotropy field along the *c* axis of hexagonal crystal, $H_{K\theta} = 2K_1/M_s \approx 17$ kOe [25]. Polycrystalline form of this material is classified as a permanent magnet because the magnetic field required to induce the magnetization is very large (i.e., large coercive force), and the resonance frequency is elevated to the millimeter wave band. Therefore, to lower the resonance frequency to the microwave range in such hexagonal ferrites, it is necessary to decrease the anisotropy field or modify the magneto-crystalline anisotropy from *c* axis to *a*–*b* plane.

When the magnetization easy axis is in the *a*–*b* plane of hexagonal structure, the ferromagnetic resonance is determined by the rotation of magnetization which is subject to the anisotropy field $H_{K\phi}$ in the preferred plane. $H_{K\phi}$ is very small compared to the *c*-axis anisotropy field $H_{K\theta}$. As a result, the resonance frequencies for single-domain particles are in the range [26]:

$$\gamma\sqrt{H_{K\theta}H_{K\phi}} \le \omega_r \le \gamma\sqrt{(H_{K\theta} + 4\pi M_s)(H_{K\phi} + 4\pi M_s)} \tag{4.70}$$

This range of resonance frequency shifts to higher frequencies at least 10 times greater as compared with the spinel ferrites. This can be seen in the frequency spectra of hexagonal Co_2Z ferrite ($Ba_3Co_2Fe_{24}O_{41}$) with a preferred plane of magnetization relative to that of cubic spinel ferrite ($NiFe_2O_4$) [27, 28]. Although the static permeability is almost equal, the dispersion of Co_2Z is substantially higher than the spinel ferrite.

4.5.4 Magnetic Relaxation

When a magnetic material is so small that it is in single domain state, the magnetization proceeds by the rotation of magnetization, as illustrated in Fig. 4.16. When the external alternating magnetic field H_{ac} is parallel to the magnetization vector M_s, the rotational force is maximum. Therefore, this is defined as the longitudinal interaction. In gyromagnetic resonance, H_{ac} is perpendicular to the magnetization vector, as illustrated in Fig. 4.14a, and it is called the transverse interaction. During the longitudinal interaction, the rotation of magnetization requires some time to reach its final equilibrium with the external field, that is called the magnetic relaxation.

In analogy to the Debye relaxation in dielectric materials, when a magnetic field H_0 is applied momentarily to a magnetic body, the magnetization reaches its final equilibrium value M_∞ over a period of time, as shown in Fig. 4.17. The increment of M with time is given by an exponential function:

$$M(t) = M_\infty\left(1 - e^{-t/\tau}\right) \tag{4.71}$$

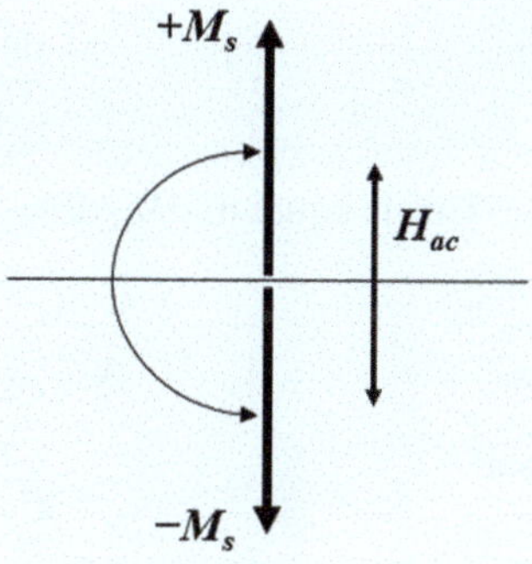

Fig. 4.16 Schematic diagram showing the longitudinal interaction between alternating magnetic drive field and magnetization vector

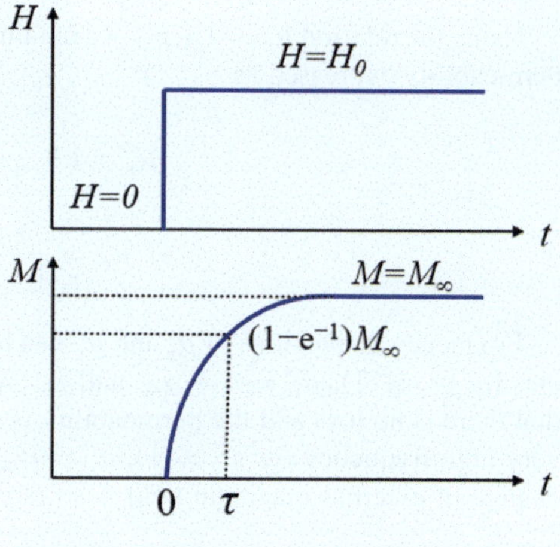

Fig. 4.17 An increase to the equilibrium value of magnetization M_∞ with the instantaneous application of the magnetic field H_0

where τ is the relaxation time, defined as the time required to take the magnetization to reach $1 - e^{-1} = 0.632$ of the equilibrium value M_∞. The shorter the τ, the faster the rate of response.

Following the procedure for dielectric relaxation described in Sect. 4.3.1 [20], we obtain the magnetic susceptibility by the relaxation process. First, differentiating Eq. (4.71), we obtain the rate of magnetization buildup:

$$\frac{dM(t)}{dt} = \frac{M_\infty e^{-t/\tau}}{\tau} \tag{4.72}$$

Substituting Eq. (4.71) into Eq. (4.72) yields

$$\frac{dM(t)}{dt} = \frac{M_\infty - M(t)}{\tau} \tag{4.73}$$

If an alternating magnetic field $H = H_0 e^{j\omega t}$ is applied, the magnetization is given by $M = \chi_m H = \chi_m H_0 e^{j\omega t}$ where χ_m is the magnetic susceptibility. At the equilibrium state, $M_\infty = \chi_\infty H_0 e^{j\omega t}$. Substituting M and M_∞ into Eq. (4.73), we obtain

$$\frac{dM(t)}{dt} = \frac{1}{\tau}(\chi_\infty - \chi_m) H_0 e^{j\omega t} \tag{4.74}$$

This is equal to $dM/dt = j\omega\chi_m H_0 e^{j\omega t}$, thus we get

$$j\omega\chi_m = \frac{1}{\tau}(\chi_\infty - \chi_m) \tag{4.75}$$

and

$$\chi_m = \frac{\chi_\infty}{1 + j\omega\tau} = \frac{\chi_\infty}{1 + \omega^2\tau^2} - j\frac{\chi_\infty\omega\tau}{1 + \omega^2\tau^2} \tag{4.76}$$

Using the relation $\mu_r = 1 + \chi_m$, we obtain the real and imaginary parts of complex permeability expressed as

$$\mu_r' = 1 + \frac{\chi_\infty}{1 + \omega^2\tau^2} \tag{4.77}$$

$$\mu_r'' = \frac{\chi_\infty\omega\tau}{1 + \omega^2\tau^2} \tag{4.78}$$

Frequency dispersions of μ_r' and μ_r'' can be sketched in Fig. 4.18. At low frequencies ($\omega^2\tau^2 \ll 1$), $\mu_r' = 1 + \chi_\infty$ and $\mu_r'' = 0$, which means the equilibrium state that there is no loss and the permeability is equal to the static permeability μ_s. At a very high frequency ($\omega^2\tau^2 \gg 1$), $\mu_r' = \mu_\infty = 1$ and $\mu_r'' = 0$. In this situation, the change of external magnetic field is so rapid that the magnetization cannot follow

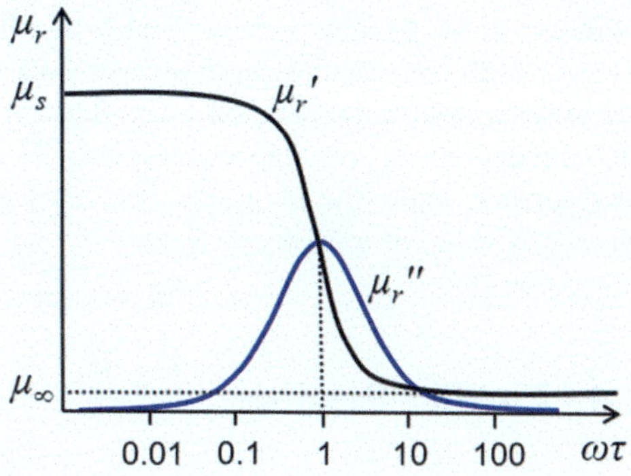

Fig. 4.18 Frequency dispersions of μ_r' and μ_r'' for the case of magnetic relaxation

it. At $\omega\tau = 1$, the loss is maximized and μ_r' is reduced to half of the initial value. This frequency is called the relaxation frequency and is inversely proportional to the relaxation time:

$$f_r = \frac{1}{2\pi\tau} \tag{4.79}$$

4.5.5 *Frequency Dependence of Magnetic Loss*

The permeability spectrum of magnetic materials is represented by the overlap of magnetic resonance and magnetic relaxation. Magnetic resonance accompanies the domain wall resonance and the gyromagnetic spin resonance. Through analyzing the damping factors, the dominant loss mechanism (resonance or relaxation) can be evaluated [29]. It was interpreted that in sintered Mn–Zn ferrites, domain wall resonance is dominant, while the gyromagnetic spin resonance or magnetization rotation is great in Ni–Zn ferrites. In the polymer composites of fine particles, however, the relaxation loss is dominant for both Mn–Zn ferrite and Ni–Zn ferrite. This is because the damping factor is large in the particulate composites, which suppress the gyromagnetic spin resonance. In addition, the reduced area of the domain walls in the fine ferrite particles reduces the contribution of domain wall resonance.

Frequency dependence of magnetic permeability and magnetic loss was discussed in Ref. [28]. The low-frequency dispersion is associated with domain wall dynamics. In the specimens of multiple domains and low anisotropy constant, the domain wall resonance is the dominant loss mechanism and a small domain wall damping allows a higher permeability. In the case of cubic ferrites, the resonance frequency of wall motion is in the range of 1~100 MHz, depending on the composition [22, 25]. Above the domain wall damping edge, the permeability of single-domain rotation is present up to the relaxation frequency which is inversely proportional to the relaxation time. An increase in the number of crystal defects (damping factor) generally results in a decrease in the magnetic relaxation time, thereby raising the relaxation frequency. At very high frequencies, domain walls are unable to follow the external field, and the magnetization mechanism takes the form of gyromagnetic resonance, which is also dependent on spin–lattice coupling. High-anisotropy materials, such as hexagonal ferrites, show the later appearance of gyromagnetic loss effects because of an increased value of anisotropy field.

4.6 Further Reading

The topics in this chapter including dielectrics, permittivity, dielectric relaxation and resonance are presented in greater detail in public domain [1, 30] and advanced textbooks [2, 19] on electronic materials. Magnetization, permeability, and magnetic resonance and relaxation are presented in greater detail in public domain [31, 32] and advanced textbooks [22, 25, 27] on magnetism and magnetic materials.

References

1. *Dielectric* (Wikipedia). https://en.wikipedia.org/wiki/Dielectric
2. S.O. Kasap, *Principles of Electronic Materials and Devices*, 4th edn. (McGraw-Hill, New York, 2018), pp. 659–678
3. *Ion Jump Polarization* (Big Chemical Encyclopedia)
4. R. Dorey, *Ceramic Thick Films for MEMS and Microdevices* (Elsevier Inc., New York, 2011), pp. 85–112
5. R.P. Jimenez, J.A. Eiras, Ferroelectrics **569**, 295 (2020)
6. G. Arlt, H. Dederichs, R. Herbiet, Ferroelectrics **74**, 37 (1987)
7. A.A. Bokov, Z.-G. Ye, J. Appl. Phys. **95**, 6347 (2004)
8. J. Erhart, Phase Trans. **77**, 989–1074 (2004)
9. G. Liu, S. Zhang, W. Jiang, W. Cao, Mater. Sci. Eng. R **89**, 1 (2015)
10. G. Arlt, U. Böttger, S. Witte, Ann. Phys. **506**, 578 (1994)
11. M. Qin, L. Zhang, H. Wu, Adv. Sci. **9**, 2105553 (2022)
12. Y. Wu, R. Shu, J. Zhang, R. Sun, Y. Chen, J. Yuan, J. Alloys Compd. **785**, 616 (2019)
13. X. Wang, M. Zhang, J. Shu, B. Wen, W. Cao, M. Cao, Carbon **184**, 136 (2021)
14. J. Feng, Y. Zong, Y. Sun, Y. Zhang, X. Yang, G. Long, Y. Wang, X. Li, X. Zheng, Chem. Eng. J. **345**, 441–451 (2018)
15. P. Liu, S. Gao, G. Zhang, Y. Huang, W. You, R. Che, Adv. Funct. Mater. **31**, 2102812 (2021)
16. L. Wang, X. Yu, X. Li, J. Zhang, M. Wang, R. Che, Chem. Eng. J. **383**, 123099 (2020)
17. Z. Gao, Y. Song, S. Zhang, D. Lan, Z. Zhao, Z. Wang, D. Zang, G. Wu, H. Wu, J. Colloid Interface Sci. **600**, 288 (2021)
18. T. Li, L. Xia, H. Yang, X. Wang, T. Zhang, X. Huang, L. Xiong, C. Qin, G. Wen, ACS Appl. Mater. Interfaces **13**, 11911 (2021)
19. S.O. Kasap, *Principles of Electronic Materials and Devices*, 4th edn. (McGraw-Hill, New York, 2018), pp. 679–690
20. G.G. Raju, Dielectric loss and relaxation—I, in *Dielectrics in Electric Fields* (CRC Press, 2016), pp. 83–90
21. *Plasma Oscillation* (Wikipedia). https://en.wikipedia.org/wiki/Plasma_oscillation
22. T.H. Noh, *Magnetic Materials* (Dooyangsa Co., Seoul, 2006), pp. 323–366 (in Korean)

23. *Landau–Lifshitz–Gilbert Equation* (Wikipedia)
24. D.M. Pozar, *Microwave Engineering*, 4th edn. (Wiley, USA, 2012), pp. 452–462
25. J. Smit, H.P.J. Wijn, *Ferrites* (Philips Technical Library, Eindhoven, 1959), pp. 202–211
26. C. Kittel, Phys. Rev. **73**, 155 (1949)
27. J. Smit, H.P.J. Wijn, *Ferrites* (Philips Technical Library, Eindhoven, 1959), pp. 268–300
28. G.F. Dionne, IEEE Trans. Magn. **39**, 3121 (2003)
29. T. Tsutaoka, J. Appl. Phys. **93**, 2789 (2003)
30. *Permittivity* (Wikipedia). https://en.wikipedia.org/wiki/Permittivity
31. *Magnetization* (Wikipedia). https://en.wikipedia.org/wiki/Magnetization
32. *Permeability (Electromagnetism)* (Wikipedia). https://en.wikipedia.org/wiki/Permeability_%28electromagnetism%29

Chapter 5
Electromagnetic Wave Absorber Design

This chapter describes the design methods for different types of EM wave absorbers based on equivalent circuit models and simulation techniques. All EM wave absorbers of various geometries, from plane to pyramid or from monolayer to multilayer with or without periodic conductive elements, can be represented as equivalent electrical circuits, allowing for the design of optimal structures with easier mathematical processing. As one of the simulation techniques, HFSS is introduced, including the entire simulation process. Starting from the conventional resonant-type absorbers (1/4 wavelength absorber, Salisbury screen, etc.), we review the latest metamaterial and frequency selection surface (FSS) absorbers with more advanced design techniques. Equivalent circuit method is introduced to determine the circuit parameters of FSS (inductance L, capacitance C) as a function of the shape and dimensions of the periodic conductor elements. Recent research on how to design ultra-wide band absorbers by multilayering FSSs with different dimensions or patterns is also included. Another design technique for wide-bandwidth absorbers describes a method for controlling the shape of a lossy material (grid-type, reverse grid-type, and pyramid-shaped absorbers). Finally, the design method for oblique incidence absorbers is included with the results of simulation and experiment for single-layer and multilayer structures.

5.1 Equivalent Circuit Model and Simulation Technique

Electromagnetic (EM) waves are composed of oscillating electric and magnetic fields, propagating through space with an electromagnetic radiant energy. When EM wave encounters other media, it is partly reflected and partly transmitted. The transmitted wave can also be attenuated (absorbed) if the medium has loss characteristics. The reflection, transmission, and absorption of EM waves will depend on the

S. Kim, *Fundamentals and Principles of Electromagnetic Wave Absorbers*, Springer Series in Advanced Microelectronics 66, https://doi.org/10.1007/978-981-97-6477-8_5

material parameters, complex permittivity ($\varepsilon_r = \varepsilon_r' - j\varepsilon_r''$) and complex permeability ($\mu_r = \mu_r' - j\mu_r''$), of the medium involved.

The reflection of EM waves at the boundary is determined by the impedance of medium. The impedance Z is defined as the ratio of electric field E and magnetic field H, and expressed as a function of the material parameters, ε_r and μ_r:

$$Z = \frac{E}{H} = Z_0\sqrt{\frac{\mu_r}{\varepsilon_r}} \tag{5.1}$$

where Z_0 is the impedance of free space (= 377 Ω). The transmitted wave may decrease exponentially in magnitude if the medium has loss characteristics (imaginary part of complex permeability and permittivity, μ_r'' and ε_r''). The attenuation increases as the loss tangent ($\tan\delta$) increases, which is defined as

$$\tan\delta = \frac{\varepsilon_r''}{\varepsilon_r'} \text{ or } \frac{\mu_r''}{\mu_r'} \tag{5.2}$$

The ideal EM wave absorber is based on two basic principles: the first is that the impedance of material should be equal to the impedance of air, so that the reflection of the EM wave at the interface does not occur. Second, the loss parameter of the material is large, so the transmitted wave must be attenuated inside the material. As such, the EM wave absorber can be regarded as an electronic device that controls the reflectivity of the EM wave below the required value by using the permittivity, permeability, and loss characteristics of the constituent medium. Wave propagation in an infinite media can be interpreted by electrical circuit theory with the advantage of easier mathematical treatment, and the tool that connects them is the transmission line. Transmission line theory is useful for the design and analysis of EM wave absorbers by replacing the medium of the constituent material parameters with an electrical circuit of hypothetical lumped elements, which is called the equivalent circuit model.

If the structure of the EM wave absorber becomes complex, it may be difficult to apply the equivalent circuit model. A typical example is an absorber consisting of a metamaterial or frequency-selective surface. This type of absorber has a periodic structure of unit cells that are repeated in two dimensions (2D) or three dimensions (3D). Simulation tools can be used to model the interaction between EM waves and the periodic structures to derive the optimized materials and geometries. The advantage of simulation technology is that it is possible to design the optimized structures and predict their performance with high accuracy before final confirming through experiments. Well-known commercial simulation tools include CST Microwave Studio [1], HFSS [2], and Comsol [3]. For example, High Frequency Structure Simulator (HFSS) is a 3D EM simulation software for designing and simulating the high-frequency electronic products such as antennas, antenna arrays, RF or microwave components, etc. Even without knowledge of esoteric electromagnetic field theory, it is possible to perform the high-speed and high-accuracy analysis simply by simulation with basic data on device geometry, material properties, and analysis frequency.

5.1.1 EM Simulation Techniques

There are several technical approaches to EM simulation. If there are problems with complex geometries, loads, and material properties, it is usually impossible to obtain an analytical mathematical solution. Here, numerical methods can be used to obtain a proper approximation which is very close to real values. The three most well-known EM simulation techniques are MoM (method of moments), FEM (finite element method), and FDTD (finite difference time domain) solutions [4–6]. For planar structures such as PCBs and RF boards, MoM provides the most efficient simulation results. For 3D structures such as microstrips, waveguides, 3D antennas, etc., FEM or FDTD is generally more suitable.

FEM is a numerical analysis method that approximates the answer to a phenomenon by dividing the area or volume of an object into small regions called elements, and is used in various ways to solve problems such as engineering and mathematical physics [7]. In FEM, instead of solving the problem for the entire object at once, the structure is divided into small areas or volumes to find the approximate solution. The solutions obtained from each small element are then combined or summed to obtain the solution of the entire system. The modeling process of dividing objects into smaller objects or units (elements) and making them into equivalent systems connected by points (nodes), boundaries, or interfaces shared by two or more elements is called the discretization or lattice division [7, 8]. In FEM, the location where the result (unknown) is calculated is commonly referred to as the node, which is the vertex of an element. Recently, this FEM has been used as one of the standard models for analyzing electromagnetic fields. In FEM, the simulation volume is typically discretized using the tetrahedral mesh cells. The value of vector field quantity (e.g., E field or H field) at a point inside each tetrahedron is interpolated from the vertices of the tetrahedron. Finite element models are analyzed through a numerical analysis process, and the results are evaluated through post-processing. For electromagnetic field problems, the output is evaluated, including the S (scattering) parameter, Y (admittance) parameter, Z (impedance) parameter, etc.

Using a similar approach, FDTD technology is often used to solve the specific EM problems. In FDTD analysis, the entire volume of simulation is typically discretized using the rectangular mesh cells. FDTD uses a time-step algorithm that updates the field values across the mesh cell time-step by time-step, providing data over a wide frequency range [5].

5.1.2 HFSS Simulation Process

The design and simulation of EM wave absorbers often uses the electromagnetic simulation tools that use FEM. Here we introduce ANSYS HFSS. HFSS is a three-dimensional (3D) analysis tool using FEM that divides the object of analysis into a mesh of tetrahedral structures, and derives the electromagnetic properties of the entire structure by successively calculating the E fields and H fields at the points of tetrahedron. The simulation process of HFSS can be divided into four parts: model design, parametric model (input of material parameters, boundary conditions, port excitations), analysis (mesh division and frequency sweep), and results output, as shown in Fig. 5.1 [9].

(1) **Model Design**

It is a step of generating layout geometry for designing the structure and pattern of the electronic device to be analyzed. Various length units can be set according to the user's convenience. Although the rectangular coordinates are usually used, various coordinate axes can be set. We can use a variety of 3D modeling tools, including drawing and manipulating, to construct the model structure. In the case of a periodic structure, only one unit cell can be designed, and an infinite repeating pattern can be applied with the appropriate boundary conditions described below.

(2) **Parametric Model**

When the modeling of structure and pattern is completed, it is necessary to specify the material of structure. Initially, it is designated as vacuum, but we can select the material that is set by default, or we can add a substance to apply. At this time, we enter the material parameters to specify the desired material.

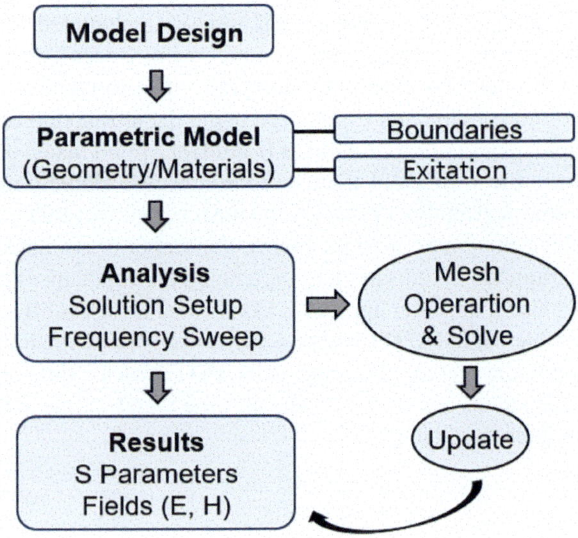

Fig. 5.1 HFSS simulation process

Setting boundary conditions is necessary to simplify the model structure and increase the mesh effect. When electromagnetic waves traveling in a certain direction meet the interface, the electric and magnetic fields of the electromagnetic waves at the boundary can be divided into normal and vertical (tangential) components. Boundary conditions determine the reflection and transmission of the vertical and horizontal components of electromagnetic waves, depending on the characteristics of the interface.

The basic boundary conditions are Perfect Electric Conductor (PEC) and Perfect Magnetic Conductor (PMC), as shown in Fig. 5.2. In PEC boundary condition, an E field is formed in the vertical direction of the surface. It is modeled like a metal plate, acting as if electromagnetic waves are completely reflected. At the PMC boundary, the H field is formed vertically and the E field is formed horizontally. The magnetic field is reflected and the electric field is transmitted. When modeling the infinitely large repeating array structures, such as frequency-selective surfaces (FSS) or metamaterials, Master–Slave boundary conditions are useful instead of PEC/PMC boundary conditions, as shown in Fig. 5.3. The Master–Slave boundary must exist in pairs on opposite sides.

The 3D tool uses a model that sets a 2D square area of a certain size at the input site and fully incidents electromagnetic waves into the cross-section. This is called a waveguide port. Wave ports are used in closed structures such as waveguides and coaxial cables, and the entire cross-section of the input is set to the port. After the plane waves are randomly emitted into this circular or square-shaped port, the electromagnetic waves will automatically converge to a specific mode. If we need to consider the simulation of large planar and periodic structures, such as frequency selective surfaces, the Floquet port is designated as an I/O port. Floquet ports can only be granted under Master–Slave boundary conditions.

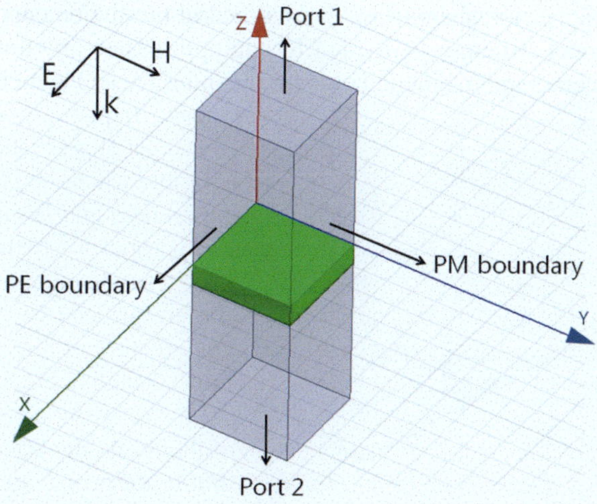

Fig. 5.2 PEC and PMC boundary conditions in HFSS simulation

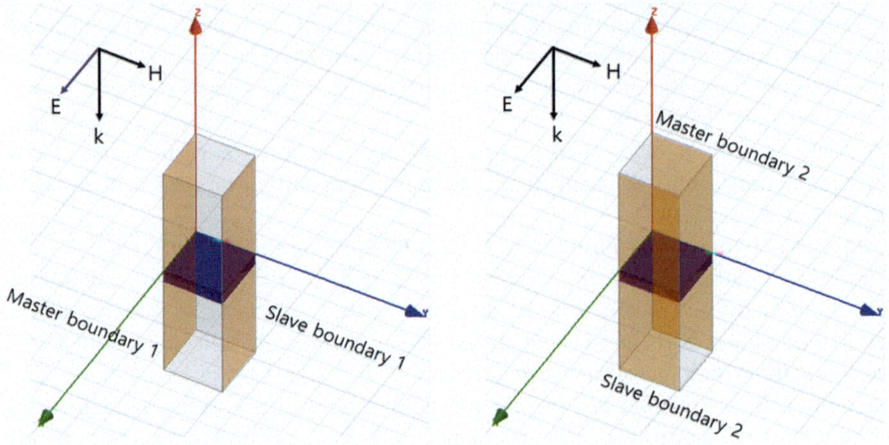

Fig. 5.3 Master–Slave boundary conditions in HFSS simulation

(3) **Analysis**

After setting up the parametric model, we enter the analysis conditions before running the program. In this step, there are two tasks: the input of adaptive conditions for mesh generation and the frequency sweep conditions for viewing the frequency-specific results.

The physical model (layout geometry) is divided into the appropriate units, and Maxwell's equations are solved for each unit, each of which is referred to as a mesh or cell [9, 10]. The field/current across the mesh cell is then approximated using a local function (often referred to as the Extension or Base function). Figure 5.4 shows an example of mesh generation on a microstrip line covered with a noise suppression sheet using the tetrahedral mesh elements in an HFSS simulation. The model structure is divided by the appropriate amount of mesh grid, and then the electric and magnetic fields for the interface and center of each mesh are calculated.

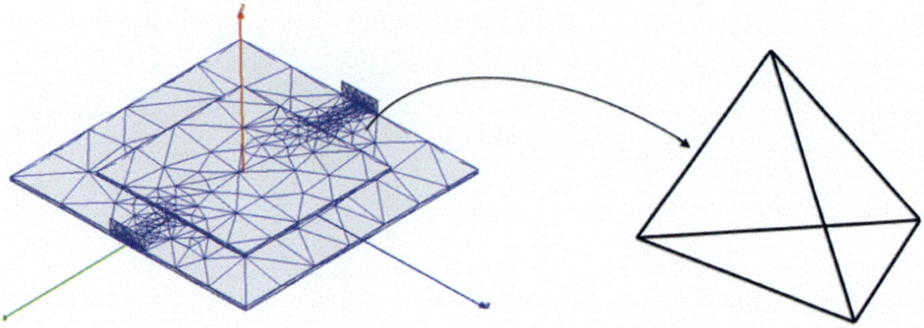

Fig. 5.4 Mesh division in microstrip line and a tetrahedral mesh element in HFSS simulation

The mesh partitioning process, called the adaptive pass, has a very decisive effect on the accuracy of analysis, so it must be carefully considered and entered. HFSS uses the concept of ΔS, which stands for the maximum fine-tuning per pass, to determine the degree of mesh splitting. When the desired ΔS value is satisfied through multiple passes (repeated execution), the mesh division is terminated and the process of acquiring the S parameters through frequency sweep begins. The iterative process of using the reference variable ΔS to create a mesh is called the adaptive process. The reason why this adaptation process is necessary in the FEM algorithm is that stable mesh partitioning is essential for accurate calculations. The adaptive solution process uses ΔS and the maximum number of passes to converge to the correct result. If convergence is reached before the maximum number of passes is performed, the solution process stops. Typically, $\Delta S \approx 0.02$ is used.

The frequency sweep calculates the distribution of the S parameter in a particular frequency band, and also includes the meaning of the reference frequency that divides the mesh. There are three types of frequency sweeps: discrete, fast, and interpolating. 'Discrete' is a method of deriving S parameters for each set frequency, so it takes a long time to obtain the correct value. 'Fast' is a method of extracting the results of the upper and lower bounds based on the results of a single frequency. It has a very short time and is suitable for narrowband analysis. 'Interpolation' is an intermediate method between 'discrete' and 'fast' and is recommended for calculating the wideband S parameters.

(4) **Results**

It is a part that shows a diagram, graph, or 2D/3D field distribution on the screen after receiving the analysis result from the solver. The performance of EM wave absorber is represented by absorbance $A(\omega)$, which is expressed as $A(\omega) = 1 - R(\omega) - T(\omega)$, where $R(\omega)$ is the reflectance and $T(\omega)$ is the transmittance. The reflectance and transmittance of infinite periodic structure are determined by simulation of one unit cell with the assignment of appropriate boundary conditions and port excitations. The simulation output is indicated by S parameters (reflection coefficient S_{11} and transmission coefficient S_{21}). Using the relation $R(\omega) = |S_{11}|^2$ and $T(\omega) = |S_{21}|^2$, power absorption can be calculated.

The simulation also provides Y (admittance) and Z (impedance) parameters. The simulation allows monitoring of the distribution of electric fields, magnetic fields, and surface currents, and provides information about the resonance mechanism of EM wave absorbers. For example, if electric currents in opposite direction are distributed on the metal elements located above and below the substrate, it is interpreted that magnetic resonance has occurred. Conversely, the parallel current distribution on both sides implies electrical resonance.

(5) **Simulation Example**

An example of HFSS simulation process is presented in Appendix A, which shows how to design, simulate, and analyze an EM wave absorber with a sample of metamaterial consisting of a front cross-conductor, a dielectric substrate, and a ground back plane.

5.2 Reflection Coefficient

The ideal EM wave absorber is to absorb the energy of incident wave at all frequencies without any reflections. In reality, however, it is impossible to obtain the perfect absorbers with a simple geometry and a reasonable thickness. In general, the reflection coefficient Γ is assigned below a specific value and the corresponding frequency band is set as the range of uses. If the allowed reflection coefficient $|\Gamma_0|$ is given in voltage (or electric field), then, in power, it becomes $|\Gamma_0|^2$. The values of $|\Gamma_0|$ and $|\Gamma_0|^2$, and the reflection coefficient in dB ($= 20\log|\Gamma_0|$) are shown in Table 5.1.

EM wave absorbers are classified according to the frequency characteristics of absorption bandwidth. Narrowband (N-type) absorbers satisfy $|\Gamma| \leq |\Gamma_0|$ in a narrow frequency range near f_0, as illustrated in Fig. 5.5a. Wideband (W-type) absorbers satisfy $|\Gamma| \leq |\Gamma_0|$ in a wide frequency range above f_L, as illustrated in Fig. 5.5b. For the N-type absorbers, a figure of merit on absorber performance can be represented by the fractional bandwidth (FBW), which is defined by the ratio of bandwidth B to the center frequency f_0. The conventional narrow-band absorbers, such as Salisbury screens [11, 12] or Dallenbach layers [12] exhibit a single absorption peak with a typical value of FBW smaller than 20%. In the W-type absorbers, the bandwidth is limited by the lowest frequency f_L, which is a theoretical value determined from the given reflectance and thickness [13]. Then the figure of merit of W-type absorbers will be given by the thickness d required for absorption divided by the wavelength λ_L corresponding to the lowest frequency f_L. It is considered that the smaller the d/λ_L, the better the absorber performance. The Jaumann absorber [14–16], one example of W-type absorbers with a multilayer structure, has a very large FBW up to 170%, but it suffers from a much bigger thickness ($\sim 0.3\lambda_L$). A physical bound on the lowest achievable frequency of absorption bandwidth is given by $f_L = c|\Gamma_0|/172d$ for a non-magnetic multilayer absorber [13], where c is the velocity of light, $|\Gamma_0|$ is the reflection coefficient in dB, and d is the total thickness of absorber. The lower the reflectance and the smaller the thickness, the higher the lowest frequency (i.e., the theoretical bandwidth is decreased).

Table 5.1 Reflection coefficient in voltage, power, and dB

Voltage $\lvert\Gamma_0\rvert$	Power $\lvert\Gamma_0\rvert^2$	$\lvert\Gamma_0\rvert$ in dB
0.316	0.1	– 10
0.1	0.01	– 20
0.03	0.0009	– 30
0.01	0.0001	– 40

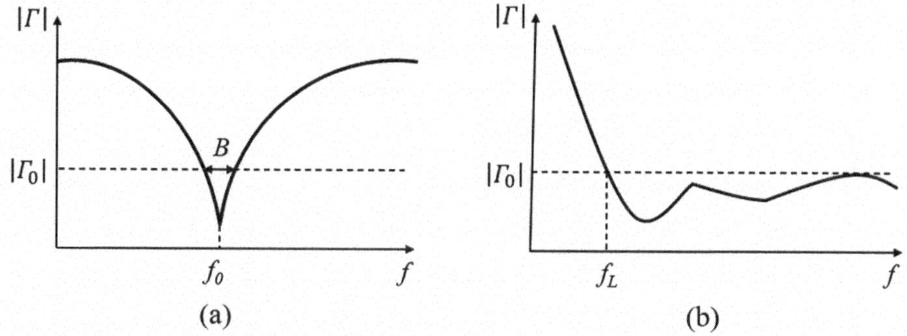

Fig. 5.5 Reflection coefficient versus frequency characteristics of EM wave absorbers: **a** narrowband (N-type) and **b** wideband (W-type)

5.3 Narrowband (N-type) Absorbers

5.3.1 1/4 Wavelength Thickness Absorbers

Because metals have a high electrical conductivity, they completely reflect EM waves. To prevent this, an EM wave absorbing material is coated or attached over the metal plate, as shown in Fig. 5.6. In this case, the principle of EM wave absorption is explained as follows.

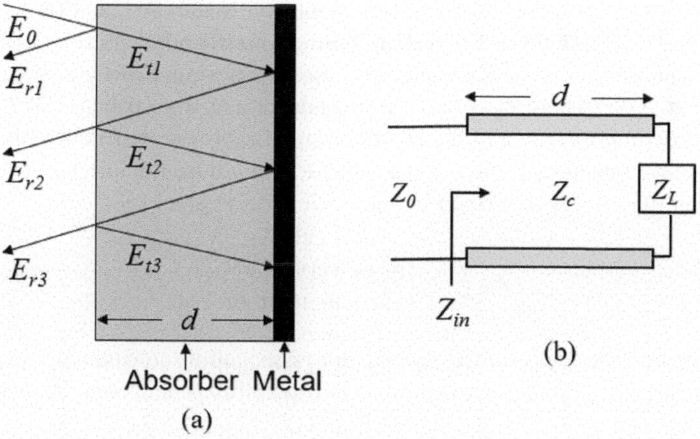

Fig. 5.6 Illustration of a 1/4 wavelength thickness EM absorber: **a** principle of operation and **b** equivalent circuit

(a) When EM wave with electric field strength E_0 is incident into the absorber whose thickness is d, a part is reflected from the surface by the impedance difference between two media (reflected wave with E_{r1}), and the rest is transmitted into the interior of absorbent layer (transmitted wave with E_{t1}).
(b) The internally transmitted waves are attenuated exponentially by the dissipation characteristics of the EM absorbing material.
(c) When the transmitted wave reaches the metal plate at the back side of EM absorber, it is fully reflected.
(d) The reflected wave is attenuated again in the EM absorption material to reach the surface of EM wave absorber.
(e) Due to the impedance mismatch at the surface, the wave is partially reflected and returned to the inside of absorbent layer (E_{t2}), and the rest is transmitted and exit outside the absorber (E_{r2}).
(f) This process is repeated.

The reflection coefficient Γ is then given by the ratio of total reflected waves to the incident wave, that is

$$\Gamma = \frac{E_r}{E_0} = \frac{\sum E_{rn}}{E_0} \tag{5.3}$$

Since the EM wave is exponentially attenuated inside the material, most of the attenuation occurs during the first-round traveling. Therefore, if there is a phase difference of 180° between reflected wave with E_{r1} and emergent wave with E_{r2}, they are canceled out, so that reflection coefficient becomes zero. When the thickness of absorbent layer is 1/4 of wavelength, this condition is satisfied, so this type of EM wave absorber is called the 1/4 wavelength thickness absorber.

An equivalent circuit for the EM wave absorber terminated by a metal plate is shown in Fig. 5.6b, where Z_0 is the wave impedance of free space (= 377 Ω), Z_c is the characteristic impedance of the absorber layer, Z_L is the impedance of load, and Z_{in} is the input impedance at the surface of absorber whose thickness is d. Using the transmission line theory described in Sect. 2.2.3, Z_{in} is given by

$$Z_{in} = Z_c\left[\frac{Z_L + Z_c \tanh \gamma d}{Z_c + Z_L \tanh \gamma d}\right] \tag{5.4}$$

The characteristic impedance Z_c and the propagation constant γ of absorbent medium are given as a function of complex permeability μ_r and complex permittivity ε_r:

$$Z_c = Z_0\sqrt{\frac{\mu_r}{\varepsilon_r}} \tag{5.5}$$

$$\gamma = j\frac{2\pi}{\lambda}\sqrt{\mu_r\varepsilon_r} \tag{5.6}$$

where λ is the wavelength in free space. Since the back side is shorted to a conductor ($Z_L = 0$), Eq. (5.4) is simplified to

$$Z_{in} = Z_c \tanh \gamma d \tag{5.7}$$

For a plane wave incident normally, the reflection coefficient Γ is given by

$$\Gamma = \frac{Z_{in} - Z_0}{Z_{in} + Z_0} \tag{5.8}$$

In order to satisfy the non-reflected condition ($\Gamma = 0$), the input impedance must be equal to the free space impedance ($Z_{in} = Z_0$). This is called the impedance matching. Using Eqs. (5.7) and (5.8), the equation that satisfies the impedance matching can be expressed as

$$1 = \sqrt{\frac{\mu_r}{\varepsilon_r}} \tanh\left(j\frac{2\pi}{\lambda}\sqrt{\varepsilon_r\mu_r}d\right) = \sqrt{\frac{\mu_r}{\varepsilon_r}} \tanh\left(j\frac{2\pi f}{c}\sqrt{\varepsilon_r\mu_r}d\right) \tag{5.9}$$

where c is the speed of light. The variables that satisfy Eq. (5.9) include frequency f, absorber thickness d, real and imaginary parts of complex permeability (μ_r', μ_r'') and complex permittivity (ε_r', ε_r''). For a given material, the frequency and thickness satisfying Eq. (5.9) are called the matching frequency f_m and the matching thickness d_m, respectively.

Figure 5.7 shows the $\mu_r' - \mu_r''$ plane that satisfies Eq. (5.9) in the range of $f \cdot d = 300 \sim 9000$ MHz mm [17–19]. Since the complex permittivity does not change greatly with frequency, in general, a constant value of dielectric loss tangent ($\tan \delta_\varepsilon$) can be used in calculation. The impedance matching solution map can be divided into two regions.

In the region of $\mu_r'' \geq \mu_r'$, a product of matching frequency and matching thickness ($f \cdot d$) does not change with μ_r' and ε_r', but mainly decreases with increasing μ_r''. Even for a higher value of $\tan \delta_\varepsilon$, $f \cdot d$ trajectory in this region does not change. In order to reduce the absorber thickness, a high magnetic loss material is required. While, in the region of $\mu_r' \geq \mu_r''$, $f \cdot d$ trajectory is dependent on all material parameters including the complex permittivity in addition to the complex permeability. In order to satisfy the impedance matching in this region, particularly, a high dielectric constant greater than 100 will be required for $f \cdot d = 300 \sim 9000$ MHz mm.

In Fig. 5.7, if $\mu_r'' \geq \mu_r'$ and $\tan \delta_\varepsilon$ is very small, the solution of the impedance matching equation can be simplified to the following equations [20, 21]:

$$\varepsilon_r' = 3\mu_r' \tag{5.10}$$

$$d/\lambda = (\pi \mu_r'')/2 \tag{5.11}$$

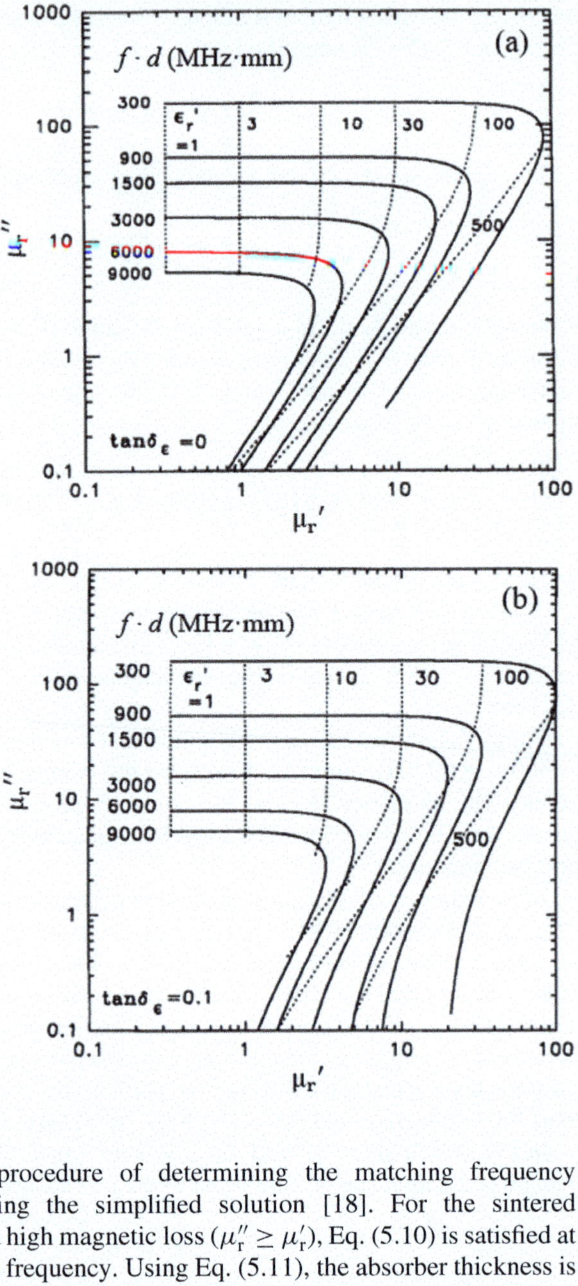

Fig. 5.7 The trajectory of frequency and thickness ($f{\cdot}d$) for the impedance matching in $\mu_r' - \mu_r''$ plane for a 1/4 wavelength thickness EM absorber: **a** $\tan\delta_\varepsilon = 0$, **b** $\tan\delta_\varepsilon = 0.1$. Reproduced with permission from Ref. [18]

Figure 5.8a shows the procedure of determining the matching frequency and matching thickness using the simplified solution [18]. For the sintered $Ni_{0.5}Zn_{0.5}Fe_2O_4$ ferrite with a high magnetic loss ($\mu_r'' \geq \mu_r'$), Eq. (5.10) is satisfied at 592 MHz and $\mu_r'' = 20$ at that frequency. Using Eq. (5.11), the absorber thickness is calculated to be $d = 4$ mm. The matching frequency and matching thickness obtained using the approximated solution are in good agreement with the result of reflection loss shown in Fig. 5.8b.

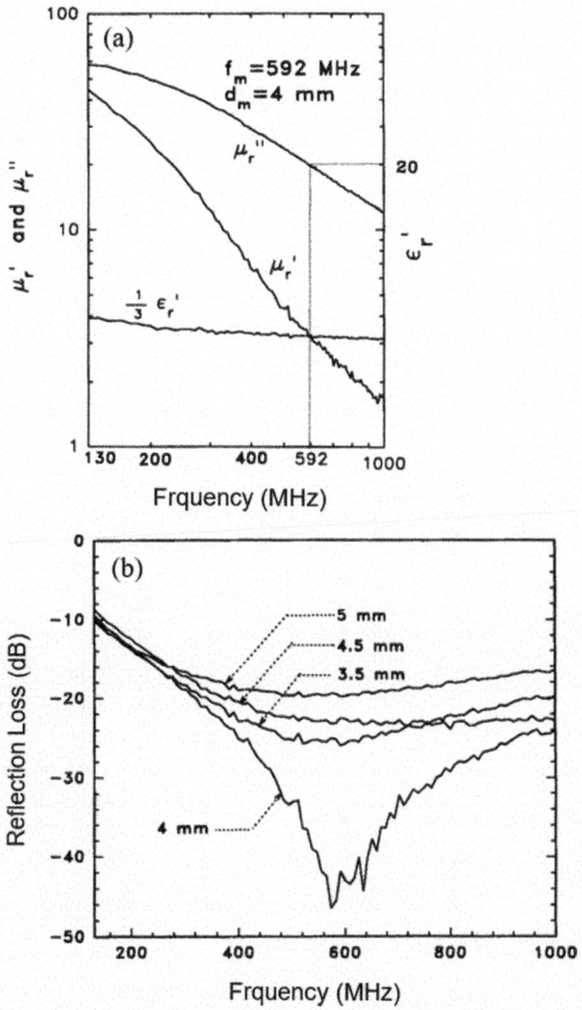

Fig. 5.8 EM wave absorption properties of sintered $Ni_{0.5}Zn_{0.5}Fe_2O_4$ ferrite: **a** material parameters for matching frequency and thickness, and **b** reflection loss with varying the plate thickness. Reproduced with permission from Ref. [18]

Figure 5.9 shows $\mu_r' \geq \mu_r''$ region to obtain the matching frequency and matching thickness using the impedance matching solution trajectory. A typical material belonging to this region is the polymer composites containing the ferrite particles as absorbent fillers [22]. Using the complex permeability and permittivity of the composite material mixed with NiZn ferrite powders with rubber matrix, the trajectory of $\mu_r' - \mu_r''$ intersects $\varepsilon_r' = 6.2$ curve at 10.4 GHz, where $f \cdot d = 42$ GHz mm. From this, we predict the matching frequency and matching thickness, $f_m = 10.4$ GHz and $d_m = 4$ mm, respectively. The reflection loss shown in Fig. 5.10 is in good agreement with the theoretical prediction.

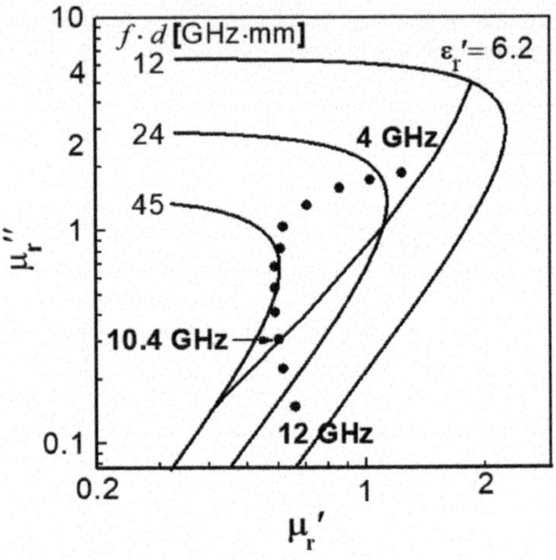

Fig. 5.9 Determination of matching frequency and matching thickness in the polymer composite of NiZn ferrite powders with material parameters (μ_r', μ_r'', ε_r') shown in figure

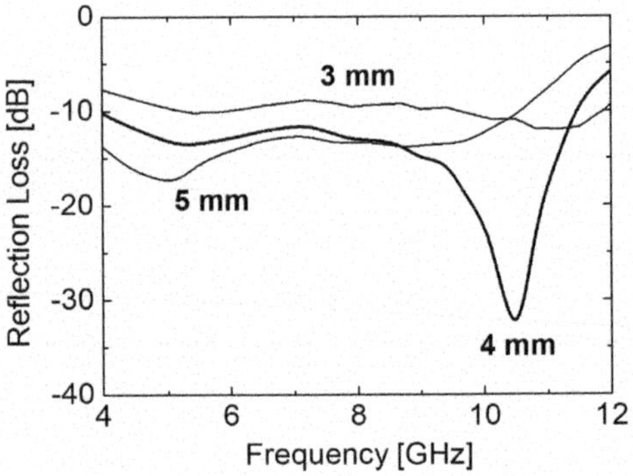

Fig. 5.10 Reflection loss determined in the composite of NiZn ferrite powder with varying the layer thickness

5.3.2 *Dielectric Absorbers*

A non-magnetic (dielectric) material has a relative permeability $\mu_r = 1$. Equation (5.9) for the impedance matching is simplified to

$$1 = \frac{1}{\sqrt{\varepsilon_r}} \tanh\left(j\frac{2\pi}{\lambda}\sqrt{\varepsilon_r}d\right) \tag{5.12}$$

The d/λ (thickness/wavelength) trajectory satisfying Eq. (5.12) for the given values of ε_r' and ε_r'' is shown in Fig. 5.11a.

Table 5.2 shows the values of ε_r' and ε_r'' that satisfy the impedance matching for a given value of d/λ. For an example of impedance matching at $d/\lambda = 0.1$, a dielectric material with $\varepsilon_r' = 6.5$ and $\varepsilon_r'' = 3.1$ is required. When the frequency is fixed at 10 GHz, the matching thickness of the absorber is given in Fig. 5.11b. It can

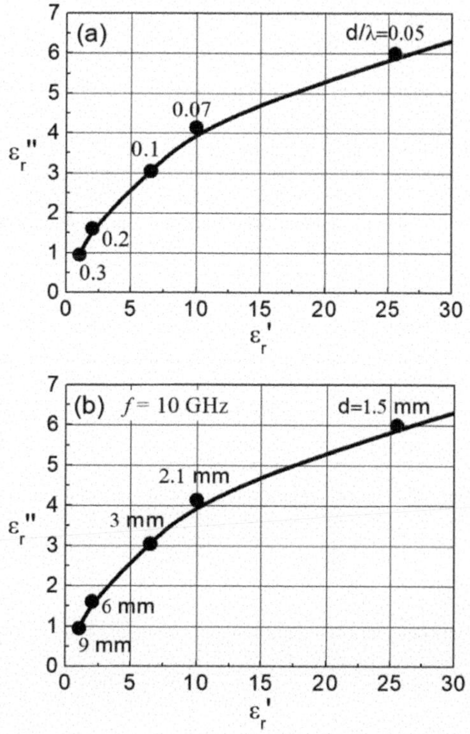

Fig. 5.11 A trajectory of **a** thickness/wavelength (d/λ) and **b** thickness d at a fixed frequency of 10 GHz ($\lambda = 30$ mm) for impedance matching in $\varepsilon_r' - \varepsilon_r''$ plane of dielectric absorbers

be seen that in order to reduce the matching thickness, it is necessary to increase the complex permittivity along the d/λ trajectory [17, 23].

Figure 5.12 shows the procedure of determining the matching frequency and matching thickness with the solution map of impedance matching. A typical material for dielectric absorbers is SiC fiber-epoxy composite whose ε_r' and ε_r'' are given in Fig. 5.12a. The trajectory of $\varepsilon_r' - \varepsilon_r''$ intersects $d/\lambda = 0.08$ at 10 GHz ($\lambda = 30$ mm), as shown in Fig. 5.12b. From this, we can predict the matching frequency and matching thickness, $f_m = 10$ GHz and $d_m = 0.08 \times 30 = 2.4$ mm, respectively. The reflection loss shown in Fig. 5.13 is in good agreement with the theoretical prediction.

Table 5.2 Complex permittivity that satisfies the impedance matching for a given value of thickness/wavelength (d/λ)

d/λ	0.03	0.05	0.07	0.1	0.2	0.3
ε_r'	70.0	25.5	10.0	6.5	2.0	1.0
ε_r''	10.0	6.0	4.1	3.1	1.6	0.95

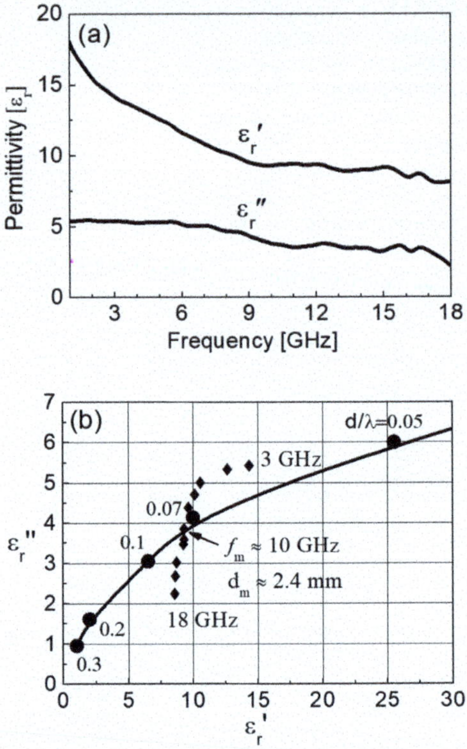

Fig. 5.12 **a** Complex permittivity of SiC fiber-epoxy composite and **b** determination of matching frequency and matching thickness in $\varepsilon_r' - \varepsilon_r''$ plane

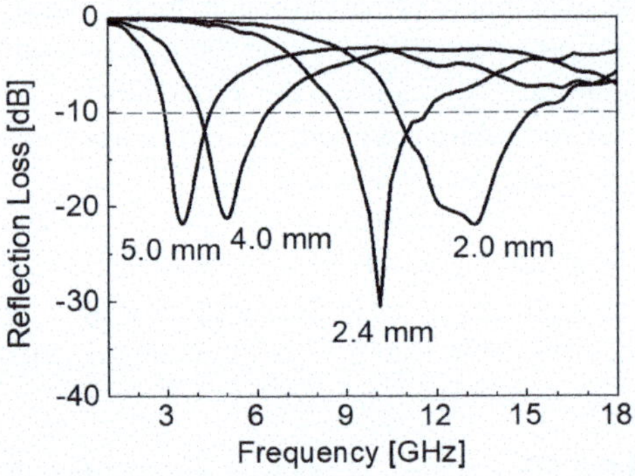

Fig. 5.13 Reflection loss determined in SiC fiber-epoxy composite with varying the thickness

5.3.3 Resistive Film Absorbers (Salisbury Screens)

N-type EM wave absorber using a resistive film is the Salisbury screen [11, 24]. As shown in Fig. 5.14a, it has a structure in which a lossless dielectric spacers of 1/4 wavelength thickness is placed in front of the metal plate and a resistive film is coated on its surface. Expressing this as an equivalent circuit is shown in Fig. 5.14b.

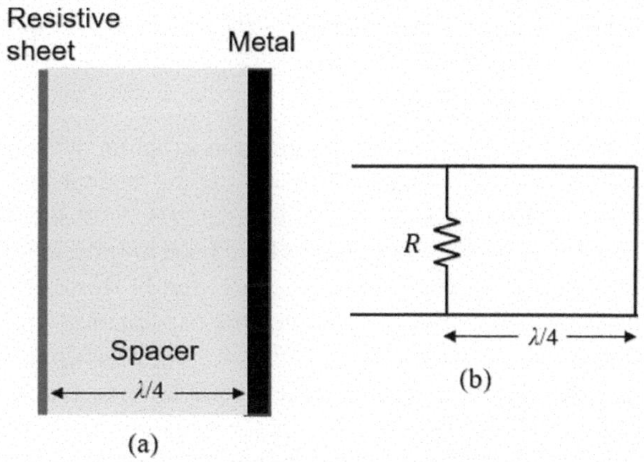

Fig. 5.14 Illustration of **a** Salisbury screen absorber and **b** equivalent circuit

The equivalent impedance Z at the surface of dielectric spacer with thickness d is determined by the parallel connection of input impedance of spacer Z_{in} and resistive film impedance Z_R equal to film resistance R, and each impedance is given by the following formula:

$$Z_{in} = Z_0\sqrt{\frac{1}{\varepsilon_r}}\tanh\left(j\frac{2\pi}{\lambda}\sqrt{\varepsilon_r}d\right) \tag{5.13}$$

$$Z_R = R \tag{5.14}$$

$$Z = \frac{Z_{in}Z_R}{(Z_{in} + Z_R)} \tag{5.15}$$

When the imaginary part of Z becomes 0 and the real part of Z becomes 377 Ω, impedance matching is achieved, and the design of a non-reflected absorber can be made. The wavelength within this material of dielectric constant ε_r is given by $\lambda/\sqrt{\varepsilon_r}$. When the thickness of substrate is equal to 1/4 wavelength ($d = (\lambda/\sqrt{\varepsilon_r})/4$), $Z_{in} = \infty$. From Eq. (5.15), $Z = R = 377\ \Omega$, it becomes a non-reflected absorber. If we control the thickness of dielectric spacer to 1/4 wavelength and the resistance of surface film to 377 Ω, a perfect EM absorber can be produced. As the frequency changes from the designed value, the reflection coefficient increases because the 1/4 wavelength thickness changes. Therefore, it exhibits narrowband absorption characteristics with an FBW value of about 25% [24].

5.3.4 Circuit Analog Absorbers and Frequency Selective Surfaces

Salisbury screen absorber is simple in structure and easy to manufacture, but it is not easy to increase the bandwidth because the resonance occurs at a specified frequency. To overcome this problem, circuit analog (CA) absorbers containing capacitance and inductance in the resistive films have been proposed [24]. Several conductor patterns, such as patch, cross, loop, etc., are used in those absorbers. The spacing between the patterned elements generates the capacitance, and the current flow through the conductors generates the inductance. A similar concept has been proposed for EM wave absorbers using the frequency selective surfaces (FSS) [25–30]. FSS structures are generally composed of metal conductor patterns and exhibit selectively bandpass or bandstop characteristics for specific frequency bands. Typical forms of FSS are patch, cross, loop (square, circular), fractal, etc., as shown in Fig. 5.15. Granting resistance to these FSSs makes it correspond to the circuit analog absorbers.

A structural diagram of the resistive FSS absorber and its equivalent circuit is shown in Fig. 5.16. This absorber contains not only resistance component but also reactive components in the sheet, making impedance control more versatile. In the strip part of conductor, an inductance L is formed, and a capacitance C is produced at the gap between the pattern elements. The FSS, including the conductor resistance R, can be represented by a series RLC equivalent circuit, as shown in Fig. 5.16c.

The equivalent circuit of the EM wave absorber with a combination of FSS and dielectric substrate is shown in Fig. 5.16b. The total admittance at the absorber surface is expressed Y_{Total} as the sum of the input admittance at the surface of substrate Y_{d} and the admittance of FSS Y_{FSS}. When the back side of dielectric substrate having a thickness d is terminated by metal, the input admittance at the dielectric surface Y_{d} is given by

$$Y_{\mathrm{d}} = -Y_0\sqrt{\frac{\varepsilon_{\mathrm{r}}}{\mu_{\mathrm{r}}}}\coth\left(j\frac{2\pi fd\sqrt{\mu_{\mathrm{r}}\varepsilon_{\mathrm{r}}}}{c}\right) = Y_0\left[\mathrm{Re}(Y_{\mathrm{d}}) + j\mathrm{Im}(Y_{\mathrm{d}})\right] \tag{5.16}$$

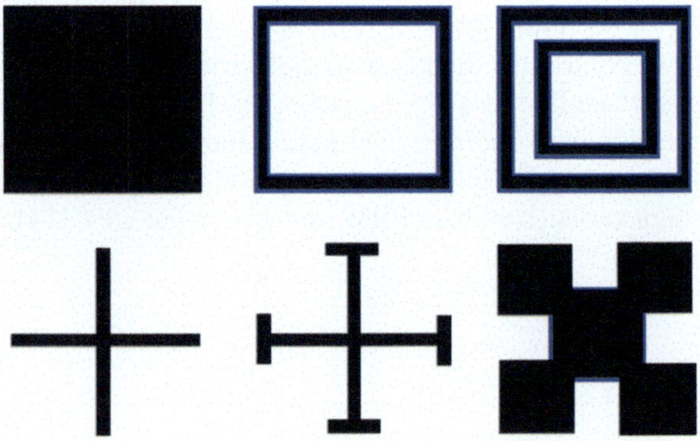

Fig. 5.15 Various FSS patterns

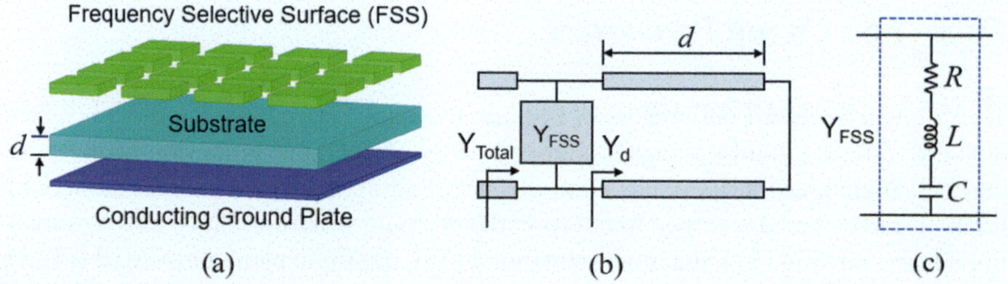

Fig. 5.16 Illustration of an FSS absorbers: **a** schematic structure, **b** equivalent circuit, and **c** series *RLC* circuit of FSS

where Y_0 is the admittance of free space ($= 1/377\ \Omega^{-1}$), ε_r and μ_r are the relative permittivity and permeability of substrate, respectively, f is the frequency, and c is the velocity of light. $\text{Re}(Y_d)$ and $\text{Im}(Y_d)$ are the real and imaginary parts of Y_d normalized by Y_0.

The admittance of FSS Y_{FSS} is expressed as a *RLC* series circuit:

$$Y_{FSS} = \frac{1}{R + j\left(\omega L - \frac{1}{\omega C}\right)} = Y_0\left[\text{Re}(Y_{FSS}) + j\text{Im}(Y_{FSS})\right] \tag{5.17}$$

where $\text{Re}(Y_{FSS})$ and $\text{Im}(Y_{FSS})$ are the real and imaginary parts of Y_{FSS} normalized by Y_0. Y_d and Y_{FSS} are in parallel connection, and the total admittance Y_{Total} is then given by

$$Y_{Total} = Y_d + Y_{FSS} = Y_0\left[\text{Re}(Y_{Total}) + j\text{Im}(Y_{Total})\right] \tag{5.18}$$

where $\text{Re}(Y_{Total})$ and $\text{Im}(Y_{Total})$ are the real and imaginary part of total admittance normalized by Y_0. In such a structure, the admittance matching conditions are given by

$$\text{Re}(Y_{Total}) = 1 \tag{5.19}$$

$$\text{Im}(Y_{Total}) = 0 \tag{5.20}$$

For a non-reflected FSS absorber, the imaginary part of Y_{Total} should be 0 and the real part should be equal to the free space admittance Y_0, which can be obtained by optimizing the FSS patterns and their dimensions, and the dielectric substrate and its thickness.

5.3.5 FSS Circuit Parameters

The equivalent circuit parameters of FSS are resistance R, inductance L, and capacitance C. These circuit parameters depend on the shape and dimension of periodic unit cell (patch, cross, loop, etc.) and the surrounding medium. Various numerical methods have been developed to interpret the periodic structures, such as the mutual impedance method [31], the moment method [32], the finite element method (FEM) [33], and the equivalent circuit method [34]. Using the simple and efficient equivalent circuit method, circuit parameters in FSS can be calculated as a function of the shape and dimensions of their elements.

(1) **Sheet Resistance**

Sheet resistance means the resistance of a square plate and is denoted by R_{sheet} or R_s, and the unit is Ω/sq or Ω/□. Also called the surface resistance, it refers to the resistance of a thin film conductor. Given the width w, thickness t, and length of the conductor l, the resistance R is expressed as

$$R = \frac{1}{\sigma}\frac{l}{wt} = \frac{1}{\sigma t}\frac{l}{w} = R_s\frac{l}{w} \tag{5.21}$$

The sheet resistance R_s equals to the resistance at $w = l$ and is given by an inverse function of the electrical conductivity σ and thickness t, that is

$$R_s = \frac{1}{\sigma t} \tag{5.22}$$

In measuring the sheet resistance, the four-probe method is most commonly used. In particular, the four-probe method is widely used to measure the surface resistance of thin films. Figure 5.17 shows an illustration of this four-probe method. A current source is connected to outer terminals through which a constant current I is supplied, and the inner terminals are connected to a device that measures the voltage V. Then the sheet resistance of thin film can be calculated using the following equation [35]:

$$R_s = \frac{\pi}{\ln(2)}\frac{V}{I} = 4.53236\frac{V}{I} \tag{5.23}$$

This equation is accurate when the material under test (MUT) is not thicker than 40% of the spacing S between the probes and the lateral size of the sample is large enough compared to S. Geometric correction factors are needed for thick and small samples [35]. The electrical resistivity ρ can be obtained from the following equation:

$$\rho = \frac{1}{\sigma} = R_s \times t \tag{5.24}$$

When a conductor pattern is periodically arranged, such as FSS, the resistance is modified by the proportion of surface area of the conductor. If we refer to the area

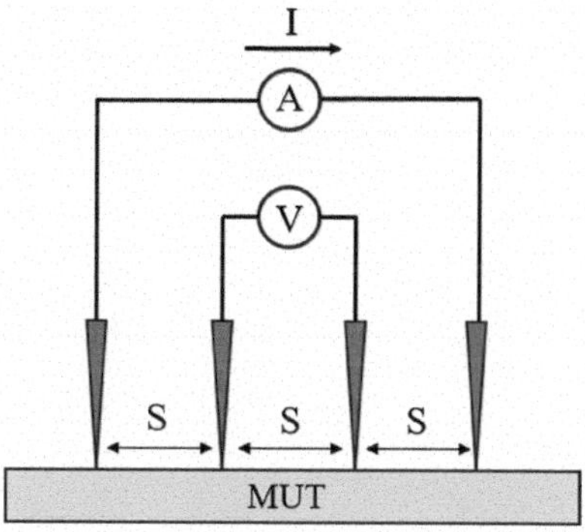

Fig. 5.17 Sheet resistance measurement by four-probe method

of unit cell A_{cell} and the area of conductor in the unit cell A_{FSS}, the FSS resistance R is given by

$$R \simeq R_s \left(\frac{A_{\text{cell}}}{A_{\text{FSS}}} \right) \tag{5.25}$$

The smaller the pattern area, the greater the FSS resistance.

(2) **Inductance, Capacitance**

The extraction of L and C of FSS by equivalent circuit model is based on the calculation formula for the periodic grating structure proposed by Marcuvitz [36]. Figure 5.18 shows a schematic representation of metal strips arranged periodically with a very thin thickness and infinite length. The repeating period is p and the width of metal strips is w. The angle of incidence of plane wave is θ. The current flow by an electric field E parallel to the metal strips creates an inductive reactance X_L, and it is represented as the equivalent circuit shown in Fig. 5.18c.

The inductive reactance X_L at the resonance frequency, normalized by the free space impedance ($Z_0 = 377\ \Omega$), is expressed as a function of FSS dimensions [36]:

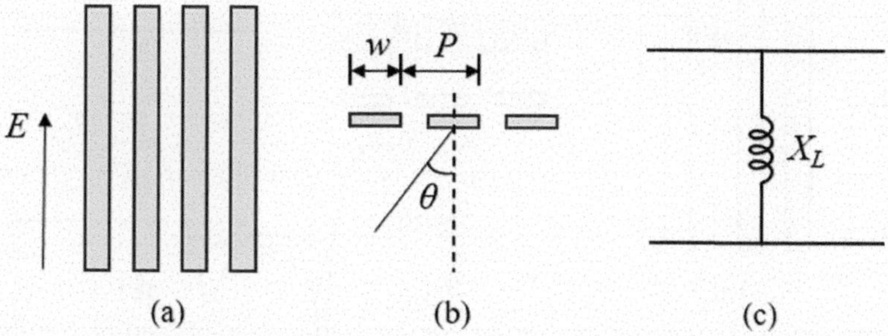

Fig. 5.18 The incidence of plane wave to the periodic metal strips to which the electric field is parallel (i.e., the magnetic field is perpendicular): **a** front view, **b** top view, and **c** equivalent circuit

$$\frac{X_L}{Z_0} = \frac{\omega L}{Z_0} = \frac{2\pi f L}{Z_0} = \frac{p\cos\theta}{\lambda}\left[\ln\left(\csc\left(\frac{\pi w}{2p}\right) + G\right)\right] \tag{5.26}$$

where ω is the angular frequency ($= 2\pi f$), λ is the wavelength at the resonance frequency, and G is a correction term given by

$$G = \frac{1}{2}\frac{(1-\beta^2)^2\left[\left(1-\frac{\beta^2}{4}\right)(A_- + A_+) + 4\beta^2 A_+ A_-\right]}{\left(1-\frac{\beta^2}{4}\right) + \beta^2\left(1+\frac{\beta^2}{2}-\frac{\beta^4}{8}\right)(A_- + A_+) + 2\beta^6 A_- A_+} \tag{5.27}$$

$$A_\pm = \frac{1}{\sqrt{1 \pm \frac{2p\sin\theta}{\lambda} - \left(\frac{p\cos\theta}{\lambda}\right)^2}} - 1 \tag{5.28}$$

$$\beta = \sin\frac{\pi w}{2p} \tag{5.29}$$

The correction factor G reflects the effects of incidence angle θ, period p, and width w of conducting strips with respect to the wavelength λ at resonance frequency, as depicted in Eq. (5.28). When θ is small (i.e., normal incidence) and the period is small relative to the wavelength ($p/\lambda \ll 1$), $G \approx 0$.

If the magnetic field H of incident plane wave is parallel to the metal strips (i.e., the electric field is perpendicular to them), as shown in Fig. 5.19, a capacitive susceptance B_C is formed, which is expressed as [36],

$$\frac{B_C}{Y_0} = \frac{\omega C}{Y_0} = \frac{2\pi f C}{Y_0} = \frac{4p\cos\theta}{\lambda}\left[\ln\left(\csc\left(\frac{\pi g}{2p}\right) + G\right)\right] \tag{5.30}$$

where Y_0 is the admittance of free space ($= 1/377\ \Omega^{-1}$), g is the spacing between two metal strips. Correction factor G is given by Eq. (5.27), where β is a function of g, which is given by [36],

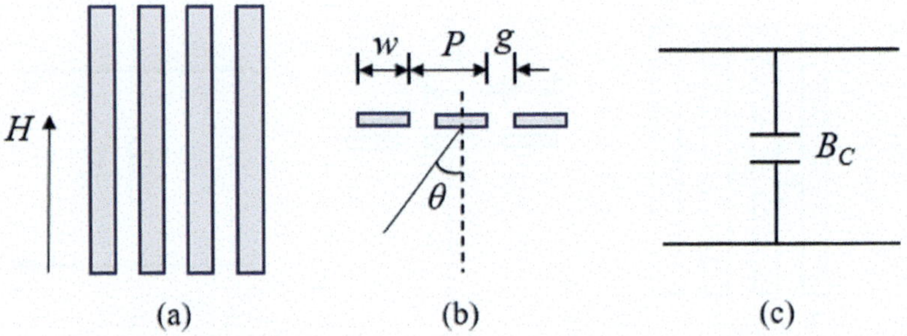

Fig. 5.19 The incidence of plane wave to the periodic metal strips to which the magnetic field is parallel: **a** front view, **b** top view, and **c** equivalent circuit

$$\beta = \sin \frac{\pi g}{2p} \tag{5.31}$$

Example 1 Square loop FSS

Figure 5.20 shows a square loop FSS (SL-FSS) placed in free space, and its geometries and equivalent circuit. For example, the dimensions of unit cell period ($p = 10$ mm), SL-FSS (length $d = 8.75$ mm, width $s = 0.625$ mm), and spacing between two SL-FSS ($g = 1.25$ mm) are given in Fig. 5.20b. Assuming the FSS conductor is perfect (resistance $R = 0$), the equivalent circuit is represented by a shunted series LC circuit, as shown in Fig. 5.20c. For this geometry, the transmission loss simulated using HFSS tool is depicted in Fig. 5.21. The band-stop behavior of transmission loss indicates this array has a single principal resonance at a frequency $f_0 = 10.2$ GHz (wavelength $\lambda = 29$ mm).

As the length of the metal strip is finite, a modification is needed in Marcuvitz's formula. The reactance and susceptance are reduced by a factor of d/p from the corresponding array of infinite strips. There are two metal strips in a unit cell parallel to the electric field, so that the effective linewidth is $w = 2s$. The inductive reactance ($X_L = \omega L$) of SL-FSS at the resonance frequency is given by the following equation [30, 34, 37–40]:

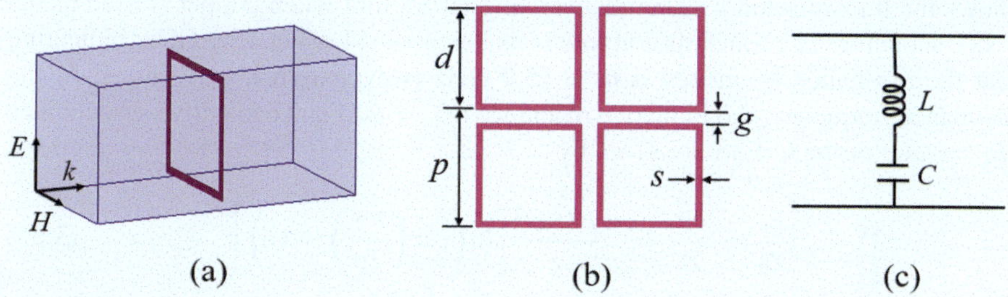

Fig. 5.20 Illustration of square loop FSS (SL-FSS): **a** simulation model, **b** geometry, and **c** equivalent circuit. Reprinted with permission from Ref. [30], copyright 2016 Elsevier

Fig. 5.21 Transmission loss of SL-FSS in free space. Reprinted with permission from Ref. [30], copyright 2016 Elsevier

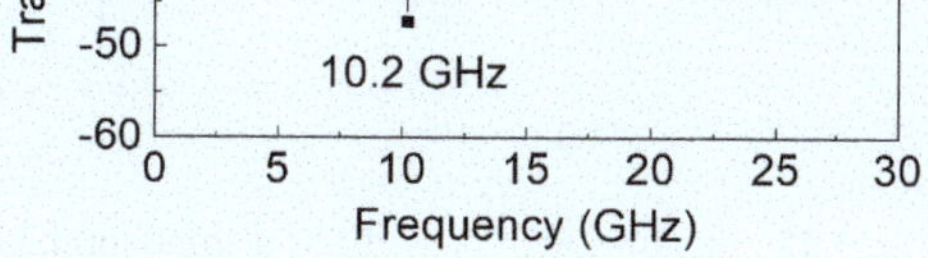

$$\frac{X_L}{Z_0} = \frac{\omega L}{Z_0} = \frac{d}{p}\frac{p\cos\theta}{\lambda}\left[\ln\left(\csc\left(\frac{\pi 2s}{2p}\right)+G\right)\right] \tag{5.32}$$

where G is given by Eq. (5.27) as the correction term, and β is given by

$$\beta = \sin\frac{\pi 2s}{2p} \tag{5.33}$$

The capacitive susceptance of SL-FSS ($B_C = \omega C$) is given by [30, 34, 37–40]

$$\frac{B_C}{Y_0} = \frac{\omega C}{Y_0} = \frac{d}{p}\frac{4p\cos\theta}{\lambda}\left[\ln\left(\csc\left(\frac{\pi g}{2p}\right)+G\right)\right] \tag{5.34}$$

where the correction terms G and β are the same as Eqs. (5.27) and (5.31), respectively. The inductance and capacitance of SL-FSS obtained using the above formula are $L = 2.965$ nH and $C = 83.45$ fF, respectively.

Example 2 Cross FSS

A similar step can be taken for the cross FSS (C-FSS). Figure 5.22 shows the C-FSS placed in free space, dimensions of the unit cell, and equivalent circuit. Given the geometry of a unit cell, $p = 10$ mm, $d = 9$ mm, $s = 0.5$ mm, $g = 1$ mm, Fig. 5.23 shows the transmission loss simulated using HFSS tool when copper is used as the FSS conductor. The transmission loss is as low as − 44 dB at 15.9 GHz, indicating that the resonance frequency is $f_0 = 15.9$ GHz (wavelength $\lambda = 19$ mm). At the resonance frequency, the inductive reactance ($X_L = \omega L$) and capacitive susceptance ($B_C = \omega C$) of the C-FSS are given by

$$\frac{X_L}{Z_0} = \frac{\omega L}{Z_0} = \frac{d}{p}\frac{p\cos\theta}{\lambda}\left[\ln\left(\csc\left(\frac{\pi s}{2p}\right)+G\right)\right] \tag{5.35}$$

$$\frac{B_C}{Y_0} = \frac{\omega C}{Y_0} = \frac{s}{p}\frac{4p\cos\theta}{\lambda}\left[\ln\left(\csc\left(\frac{\pi g}{2p}\right)+G\right)\right] \tag{5.36}$$

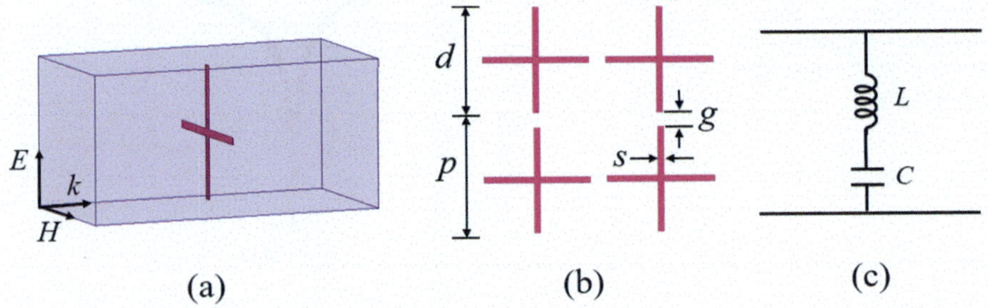

Fig. 5.22 Illustration of cross FSS: **a** simulation model, **b** geometry, and **c** equivalent circuit

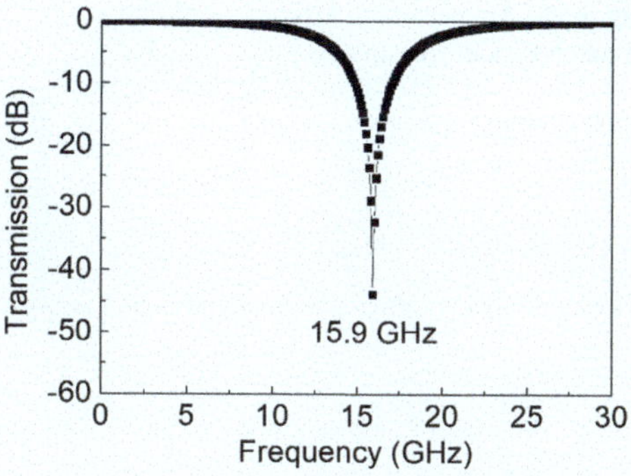

Fig. 5.23 Transmission loss of cross FSS (C-FSS) in free space

Using the correction term G given by Eq. (5.27), the inductance and capacitance can be calculated to be $L = 4.469$ nH and $C = 21.28$ fF, respectively. It is to be noted that the capacitance is significantly reduced due to a small width of C-FSS conductor.

Example 3 Patch FSS

Figure 5.24 shows the patch FSS (P-FSS) placed in free space, its dimensions, and equivalent circuit. The dimensions are given in Fig. 5.24b, where the unit cell period $p = 10$ mm, patch size $d = 9$ mm, the gap between patches $g = 1$ mm. For this geometry, simulation result is presented in Fig. 5.25, showing a single principal resonance at a frequency $f_0 = 29.9$ GHz (wavelength $\lambda = 10$ mm).

The capacitive susceptance ($B_C = \omega C$) of P-FSS at the resonance frequency is given by

$$\frac{B_C}{Y_0} = \frac{\omega C}{Y_0} = \frac{d}{p}\frac{4p\cos\theta}{\lambda}\left[\ln\left(\csc\left(\frac{\pi g}{2p}\right) + G\right)\right] \tag{5.37}$$

Using Eq. (5.27) of correction term G, the capacitance can be calculated to be $C = 159.4$ fF. The equation for the parallel-plate capacitor is given by [41]:

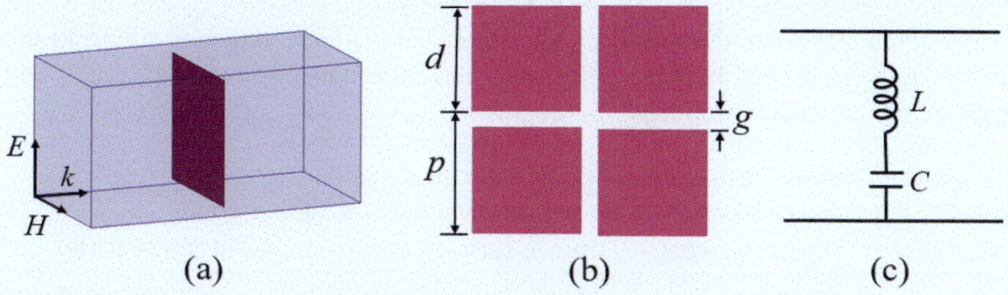

Fig. 5.24 Illustration of patch FSS: **a** simulation model, **b** geometry, and **c** equivalent circuit

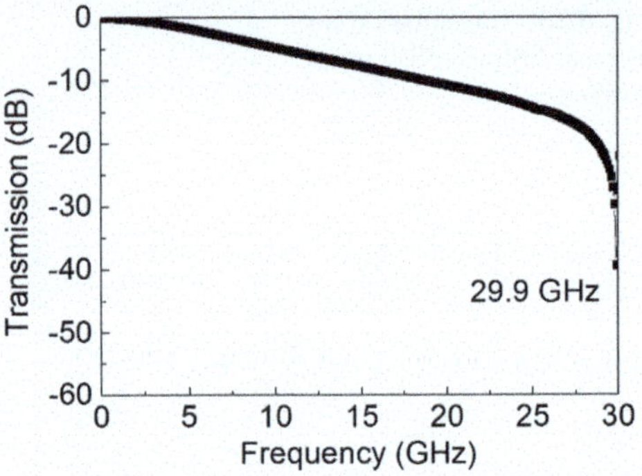

Fig. 5.25 Transmission loss of patch FSS in free space

$$C = \varepsilon_0 \varepsilon_r \frac{dt}{g} \text{ (F)} \tag{5.38}$$

where ε_0 is the permittivity of free space and ε_r is the relative permittivity of media. For the P-FSS with dimensions of $d = 9$ mm, $g = 1$ mm, and thickness $t = 0.02$ mm, the capacitance is calculated to be $C = 159.3$ fF, in good agreement with the result of equivalent circuit model.

On the other hand, there is a problem when calculating the inductance using the equivalent circuit model, that assumes that the linewidth of the FSS conductor is sufficiently small relative to the period. In the case of the P-FSS above, however, the line width is close to the period. Instead, we can calculate the inductance from the resonance frequency f_0:

$$L = \frac{1}{C(2\pi f_0)^2} \tag{5.39}$$

The inductance of P-FSS obtained using Eq. (5.39) is $L = 0.176$ nH. Because the patch has a wide side, the capacitance is large, but the inductance is very small.

In summary, Table 5.3 shows the values of capacitance, inductance, and resonance frequency of the three FSSs above with an equal period ($p = 10$ mm). In the case of patch FSS, the capacitance is large, but the inductance is very small. Thus, it shows the highest resonance frequency. Conversely, in the case of cross FSS with a narrow conductor strip, the inductance is large, but the capacitance is small. In the case of square loop FSS, both the inductance and capacitance show large values, and therefore the resonance frequency is the lowest.

Table 5.3 Capacitance, inductance, and resonance frequencies of three-type FSSs

FSS pattern	Capacitance (fF)	Inductance (nH)	Resonance frequency (GHz)
Square loop	83.5	2.965	10.2
Cross	21.3	4.469	15.6
Patch	159.4	0.176	29.9

5.3.6 *Design of FSS Absorbers*

(1) **Circuit Model**

FSS absorber is composed of dielectric substrate of which back side is terminated by a metal and the front surface is attached by FSS. The schematic structure and equivalent circuit of the FSS absorber is shown in Fig. 5.16. With knowledge of material parameters (ε_r, μ_r) and thickness d of dielectric substrate, the input admittance of substrate (Y_d) can be calculated using Eq. (5.16). FSS circuit parameters (R, L, C) can be obtained from the surface resistance of FSS and by using the equivalent circuit model, which make it possible to calculate the complex admittance of FSS (Y_{FSS}) by using Eq. (5.17). Parallel connection of Y_d and Y_{FSS} yields the total equivalent admittance Y_{Total}, and the reflection loss (RL) of the FSS absorber can be calculated by

$$\mathrm{RL} = 20\log|\Gamma| = 20\log\left|\frac{Y_0 - Y_{Total}}{Y_0 + Y_{Total}}\right| \tag{5.40}$$

where Γ is the reflection coefficient.

An example describes the non-reflected design of an EM wave absorber consisting of SL-FSS [30], which is illustrated in Fig. 5.26. For simplicity, the spacer material is assumed to be lossless air ($\varepsilon_r = 1$, $\mu_r = 1$), and its thickness is adjusted to 1/4 wavelength (λ/4) corresponding to the resonance frequency of FSS. The size of unit cell is $p = 10$ mm, and the length and width of SL-FSS conductor are $d = 8.75$ mm and $s = 0.625$ mm, respectively.

The inductance and capacitance of the SL-FSS calculated from the equivalent circuit model is $L = 2.965$ nH, $C = 83.5$ fF, and then the resonance frequency is calculated to be $f_0 = 10.2$ GHz. Figure 5.27a shows the input admittance Y_d of a short-circuited air spacer whose thickness is $\lambda/4$ (= 7.3 mm) calculated from the resonance frequency of the SL-FSS. The admittance value is normalized by free space admittance Y_0 (= 1/377 Ω^{-1}), $Y_d/Y_0 = \mathrm{Re}(Y_d) + j\mathrm{Im}(Y_d)$. At $f_0 = 10.2$ GHz,

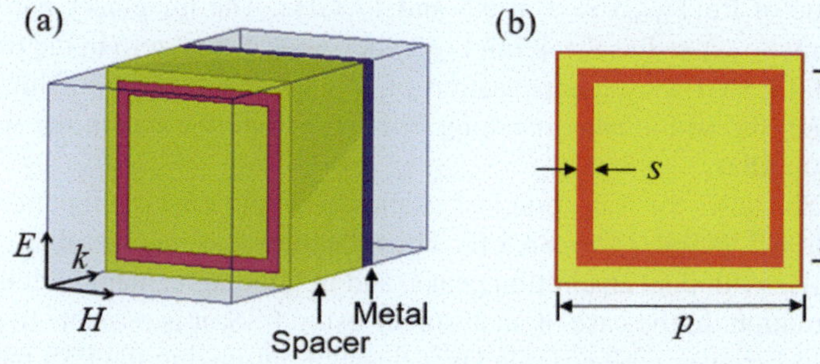

Fig. 5.26 A representation of SL-FSS absorber: **a** schematic structure and **b** dimensions of FSS ($p = 10$ mm, $d = 8.75$ mm, $s = 0.625$ mm). Reprinted with permission from Ref. [30], copyright 2016 Elsevier

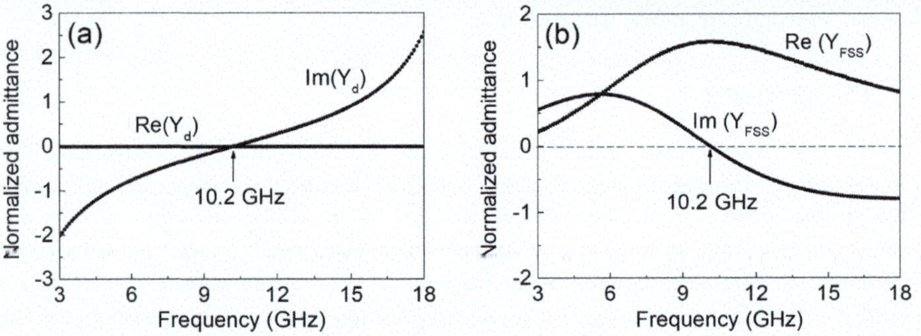

Fig. 5.27 Admittance of SL-FSS absorbers normalized by free space admittance: **a** grounded air spacer with $\lambda/4$ thickness and **b** SL-FSS with surface resistance of 26 Ω/sq. Reprinted with permission from Ref. [30], copyright 2016 Elsevier

the imaginary part $\mathrm{Im}(Y_\mathrm{d}) = 0$ due to the high-impedance surface characteristics at $\lambda/4$ thickness, and the real part $\mathrm{Re}(Y_\mathrm{d}) = 0$ due to the lossless medium of air. Below f_0, $\mathrm{Im}(Y_\mathrm{d})$ has a negative value (inductive circuit). Above f_0, $\mathrm{Im}(Y_\mathrm{d})$ has a positive value (capacitive circuit).

Figure 5.27b shows the complex admittance of SL-FSS normalized by free space admittance, $Y_\mathrm{FSS}/Y_0 = \mathrm{Re}(Y_\mathrm{FSS}) + j\mathrm{Im}(Y_\mathrm{FSS})$. The surface resistance of FSS conductor is $R_s = 26$ Ω/sq. The FSS circuit resistance is then obtained from the ratio of effective area of current flow ($= 2sd$) to unit cell area ($= p^2$), calculated to be $R = 236$ Ω by using Eq. (5.25). $\mathrm{Re}(Y_\mathrm{FSS})$ shows a broad dispersion with a peak value at the resonance frequency ($f_0 = 10.2$ GHz) due to the resistive nature of FSS. With respect to the resonance frequency, the SL-FSS is capacitive at lower frequencies and inductive at higher frequencies, exhibiting an inverse frequency dispersion to that of the grounded substrate.

Figure 5.28a presents the normalized total admittance, $Y_\mathrm{Total}/Y_0 = \mathrm{Re}(Y_\mathrm{Total}) + j\mathrm{Im}(Y_\mathrm{Total})$, that is obtained by parallel circuit connection of Y_d and Y_FSS. The real part of total admittance is equal to the FSS resistance component ($\mathrm{Re}(Y_\mathrm{Total}) = \mathrm{Re}(Y_\mathrm{d}) + \mathrm{Re}(Y_\mathrm{FSS}) = \mathrm{Re}(Y_\mathrm{FSS})$), and shows a broad peak with a matching value of $\mathrm{Re}(Y_\mathrm{Total}) = 1$ at 6.5 and 17 GHz. The imaginary part of total admittance ($\mathrm{Im}(Y_\mathrm{Total}) = \mathrm{Im}(Y_\mathrm{d}) + \mathrm{Im}(Y_\mathrm{FSS})$) becomes nearly zero in the frequency region of 6.0–14.3 GHz by compensation of $\mathrm{Im}(Y_\mathrm{d})$ and $\mathrm{Im}(Y_\mathrm{FSS})$. Thus, in this frequency band, the admittance matching is satisfied and we obtain the wideband absorption properties.

Figure 5.28b shows the reflection loss calculated using the circuit parameters of $R = 236$ Ω, $L = 2.965$ nH, $C = 83.5$ fF. The reflection loss is lower than − 10 dB in 5.0–16.8 GHz with dual absorption peaks, and in good agreement with the result of HFSS simulation. In the case of an absorber using FSS, it is possible to design a wideband absorber even with a single layer through optimizing the FSS pattern, its dimension, and the surface resistance of FSS conductor.

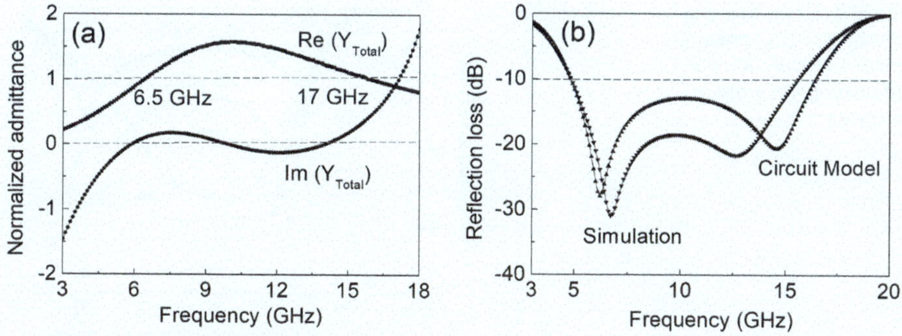

Fig. 5.28 **a** Total admittance of SL-FSS absorbers and **b** reflection loss. Reprinted with permission from Ref. [30], copyright 2016 Elsevier

(2) Surface Resistance Effect

Figure 5.29 shows the normalized admittance with variation of surface resistance of SL-FSS conductor. As the surface resistance increases, dispersion peak of real part Re(Y_{FSS}) spreads with a reduction of peak value, as shown in Fig. 5.29a. At $R_s = 30$ Ω/sq, the FSS resistance (R) corrected by the effective area of current flow approaches to the free space impedance at the resonance frequency (10.2 GHz). With respect to the resonance frequency, which is invariant with surface resistance, the SL-FSS is capacitive at lower frequencies and inductive at higher frequencies, as shown in Fig. 5.29b. Frequency dispersion of Im(Y_{FSS}) becomes more dispersive with lowering of peak value as the surface resistance increases.

Figure 5.30 shows the reflection loss with varying the surface resistance of FSS. The reflection loss decreases and bandwidth broadens with increasing surface resistance from 10 to 25 Ω/sq. At $R_s = 20$ Ω/sq and $R_s = 25$ Ω/sq, a broad bandwidth is predicted with dual peaks of absorption at the frequencies of admittance matching, as explained earlier. At a higher surface resistance above 30 Ω/sq, a single peak of

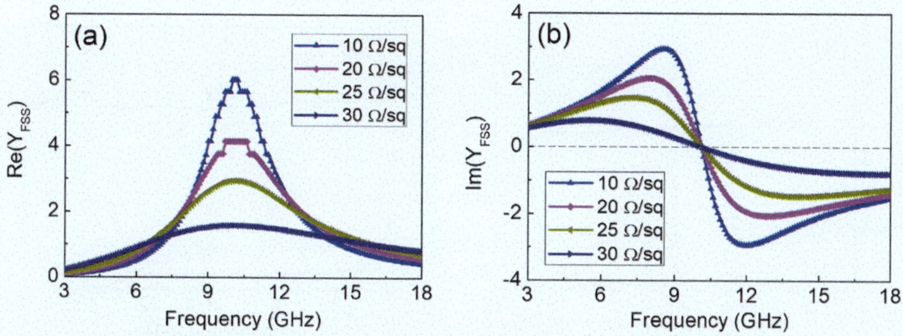

Fig. 5.29 Normalized admittance of the SL-FSS with varying the surface resistance: **a** real part and **b** imaginary part

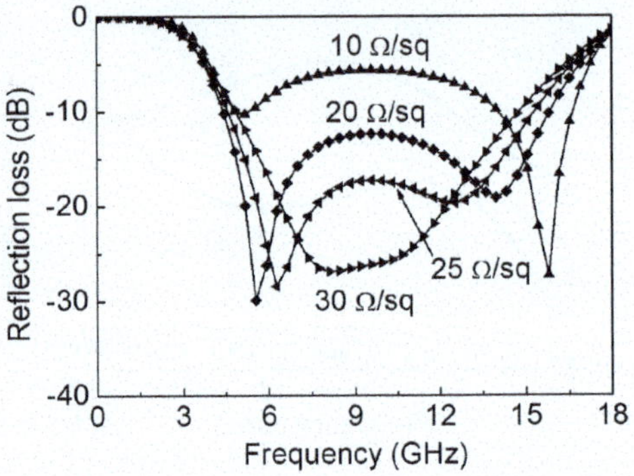

Fig. 5.30 Reflection loss determined in the grounded air substrate with SL-FSS with varying the surface resistance

absorption is predicted in the vicinity of resonance frequency (10.2 GHz), at which the FSS resistance is close to the free space impedance.

(3) **Substrate Effect**

The input admittance of substrate is dependent on its dielectric constant and dielectric loss. One way to increase the dielectric permittivity of substrate is to use the carbon black in polymer matrix. Figure 5.31 shows the frequency dispersion of complex permittivity ($\varepsilon_r = \varepsilon_r' - j\varepsilon_r''$) measured in the rubber composites of carbon black. Nearly constant values of real (ε_r') and imaginary value (ε_r'') of complex permittivity with frequency are observed. Both ε_r' and ε_r'' increase with increasing the amount of carbon black ($\varepsilon_r' \approx 5$, $\varepsilon_r'' \approx 0$ for 10 vol.% and $\varepsilon_r' \approx 15$, $\varepsilon_r'' \approx 2$ for 20 vol.%), as shown in Fig. 5.31.

The effect of dielectric permittivity of substrate on the admittance of SL-FSS absorbers is shown in Fig. 5.32. With respect to the air spacer, imaginary part of input admittance of substrate ($\mathrm{Im}(Y_d)$) increases with the use of high-permittivity substrate, as shown in Fig. 5.32a. After summing with the imaginary part of FSS admittance ($\mathrm{Im}(Y_{FSS})$) shown in Fig. 5.32b, the use of high-permittivity substrate

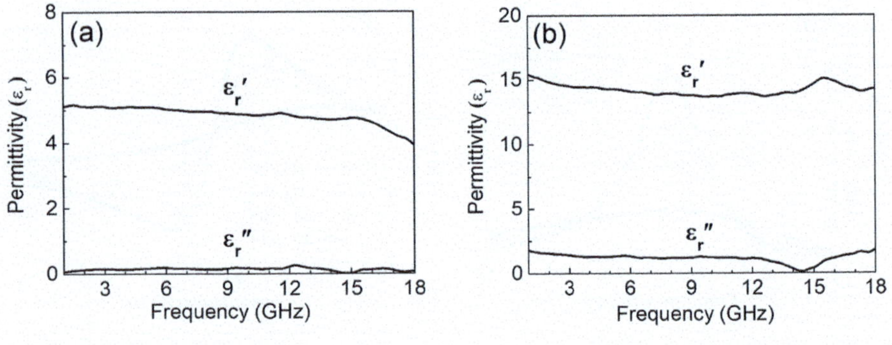

Fig. 5.31 Complex permittivity of polymer composites containing carbon black with increasing its volume fraction: **a** 10 vol.% and **b** 20 vol.%

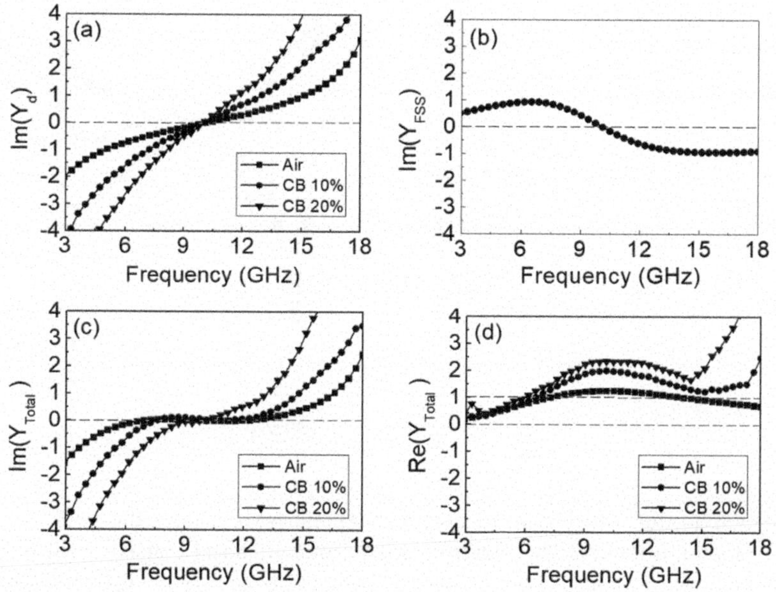

Fig. 5.32 Normalized admittance of SL-FSS absorbers with varying the substrate: **a** Im(Y_d), **b** Im(Y_{FSS}), **c** Im(Y_{Total}), **d** Re(Y_{Total})

reduces the frequency range of Im(Y_{Total}) = 0, as shown in Fig. 5.32c, which reduces the absorption bandwidth of SL-FSS absorber. The dielectric loss of carbon black composites increases the real part of total admittance (Re(Y_{Total})) in addition to the FSS resistance, as shown in Fig. 5.32d, which increases the reflection coefficient.

Figure 5.33 clearly shows that the use of high-permittivity substrates reduces the absorption bandwidth and increases the wave reflection. For the low-permittivity substrate similar to air, rohacell or polyethylene foam can be used, whose dielectric constant is about 1.1 and dissipation factor is as low as 0.0007 at GHz frequencies. In particular, rohacell has excellent mechanical properties, high thermal resistance, dimensional stability, solvent resistance, and a low coefficient of heat conductivity, which makes it highly potential in the practical application to the substrate of broadband FSS absorbers.

5.3.7 Metamaterial Absorbers

(1) **State of Art**

A metamaterial refers to a material produced by periodically placing the patterned components of an artificial conductor on a dielectric substrate. Since the size of unit cell is much bigger than the molecular size of material itself, the application of macroscopic Maxwell equation is valid. The structure of its mesoscopic scale that

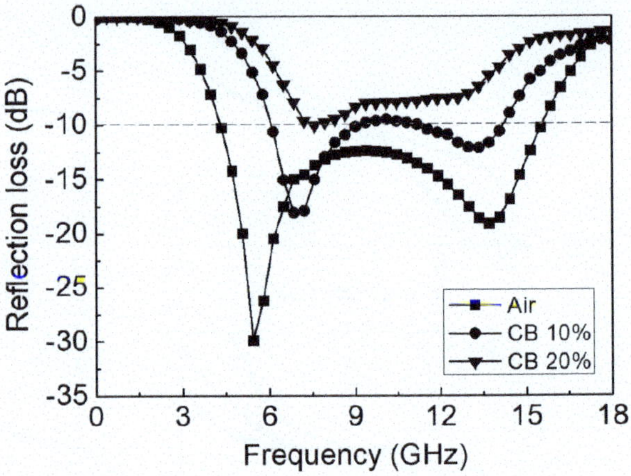

Fig. 5.33 Reflection loss determined in the SL-FSS absorbers with increasing the complex permittivity of the substrate

is smaller than the wavelength of electromagnetic wave allows the application of effective medium theory. The electromagnetic metamaterial consisting of the sub-wavelength unit elements is represented by dielectric permittivity $\varepsilon(\omega)$ and magnetic permeability $\mu(\omega)$. Usually, the metamaterial is composed of a conductor with a high electrical conductivity such as copper, gold, or silver in various patterns on a dielectric substrate. Split rings that control magnetic properties and thin wires that regulate electrical properties are the representative conductor patterns [42, 43].

The metamaterial absorber first proposed by Landy et al. [44] consists of two metal layers on top and bottom side of a dielectric layer. The upper conductor is constructed in the form of an electric ring resonator (ERR) with gaps at both ends, and the lower conductor takes the form of a cut wire (CW) and has an electrical reaction to the external electric field. Magnetic coupling occurs by antiparallel currents between the central conductor of upper ERR and the lower CW. The time-varying magnetic field of incident wave induces an antiparallel current in the upper and lower conductor pairs, and generates a magnetic response of Lorentzian resonance. Based on the above resonating structure, the control of electric and magnetic reactions is possible and the perfect absorption can be obtained in the vicinity of the magnetic resonance frequency. The absorption of this structure is as high as 99% at 11.48 GHz by simulation.

The basic structure of the metamaterial absorber is the same as the FSS absorber shown in Fig. 5.16a, where it consists of two metal conductors on the upper and lower ground planes of the dielectric substrate. However, in the case of FSS absorbers, the coupling between the two conductors is very weak because the thickness of the dielectric substrate is fairly large, equal to 1/4 wavelength ($\lambda/4$). However, in the case of metamaterial absorbers, the thickness of the substrate is so small that in addition to the inductance and capacitance of each metal conductor itself, the coupling between the upper and lower conductors must be considered. In particular, when the thickness of the substrate becomes smaller than $0.3p$ ($d \leq 0.3p$, p: unit cell size), the capacitance between the two conductors increases exponentially under the influence of the evanescent Floquet mode [45]. This is the main contribution to the thinning of perfect metamaterial absorbers. Conversely, in a thicker FSS absorber ($d \geq 0.3p$), the capacitance between the two conductors can be ignored.

An increase in capacitance leads to a decrease in the resonance frequency, which leads to a decrease in the thickness of absorber. The wavelength corresponding to the resonance frequency of the metamaterial is proportional to the period of the resonator, and is approximately $\lambda_0 \sim 2p$. The initial experimental design for the metamaterial absorber showed a thickness of $\lambda_0/35$ [44], and the later experiments reduced it to the level of $\lambda_0/40$ [46] and $\lambda_0/75$ [47]. The resonance frequency of the metamaterial can be controlled through simple adjustment of the unit cell period, which allows the absorption band to be controlled over a wide frequency range. Metamaterial absorber research initially began in the microwave region and expanded to the terahertz (THz) [48], mid-infrared (MIR) [49], near-infrared (NIR) [50], and visible light regions [51].

(2) **Circuit Model**

The operation of the metamaterial absorber can be explained by the equivalent circuit of the FSS absorber shown in Fig. 5.16b, taking into account the additional and critical effect of the electrical coupling between FSS conductor and ground plane. The reflection loss is determined by total equivalent impedance of high-impedance surface (Z_H), which is a parallel circuit of input impedance of grounded substrate (Z_d) and complex impedance of FSS (Z_{FSS}), which are given by

$$\frac{1}{Z_H} = \frac{1}{Z_d} + \frac{1}{Z_{FSS}} = \frac{1}{Z_0[\mathrm{Re}(Z_H) + j\mathrm{Im}(Z_H)]} \tag{5.41}$$

$$Z_d = \frac{Z_0}{\sqrt{\varepsilon_r}} \tanh\left(j\frac{2\pi d}{\lambda}\sqrt{\varepsilon_r}\right) = Z_0[\mathrm{Re}(Z_d) + j\mathrm{Im}(Z_d)] \tag{5.42}$$

$$Z_{FSS} = R + j\left(\omega L - \frac{1}{\omega C}\right) = Z_0[\mathrm{Re}(Z_{FSS}) + j\mathrm{Im}(Z_{FSS})] \tag{5.43}$$

where, $\mathrm{Re}(Z_H) + j\mathrm{Im}(Z_H)$, $\mathrm{Re}(Z_d) + j\mathrm{Im}(Z_d)$, and $\mathrm{Re}(Z_{FSS}) + j\mathrm{Im}(Z_{FSS})$ are the complex impedances of high-impedance surface, grounded dielectric substrate, and FSS, respectively, normalized by free-space impedance Z_0 (= 377 Ω). λ is the wavelength in free space and R is the resistance part of Z_{FSS}. In Eq. (5.43), capacitance C includes the coupling effect between FSS conductor and ground plane, which increases rapidly as the substrate thickness decreases [45]. Satisfying the impedance matching conditions, $\mathrm{Re}(Z_H) = 1$ and $\mathrm{Im}(Z_H) = 0$, a perfect non-reflected absorber can be obtained.

FR4 with a very small dielectric loss is usually used as a substrate in metamaterial absorbers. The complex permittivity of FR4 is $\varepsilon_r = 4.1 - j0.03$, and the loss tangent is very small, $\tan\delta = 0.007$. Since the dielectric loss is small, the real part of the input impedance of grounded substrate is $\mathrm{Re}(Z_d) \approx 0$. With respect to the first resonance frequency (corresponding to the $\lambda/4$ thickness), the imaginary part exhibits inductive circuit characteristics ($\text{Im}(Z_d) \geq 0$) at lower frequencies and capacitive circuit characteristics ($\text{Im}(Z_d) \leq 0$) at higher frequencies.

The real and imaginary parts of the complex impedance of FSS depend on the pattern and dimensions of FSS. In addition, due to the electrical coupling between the FSS and the ground plane, Z_{FSS} is also affected by the dielectric constant, dielectric loss, and thickness of the substrate. At the resonance frequency of FSS, $\text{Im}(Z_{FSS}) = 0$ and $\text{Re}(Z_{FSS})$ form a peak. With respect to the resonance frequency, it may exhibit the inductive-circuit characteristics ($\text{Im}(Z_{FSS}) \geq 0$) at lower frequencies and the capacitive-circuit characteristics ($\text{Im}(Z_{FSS}) \leq 0$) at higher frequencies, which will be explained later. Thus, above the resonance frequency of FSS, the inductive $\text{Im}(Z_d)$ and capacitive $\text{Im}(Z_{FSS})$ can be canceled and get close to the impedance matching condition, $\text{Im}(Z_H) = 0$. Ohmic loss is negligible when the FSS conductor is perfect. Then $\text{Re}(Z_{FSS})$ will be dependent on the thickness and complex permittivity of substrate ($\varepsilon_r', \varepsilon_r''$) [45]. $\text{Re}(Z_{FSS})$ peak increases in proportion to the thickness of substrate. Even in FR4 substrate, which typically have a small value of ε_r'', $\text{Re}(Z_{FSS})$ can have large values at resonance frequency and therefore can be close to impedance matching condition. This is a salient and notable feature of metamaterial absorbers.

A typical example of a non-reflective design is a metamaterial absorber consisting of a front split cut wire (SCW) and a rear ground plane, as shown in Fig. 5.34a. Through the dimensional change of the SCW, the effective permittivity and permeability can be controlled, and accordingly, the propagation properties of the composite structure can be controlled. As shown in Fig. 5.34b, the variables are size of unit cell (a, b), width (w) and length (s) of SCW, gap (g) between SCWs, and thickness of dielectric substrate (d). The substrate is FR4 with a complex permittivity $\varepsilon_r = 4.1 - j0.03$. The metal conductor is copper (electrical conductivity $\sigma = 5.8 \times 10^7$ S/m, thickness = 0.035 mm).

Figure 5.35 shows the simulation results of reflection loss for a unit cell with dimensions of a = 16 mm, b = 32 mm, w = 8 mm, s = 14.5 mm, g = 1 mm, d = 0.53 mm. The lowest reflection loss is − 27 dB at 4.97 GHz. These results are in good agreement with the calculations and experimental results reported in Refs. [52, 53], where the back side was partially covered by cut wire (CW) conductor. Almost the same result was obtained when the back side was completely covered

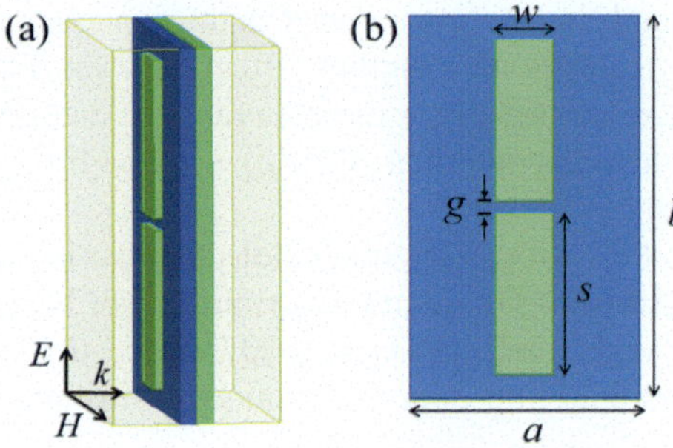

Fig. 5.34 A metamaterial absorber consisting of split cut wire (SCW) at front surface and ground plane at back surface: **a** schematic structure and **b** dimension of SCW

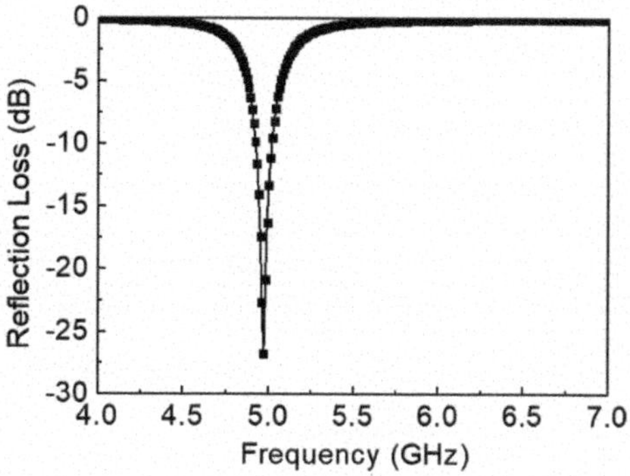

Fig. 5.35 Simulation results of reflection loss for SCW metamaterial absorber. Reproduced from Ref. [52] with the permission of Springer Nature

by metal (ground plane). Therefore, the equivalent circuit of FSS absorber shown in Fig. 5.16b can be used in the analysis of the electromagnetic behavior of the SCW metamaterial absorber.

Figure 5.36 shows the simulation results of normalized Im(Z_d), Im(Z_{FSS}), and Re(Z_{FSS}). The input impedance of substrate is inductive (Im(Z_d) ≥ 0) below the first resonance frequency of the grounded FR4 substrate, which is about 5.7 GHz [52]. With respect to the resonance frequency of FSS ($f_r = 4.85$ GHz), the imaginary part of the complex impedance of FSS is inductive (Im(Z_{FSS}) ≥ 0) at lower frequencies and capacitive (Im(Z_{FSS}) ≤ 0) at higher frequencies. Thus, at the frequency of about 5 GHz, the imaginary part of the total impedance is Im(Z_H) $= 0$ by compensation of inductive Im(Z_d) and capacitive Im(Z_{FSS}), as shown in Fig. 5.36a. This frequency is the impedance matching frequency denoted by f_m. At the resonance frequency of FSS, Re(Z_{FSS}) shows a peak (Fig. 5.36b). At the matching frequency ($f_m \approx 5$ GHz), the real part of total impedance is equal to the free-space impedance (Re(Z_H) $\approx$ Re(Z_{FSS}) $=$ 1), resulting in an impedance matching and full power absorption, which is in good agreement with the simulated reflection loss (Fig. 5.35). The present analysis technique based on the circuit model can be applied to other structures of metamaterial absorbers.

The design variables are dimensions of SCW, thickness and dielectric constant of substrate. Figure 5.37 shows the normalized complex impedance of the SCW absorber with varying the substrate thickness. As the thickness of substrate increases, FSS resonance frequency shifts to a lower frequency. For a small thickness ($d =$ 0.1 mm), impedance matching is difficult to obtain, because |Im(Z_d)| > |Im(Z_{FSS})|, as shown in Fig. 5.37a. Re(Z_{FSS}) peak increases in proportion to the substrate thickness, as shown in Fig. 5.37b. For a large substrate thickness (d = 1.0 mm), Im(Z_d) + Im(Z_{FSS}) $\simeq 0$, but Re(Z_{FSS}) < 1 at 5 GHz, which does not satisfy the impedance matching.

Figure 5.38 shows the normalized complex impedance with increase in dielectric loss of substrate ε_r''. As ε_r'' increases, the resonance frequency of the FSS tends to increase slightly, and the Re(Z_{FSS}) peak decreases and spreads. Even though the

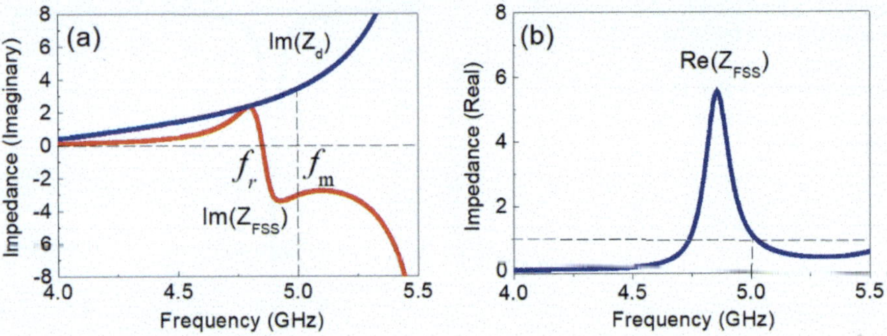

Fig. 5.36 Normalized impedance of SCW metamaterial absorbers: **a** imaginary part of the input impedance of substrate ($\mathrm{Im}(Z_d)$) and complex impedance of SCW ($\mathrm{Im}(Z_{FSS})$), **b** real part of the complex impedance of SCW ($\mathrm{Re}(Z_{FSS})$)

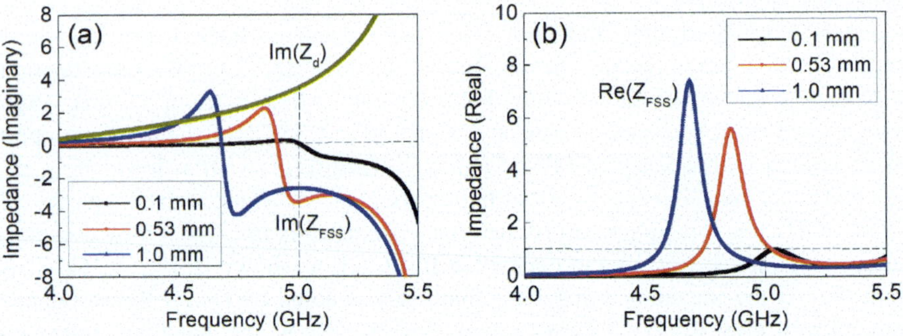

Fig. 5.37 Normalized impedance of SCW absorbers with varying the substrate thickness: **a** imaginary part, **b** real part. Reproduced from Ref. [52] with the permission of Springer Nature

dielectric loss is so small ($\varepsilon_r'' = 0.03$), $\mathrm{Re}(Z_{FSS}) = 1$ at 5 GHz and realizes the impedance matching, as shown in Fig. 5.38b. If ε_r'' is greater or less than this optimum value, $\mathrm{Re}(Z_{FSS})$ deviates from the impedance matching value ($\mathrm{Re}(Z_{FSS}) = 1$) at 5 GHz.

From the impedance analysis, it is found that the optimum thickness is $d = 0.53$ mm and the optimum value of dielectric loss is $\varepsilon_r'' = 0.03$, with a minimum reflection loss of − 27 dB at 4.97 GHz, as depicted in Fig. 5.35. Deviating from this value, the reflection loss increases, very consistent with the impedance analysis based on the equivalent circuit model [52].

In a metamaterial structure consisting of two conductors with a dielectric substrate between them, the resonance mode is divided into electric resonance and magnetic resonance, depending on the distribution of current flows through the conductors [54]. The time-varying magnetic field H between two conductors induces a circulating current in the conductor pair, which can cause a resonance with the external magnetic field. This is called the magnetic resonance. Figure 5.39a shows the current

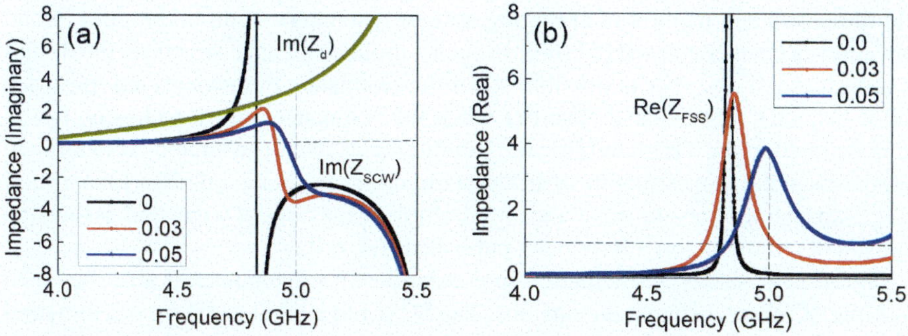

Fig. 5.38 Normalized impedance of SCW absorbers with varying the dielectric loss of substrate: **a** imaginary part, **b** real part. Reproduced from Ref. [52] with the permission of Springer Nature

distribution in SCW and ground plane at 4.97 GHz. The current flowing through front SCW and back ground plane is antiparallel, which means a magnetic response to the external oscillating *H* field. Thus, the resonance at 4.97 GHz is magnetic resonance, and it shows the Lorentz-type permeability spectrum [54, 55].

At the magnetic resonance frequency, the equivalent circuit can be represented by Fig. 5.39b, where C_m is the capacitance between SCW and ground plane and C_e is the capacitance that occurs at the gap between SCWs. The magnetic resonance frequency is given by $f_m = 1/\left(2\pi\sqrt{L(C_m + C_e)}\right)$ and is mainly controlled by C_m resulting from the electric coupling between two front and ground conductors comprising the metamaterial absorber [56].

(3) **Multiple Resonators**

Since the absorption principle of metamaterial absorbers is based on impedance matching by *LC* resonance, absorption is limited to a narrow frequency band, which is the main disadvantage of metamaterial absorbers. The acquisition of multi-band or broadband absorption is a major challenge in metamaterial absorber research. One of the methods that has been tried to achieve this is to arrange multiple resonators in a unit cell. Since the frequency response of resonators depends on their geometry

Fig. 5.39 **a** Side view of current distributions at SCW and ground plane at 4.97 GHz and **b** equivalent circuit

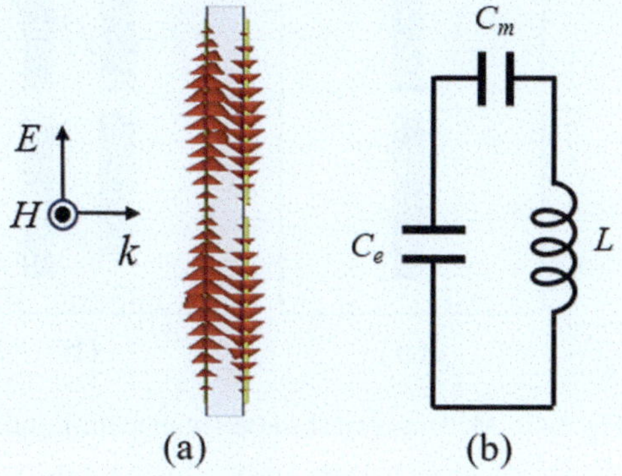

and dimensions, a method of arranging several resonators of different sizes in one unit cell has been attempted [57–60]. In such a multi-resonance structure, the design of a multi-band absorber is possible when the resonance frequencies are far apart, and it becomes a broadband absorber when the resonances overlap closely. It was demonstrated in the THz band [61, 62]. Studies have also been attempted to arrange resonators of the same structure in different directions to obtain multiband absorption [63]. Expansion of absorption bands by multiple arrays of cut wires and associated equivalent circuit analysis have been published [64, 65].

Figure 5.40 shows a schematic representation of a metamaterial absorber with multiple SCWs of different dimensions. The SCW has a fixed width ($w = 2$ mm) and different length ($s = 12.5$–15 mm). Figure 5.41 shows the multiple absorption peaks as the number of SCWs increases in a unit cell. For a single SCW on a grounded FR4 substrate (thickness = 0.5 mm), the power absorption represents a narrow single peak of 96% at 5 GHz (Fig. 5.35). The arrangement of 2 SCWs with different length results in the absorption peaks at two frequencies, as shown in Fig. 5.41a. Multi-absorption peaks are more clearly observed in the samples of 3 SCWs, as shown in Fig. 5.41b.

The multiband absorption can also be found in a metamaterial absorber composed of a split ring resonator (SRR) on a grounded dielectric substrate, which is illustrated in Fig. 5.42 [66, 67]. Power absorption in those metamaterial absorbers is due to the magnetic resonance resulting from antiparallel currents between the SRR and the ground plane. With respect to the normal ring resonator (Fig. 5.42a), the SRR has the gap and exhibits a different resonating behavior depending on the position. For perpendicular polarization with the electric field E perpendicular to the SRR gap, multiple resonances are possible at the positions of a different resonator length. For the single SRR (Fig. 5.42b), two absorption peaks are observed in reflection loss, as shown in Fig. 5.43b [66]. The first absorption at the lower frequency is due to the magnetic resonance at the wire part of SRR. The second absorption at the higher frequency is resulted from the SRR gap, which decreases the wire length and

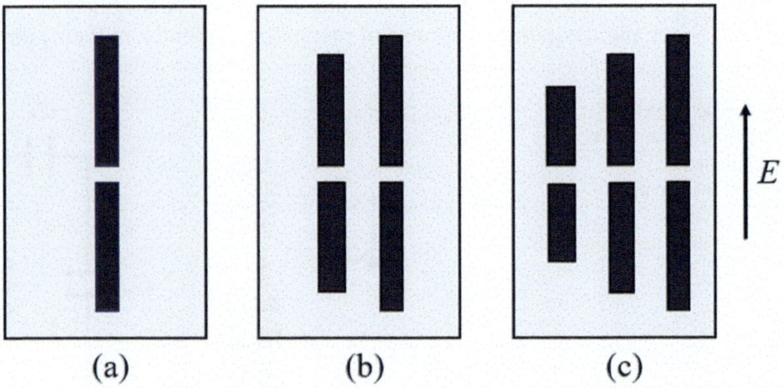

Fig. 5.40 Metamaterial absorbers with multi-resonance modes: **a** 1 SCW, **b** 2 SCWs, **c** 3 SCWs

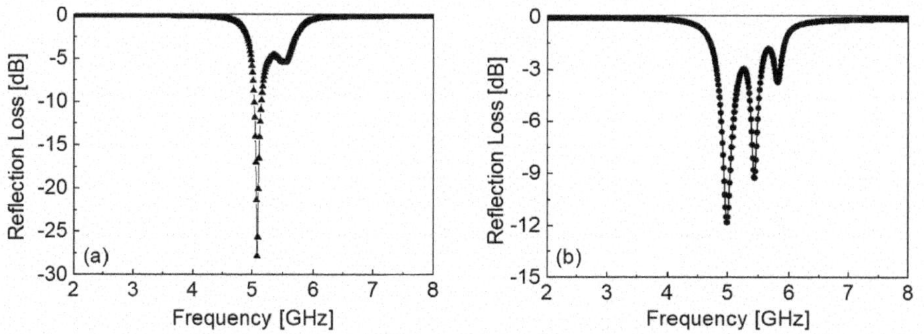

Fig. 5.41 Multiband reflection loss of the metamaterial absorbers with increasing resonators with different dimensions: **a** 2 SCWs, **b** 3 SCWs

reduces the inductance of the resonator. Extending to the double SRRs (Fig. 5.42c), three minima are observed in reflection loss, as shown in Fig. 5.43c: one is from the wire part of outer SRR and the others are from the wire part of the inner SRR with and without gap, respectively [67]. Due to the decreased inductance at the shorter resonating part, the magnetic resonance occurs at higher frequencies.

For the parallel polarization with the electric field parallel to the SRR gap, there is no effect of SRR gap, which reduces the number of absorption peaks. For the single SRR with parallel polarization, a single absorption peak is observed, corresponding to that of normal ring resonator (Fig. 5.43a). It is evident that the magnetic resonance with different values of inductance and capacitance, depending on the position in SRR, results in the multiple absorption at different resonating frequencies.

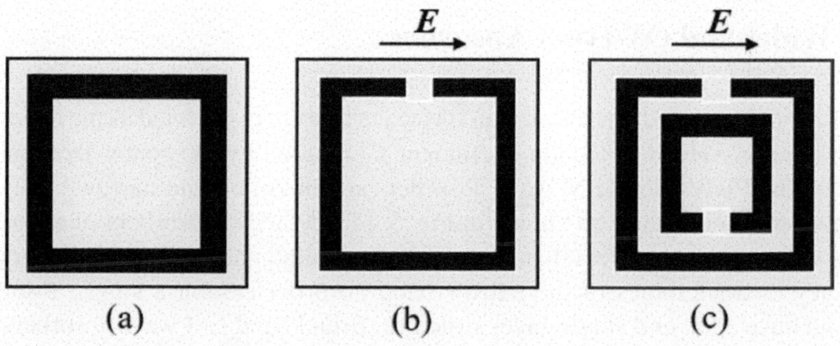

Fig. 5.42 Schematic description of SRR on grounded substrate for perpendicular polarization: **a** normal ring resonator, **b** single SRR, **c** double SRRs

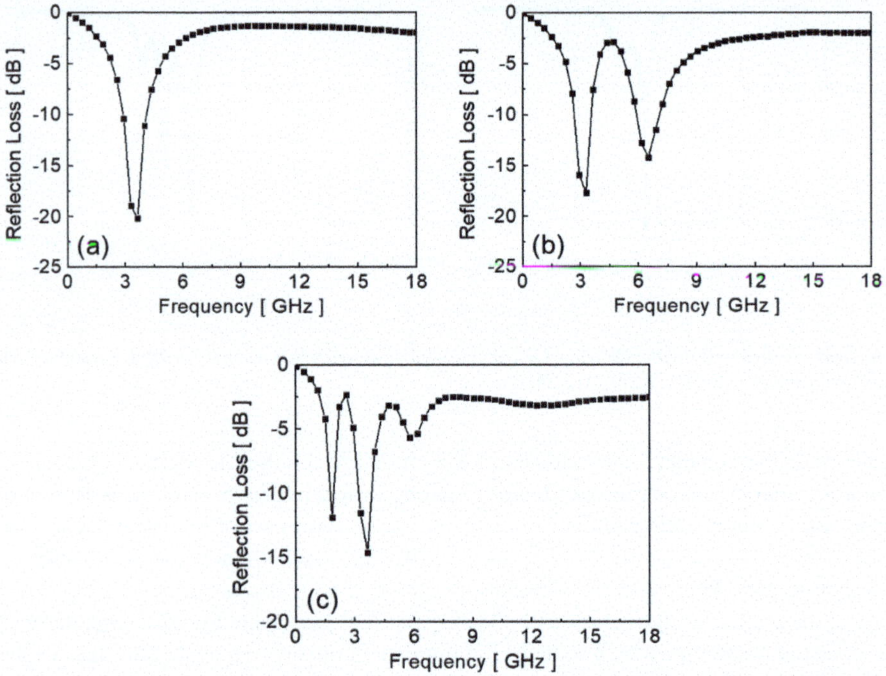

Fig. 5.43 Reflection loss determined in the metamaterial absorbers of **a** normal ring resonator, **b** single SRR, and **c** double SRRs on the grounded polymer composite of carbonyl iron powders for the perpendicular polarization. Reproduced from Ref. [66] with the permission of Springer Nature and Ref. [67] with CC-BY license

5.4 Wideband (W-type) Absorbers

The fractional bandwidth (FBW) of an EM wave absorber is defined as the bandwidth B with a target-valued reflection coefficient Γ_0 divided by the center frequency f_0. Based on the FBW value, EM wave absorbers are classified into narrow band, wide band, and ultrawide band, as shown in Fig. 5.44. EM wave absorbers of about 10% or less, or in some cases, less than 20% FBW, are called narrow band absorbers. The frequency characteristics of the narrow band absorber exhibit a single absorption peak and have a flat and single-layer structure. Broad band EM wave absorbers have an FBW value greater than 20% and generally have a multi-layer structure. The ultrawide band EM absorbers exhibit the frequency characteristics that the reflection coefficient is below Γ_0 all over the frequencies above the lowest limit f_L. Thus, the frequency bandwidth will be infinite, but there is an upper bound f_H caused by the generation of grating lobe or harmonic wave of the fundamental resonance, which will be explained later. In order to implement the ultrawide band EM absorbers, a design method is required that can simultaneously achieve a decrease in the lower limit frequency f_L and an increase in the upper limit frequency f_H.

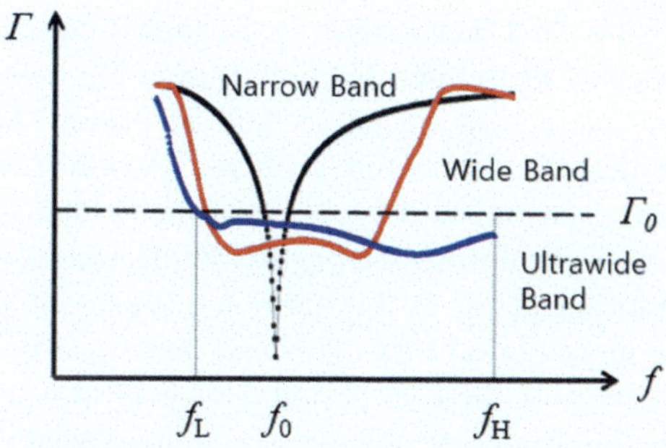

Fig. 5.44 Classification of EM wave absorbers by absorption bandwidth

5.4.1 *Lower and Upper Bounds of Absorption Band*

In an EM wave absorber consisting of a planar multilayer medium, the bandwidth expressed as a wavelength ($\Delta\lambda = \lambda_L - \lambda_H$) is a function of the absorber thickness and the reflection coefficient Γ, and the relationship between these variables is given by the theoretical formula proposed in Ref. [13]:

$$|\ln\Gamma|(\lambda_L - \lambda_H) \prec 2\pi^2 \sum_i \mu_{s,i} d_i \tag{5.44}$$

where $\mu_{s,i}$ is the permeability and d_i is the thickness of the ith layer. Assuming that the medium is non-magnetic and the absorption bandwidth is very wide ($\lambda_L - \lambda_H \simeq \lambda_L$), Eq. (5.44) is simplified to

$$\lambda_L|\Gamma| \leq 172\,d \tag{5.45}$$

It can be seen that the bandwidth in which the reflection coefficient Γ is below a specified value is inversely proportional to the absorber thickness. Conversely, given the target of reflection coefficient and bandwidth, the lower limit of the absorber thickness can be predicted. For example, at $\Gamma = -10$ dB, if the target frequency band is $f = 3$–30 GHz ($\lambda_L - \lambda_H = 100 - 10 = 90$ mm), the lower limit of the theoretical absorber thickness is approximately $d = 5.2$ mm.

The frequency selective surface (FSS), which is composed of a periodic conductor array, has resistance R, inductance of the conductor L, and capacitance C from the spacing between the conductors. Absorption frequency band can be adjusted by these circuit parameters. When the absorber is constructed by placing the FSS on a grounded dielectric substrate, the center frequency of the absorption band corresponds to the resonance frequency of FSS ($f_r = 1/2\pi\sqrt{LC}$). Therefore, in order to extend the lower bound of absorption frequency band (i.e., increase λ_L), the introduction of FSS with large inductance and capacitance is essential. The simplest way for this achievement is to enlarge the FSS period (the size of unit cell), but the problem arises that the upper frequencies are limited by the grating lobe or the harmonic appearance of fundamental resonance.

The first factor limiting the high-frequency bandwidth is the grating lobe, which appears at frequencies higher than the lowest cut-off frequency. This frequency is inversely proportional to the FSS period [68, 69]. Reducing the spatial FSS period is the solution to this problem. However, as the period decreases, the inductance and capacitance decrease, and the resonance frequency of the FSS increases. As a result, the absorption bandwidth decreases in the low-frequency region. The second limiting factor is the harmonic waves of fundamental *RLC* resonance that occur in large-period FSS. This can be solved by increasing the resonance frequency by decreasing the FSS period. However, it is accompanied by the problem of reducing the low-frequency band. In order to expand the absorption bandwidth, efforts must be made to expand the lower and upper limits at the same time, which will be discussed later.

5.4.2 Multilayer Absorbers

A microwave absorbing structure containing N layers is illustrated in Fig. 5.45a. The planar ith layer is characterized by the thickness d_i, and the complex permittivity $\varepsilon_{r,i}$ and the complex permeability $\mu_{r,i}$. The incident media is free space and the last layer is terminated by a metal (ground plane). An equivalent circuit for this multilayer structure can be represented in Fig. 5.45b.

The input impedance $Z_{in,i}$ at the surface of the ith layer is the wave impedance of the layered structure including i layers to ground plane (with a load impedance $Z_L = 0$). $Z_{in,N}$ is, therefore, the input impedance of the total N-layers structure, which can be calculated by the following recurrence formula:

$$Z_{in,1} = Z_{c,1} \tanh \gamma_1 d_1 \tag{5.46}$$

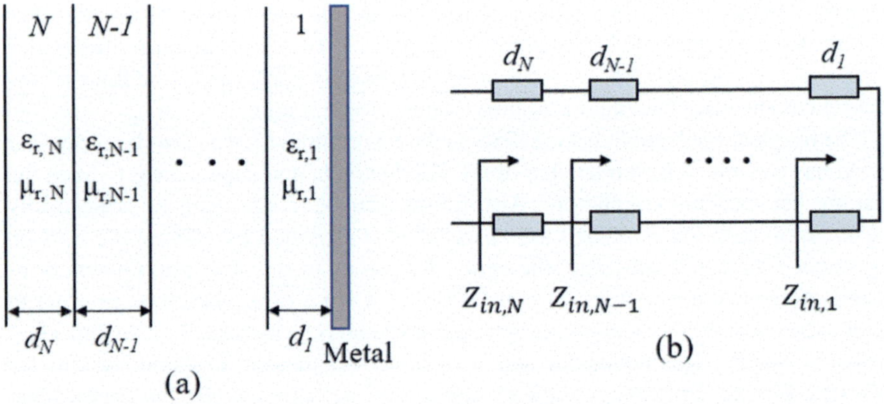

Fig. 5.45 Schematic representation of **a** N-layered microwave absorber and **b** equivalent circuit

$$Z_{\mathrm{in},i} = Z_{\mathrm{c},i}\frac{Z_{\mathrm{in},i-1} + Z_{\mathrm{c},i}\tanh\gamma_i d_i}{Z_{\mathrm{c},i} + Z_{\mathrm{in},i-1}\tanh\gamma_i d_i} \quad (i = 2 \sim N) \tag{5.47}$$

where $Z_{\mathrm{c},i}$ and γ_i are the characteristic impedance and propagation constant of the ith layer, and given by

$$Z_{\mathrm{c},i} = Z_0\sqrt{\frac{\mu_{\mathrm{r},i}}{\varepsilon_{\mathrm{r},i}}} \tag{5.48}$$

$$\gamma_i = j\frac{2\pi}{\lambda}\sqrt{\mu_{\mathrm{r},i}\,\varepsilon_{\mathrm{r},i}} \tag{5.49}$$

where Z_0, λ is the wave impedance and wavelength in free space, respectively. Reflection coefficient of the N-layers media is given by

$$\Gamma = \frac{Z_{\mathrm{in},N} - Z_0}{Z_{\mathrm{in},N} + Z_0} \tag{5.50}$$

Reflection loss (RL) in decibels is then calculated using

$$\mathrm{RL} = 20\log|\Gamma| \tag{5.51}$$

Figure 5.46 shows an example of reflection loss determined in the multilayer absorber [70]. Two types of polymer composites are used: the iron flake particles-rubber composite (designated IFC) is arranged as the 1st layer and the carbon black-rubber composite (designated CBC) is used as the 2nd layer. IFC has high values of magnetic permeability and dielectric permittivity, as shown in Fig. 5.46a, which are due to the ferromagnetic resonance of high-permeability iron particles and space charge polarization between the metal flakes dispersed in rubber matrix. In CBC, low dielectric permittivity and dielectric loss ($\varepsilon_{\mathrm{r}}' = 4.6$, $\varepsilon_{\mathrm{r}}'' = 0.3$) are observed, as shown in Fig. 5.46b. Due to the non-magnetic property of carbon black and rubber matrix, $\mu_{\mathrm{r}}' = 1$ and $\mu_{\mathrm{r}}'' = 0$. Figure 5.46c shows the reflection loss for normal incidence of plane wave determined in the single-layer composite of iron flake particles terminated by metal. A high value of reflection loss (greater than – 1 dB) is predicted even though the composite has a high magnetic and dielectric loss. Most of the power (about 80%) is reflected due to its high value of permeability and permittivity. Impedance-matching layer is required to minimize the wave reflection from such a highly reflecting surface. Carbon black composite of low dielectric constant is attached on the reflecting surface. Figure 5.46d shows the reflection loss determined in the two-layered composite laminate (CBC/IFC/metal) using the recurrence formula. At the frequency of 6 GHz, the minimum reflection loss (– 31 dB) is obtained at the thickness of 7 mm of CBC and 1 mm of IFC. This shows that by combining of absorbing layer of IFC (high magnetic and dielectric loss) and impedance-matching layer of CBC (low dielectric constant), a good absorbing structure could be designed.

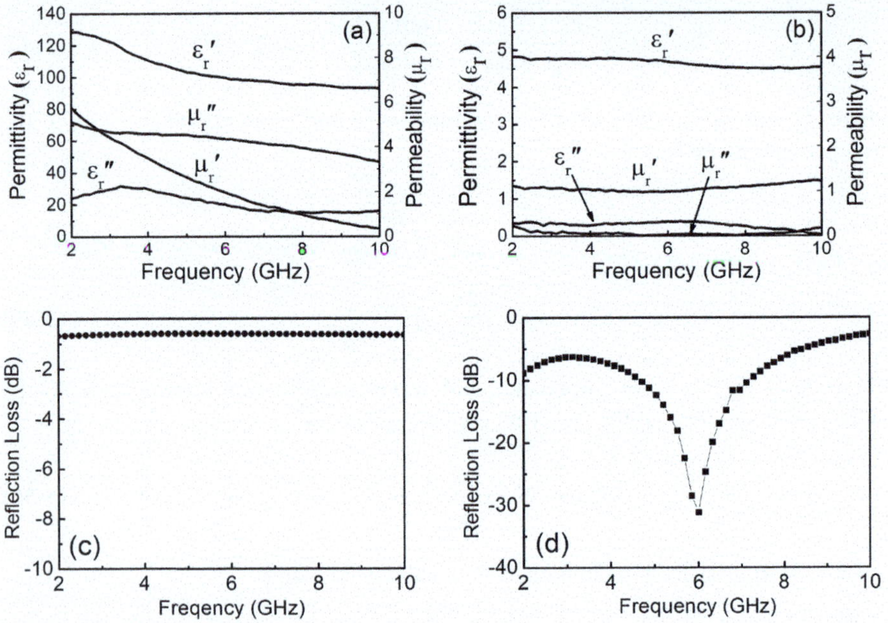

Fig. 5.46 Complex permittivity and permeability determined in the composites of **a** iron flake particles (IFC) and **b** carbon black (CBC), and reflection loss determined in **c** single-layer of IFC and **d** two-layer composites laminate (CBC/IFC). Reprinted with permission from Ref. [70], copyright 2010 Elsevier

Studies have also been conducted to increase the bandwidth by multilayering U-type hexagonal ferrites ($Ba_4(Co_{1-3x}Cr_x)_2Fe_{36}O_{60}$) with different CoCr substitutions [71]. With respect to the monolayer absorber that has an effective bandwidth of 1.5 GHz (with respect to − 10 dB reflection loss), the bandwidth is increased to 3.9 GHz by increasing the number of layers to 4. Likewise, Four-layers design with hexagonal ferrites of $BaCo_{\delta}Ti_{\delta}Fe_{12-2\delta}O_{19}$ composition with different resonance frequencies produced a wideband absorber with 10 dB bandwidth of 5–20 GHz [72].

Broadband and better reflectivity performance has been achieved by using the dielectric grading of rice-husk materials [73]. The absorber was fabricated using multiple layers of rice-husk materials with increasing dielectric loss along the direction of incident wave propagation. The proposed absorber demonstrated good performance (< − 10 dB) for normal and 60° oblique incident angle over the frequency range of 4–8 GHz.

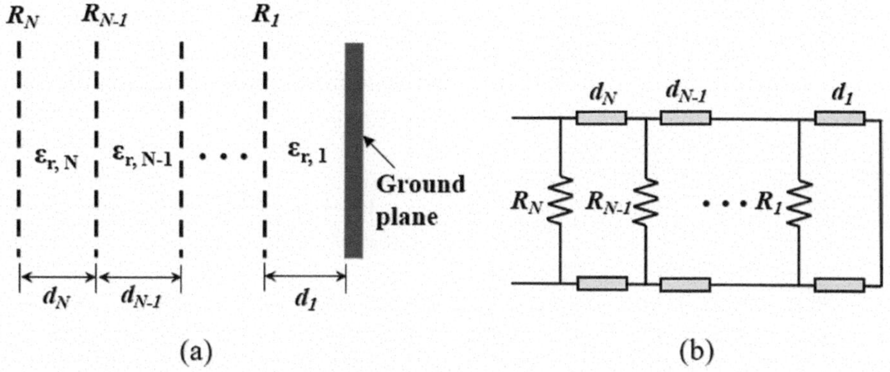

Fig. 5.47 Schematic description of **a** N-layered Jaumann absorber and **b** equivalent circuit

5.4.3 *Jaumann Absorbers*

Jaumann absorbers have a multilayer structure consisting of resistive sheets separated by low-loss dielectrics, which extends the design concept of Salisbury screens to multiple layers to increase the absorption bandwidth. In order to obtain optimal performance, the sheet resistance of each layer R_n ($n = 1 \sim N$) must be properly controlled, and the spacer takes the thickness of 1/4 wavelength at the center frequency of the operating band, as shown in Fig. 5.47. The greater the number of resistive sheets, the larger the absorption bandwidth. When the number of resistive sheets is N, the fractional bandwidth (bandwidth divided by the center frequency) increases approximately N times compared to a single-layer Salisbury screen, while the overall thickness increases N times [74].

In the practical design of Jaumann absorbers, the optimization process becomes more complex because of the increased number of parameters as the number of layers increases. A variety of solutions and optimization techniques were used in the design of Jaumann absorbers, including maximally flat design (binomial expansion), Tschebyscheff (equal-ripple) design, gradient methods, genetic algorithms, and numerical methods (FEM, FDTD) [14, 75]. For an example of ultrawide band Jaumann absorber, the 20 dB absorption bandwidth was designed to be 2.95–35.2 GHz with eight resistive sheets and total thickness of 31.4 mm [69]. Jaumann absorber is favored in bandwidth requirement, but the large thickness may disqualify its practical applications.

5.4.4 *Multilayer FSS Absorbers*

The multilayer FSS absorber consists of a structure in which various patterns or geometries of FSS on a dielectric substrate are arranged in multiple layers. As described in Sect. 5.3.6, the center frequency of the absorption band in a single-layer FSS absorber is the same as the resonance frequency f_r of the FSS [30]. Since f_r is inversely proportional to the inductance L and the capacitance C of FSS ($f_r = 1/2\pi\sqrt{LC}$), it is possible to increase the absorption bandwidth through multi-layer lamination of FSSs with different L and C values. Generally, patch FSS with wide sides has a small inductance and a large capacitance. Conversely, in the case of a narrow strip of crossed FSS, the inductance is large and the capacitance is small. The square loop FSS has a large inductance and capacitance, so it shows the lowest f_r. At a given FSS pattern, the LC resonance frequency depends on the period of FSS; the shorter the period of FSS, the greater the LC resonance frequency. Therefore, the main design variables of the multilayer FSS absorber are the FSS pattern and its dimensions, and optimization is required for the maximum bandwidth with respect to the surface resistance of the FSS conductors, the substrate materials and their thickness, and the stacking sequence of the FSS [68, 76].

(1) **FSS Period**

As an example of multilayer FSS absorber, consider a planar EM wave absorber consisting of two square-loop FSSs (designated SL-FSS) with different periods, which is illustrated in Fig. 5.48a. In the double-layered SL-FSS absorber, FSS1 is placed on the ground substrate of thickness t_1 and FSS2 is placed on the second substrate of thickness t_2. For simplicity, the substrate is assumed to be air. The equivalent circuit can be represented in Fig. 5.48b. The circuit parameters of two resistive SL-FSSs are the resistance (R_1, R_2), inductance (L_1, L_2), and capacitance (C_1, C_2).

The reflection loss of the double-layered FSS absorber is determined by the total impedance of high-impedance surface (Z_1, Z_2), which is a parallel circuit of input impedance of substrate ($Z_{in,1}$, $Z_{in,2}$) and complex impedance of FSS ($Z_{FSS,1}$, $Z_{FSS,2}$). They are given by the following equations for the air substrate with a characteristic impedance Z_0 (= 377 Ω) and a propagation constant $\gamma = j(2\pi/\lambda)$:

$$Z_{in,1} = Z_0 \tanh\left(j\frac{2\pi t_1}{\lambda}\right) \tag{5.52}$$

$$Z_{FSS,1} = R_1 + j\left(\omega L_1 - \frac{1}{\omega C_1}\right) \tag{5.53}$$

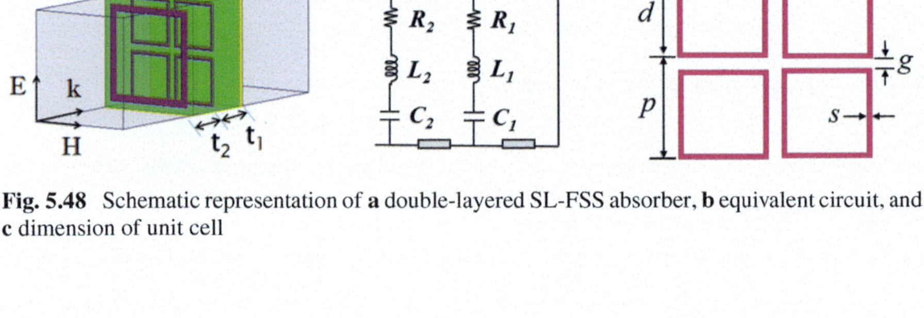

Fig. 5.48 Schematic representation of **a** double-layered SL-FSS absorber, **b** equivalent circuit, and **c** dimension of unit cell

$$Z_1 = \frac{Z_{\text{in},1} Z_{\text{FSS},1}}{Z_{\text{in},1} + Z_{\text{FSS},1}} \tag{5.54}$$

$$Z_{\text{in},2} = Z_0 \frac{Z_1 + Z_0 \tanh\left(j\frac{2\pi t_2}{\lambda}\right)}{Z_0 + Z_1 \tanh\left(j\frac{2\pi t_2}{\lambda}\right)} \tag{5.55}$$

$$Z_{\text{FSS},2} = R_2 + j\left(\omega L_2 - \frac{1}{\omega C_2}\right) \tag{5.56}$$

$$Z_2 = \frac{Z_{\text{in},2} Z_{\text{FSS},2}}{Z_{\text{in},2} + Z_{\text{FSS},2}} \tag{5.57}$$

The reflection loss (RL) can be calculated using

$$RL = 20 \log \left| \frac{Z_2 - Z_0}{Z_2 + Z_0} \right| \tag{5.58}$$

The dimension of SL-FSS is shown in Fig. 5.48c, including the unit cell period p, the length d and width s of conductor, and the spacing between two FSSs g. Now we explain the design procedure with the geometries of FSS1 and FSS2 given in Table 5.4. The inductance and capacitance are calculated to be $L_1 = 1.1$ nH and $C_1 = 29.3$ fF for FSS1, and $L_2 = 3.0$ nH and $C_2 = 83.0$ fF for FSS2, using the equivalent circuit model explained in Sect. 5.3.5 [30]. As the unit cell size increases, both L and C increase due to the increased length of FSS conductor. The resonance frequency calculated from L and C values is $f_r = 23.0$ GHz for FSS1 and $f_r = 9.6$ GHz for FSS2. The thickness corresponding to 1/4 wavelength ($\lambda/4$) at the resonance frequency is $t = 3.3$ mm for FSS1 and $t = 7.8$ mm for FSS2.

Figure 5.49 shows the results of reflection loss. Initially, the previously determined L and C values and the optimized surface resistance R_s and thickness t for the single-layer absorber of SL-FSS1 and SL-FSS2 (i.e., $R_{s,1} = 25\ \Omega/\text{sq}$, $R_{s,2} = 10\ \Omega/\text{sq}$, $t_1 = 3.3$ mm, $t_1 + t_2 = 7.8$ mm) were used for the calculation. The 10 dB bandwidth of each single-layer SL-FSS1 and SL-FSS2 absorber is 9–36 GHz and 4.3–15.6 GHz, respectively. The center frequency of absorption band is equal to the resonance frequency of each SL-FSS. When SL-FSS1 and SL-FSS2 are double-layered with the same design parameters, the bandwidth is broadened, but there is a frequency region (7–20 GHz) with reflection loss greater than − 10 dB, as shown in Fig. 5.49.

Table 5.4 Dimensions, circuit parameters, and resonance frequency (wavelength) of two square-loop FSSs with different period used for absorber design

FSS	Dimension (mm)				Circuit parameters		*LC* resonance	
	p	d	s	g	L (nH)	C (fF)	f_r (GHz)	$\lambda/4$ (mm)
FSS1	3.75	3.5	0.3	0.25	1.1	29.3	23.0	3.3
FSS2	7.5	7.0	0.3	0.5	3.0	83.0	9.6	7.8

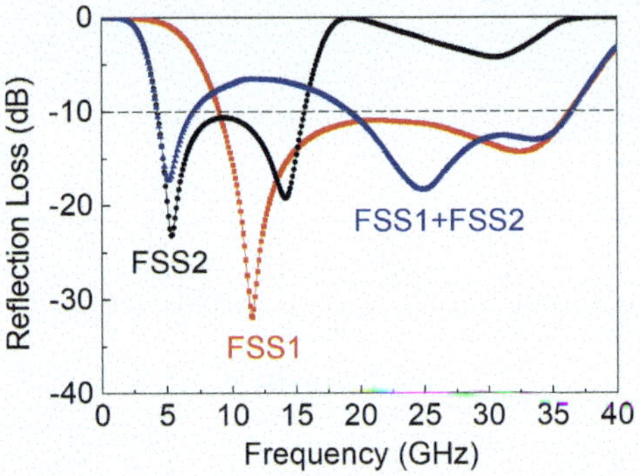

Fig. 5.49 Reflection loss determined in the single- and double-layered SL-FSS absorber with design parameters presented in Table 5.4

In Fig. 5.49, the low-frequency absorption peak of the double-layered absorber almost coincides with that of SL-FSS2 absorber, which means that the wave reflection from SL-FSS1 is negligible in this low-frequency region, as illustrated in Fig. 5.50a. At high frequencies above 30 GHz, we observe a nearly identical reflection loss as that of SL-FSS1 absorber, implying that wave reflection from SL-FSS2 is very small, as illustrated in Fig. 5.50c. In the middle-frequency region, multiple reflections occur from both SL-FSS1 and SL-FSS2, as shown in Fig. 5.50b, which raises the reflection loss, especially at the band where two absorption peaks of each single-layer absorber overlap. This requires the modification of design parameters.

Further optimization of the reflection loss was conducted through varying the surface resistance of the SL-FSS conductors and the total thickness of the absorber, as depicted in Fig. 5.51 [76]. With increasing the surface resistance of FSS2 from $R_{s,2} = 10\ \Omega/\text{sq}$ to $R_{s,2} = 25\ \Omega/\text{sq}$ at the thickness of $t_1 = 3.3$ mm and $t_2 = 4.5$ mm, the reflection loss could be decreased to less than $-$ 10 dB in the frequency range of 4.6–36.3 GHz, as shown in Fig. 5.51a. At the modified surface resistance ($R_{s,1} = 25\ \Omega/\text{sq}$ and $R_{s,2} = 25\ \Omega/\text{sq}$), reducing the total thickness from 7.8 to 6.3 mm (with the first layer thickness $t_1 = 3.3$ mm) resulted in the widest bandwidth (5.9–38.2 GHz for $-$ 10 dB reflection loss), as shown in Fig. 5.51b. The result of the reflection loss

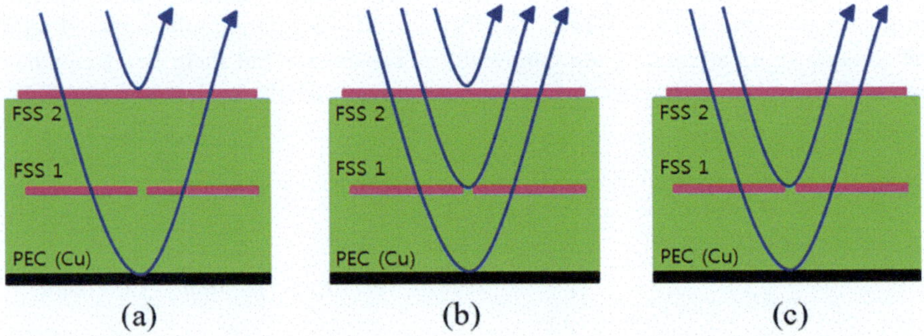

Fig. 5.50 Frequency dependence of wave reflection in the double-layer SL-FSS absorber: **a** 1–7 GHz, **b** 7–30 GHz, **c** 30–40 GHz

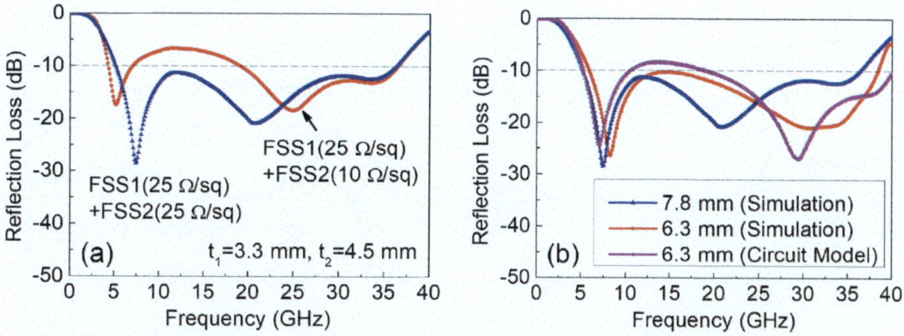

Fig. 5.51 Simulated reflection loss determined in the double-layered SL-FSS absorber with variations in the design parameters: **a** the surface resistance (R_s) of SL-FSSs and **b** the total substrate thickness ($t_1 + t_2$). Reproduced from Ref. [76] with the permission of John Wiley and Sons

calculated from the circuit model of transmission line is also provided in Fig. 5.51b, which exhibits almost identical values of the absorption bandwidth and the reflection loss as those determined using HFSS simulation.

The admittance analysis gives a better insight on the effect of surface resistance of SL-FSS on the reflection loss and the bandwidth of double-layered absorber. Figure 5.52 presents the total admittance (normalized by free-space admittance) at the surface of the double-layered absorber calculated from the equivalent circuit (Fig. 5.48b) before and after optimization. For the absorber with $R_{s,1} = 25$ Ω/sq, $R_{s,2} = 10$ Ω/sq, $t_1 = 3.3$ mm, and $t_2 = 4.5$ mm, the real part of admittance is much higher than matching value especially in the frequency region of 7–15 GHz (Fig. 5.52a), which causes higher reflection loss above − 10 dB (Fig. 5.51a). Surface resistance needs to be raised for suppressing the real part of admittance, particularly for SL-FSS2 whose frequency response is dominant at low frequencies. Through increasing the surface resistance of SL-FSS2 from $R_{s,2} = 10$ Ω/sq to $R_{s,2} = 25$ Ω/sq, the real part of admittance was lowered to the matching value (Fig. 5.52b), and the absorption properties were greatly improved.

Increasing the FSS period can shift the absorption band to lower frequencies. Figure 5.53 shows the reflection loss when the period of SL-FSS1 and SL-FSS2 are increased to 5 mm and 10 mm, respectively. Under optimal design conditions ($R_{s,1} = 30$ Ω/sq, $R_{s,2} = 10$ Ω/sq, $t_1 = 3.3$ mm, $t_2 = 5.0$ mm), the lower bound of absorption bandwidth extends to $f_L = 3.5$ GHz. On the other hand, the upper bound of bandwidth is limited to $f_H = 30$ GHz ($\lambda = 10$ mm), above which the period is greater than the wavelength ($p > \lambda$), resulting in a grating lobe or high-frequency harmonics of resonating FSS [68, 69].

(2) **FSS Patterns**

Another approach for designing the ultrawide bandwidth absorbers is the combination of FSSs with different patterns [68]. As explained in Sect. 5.3.5, the values of capacitance, inductance, and thus the resonance frequency can be more efficiently controlled through selecting the appropriate FSS patterns. In the case of square

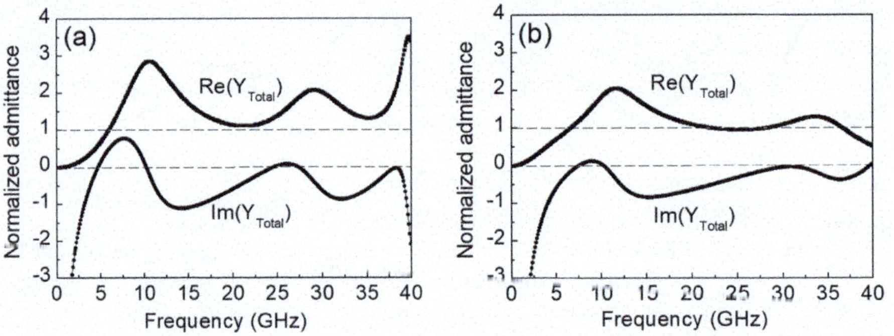

Fig. 5.52 Real and imaginary part of normalized admittance by free-space admittance of double-layered absorber with surface resistance of SL-FSS: **a** $R_{s,1} = 25\ \Omega/\text{sq}$, $R_{s,2} = 10\ \Omega/\text{sq}$, **b** $R_{s,1} = 25\ \Omega/\text{sq}$, $R_{s,2} = 25\ \Omega/\text{sq}$

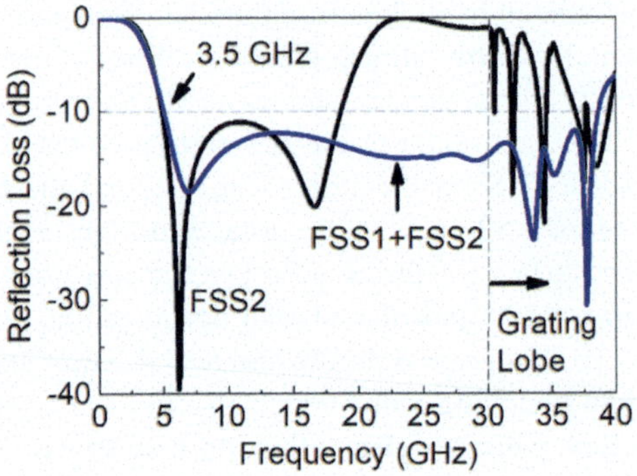

Fig. 5.53 Reflection loss determined in SL-FSS absorbers with larger periods, showing the improvement in low-frequency bandwidth and the high-frequency grating lobe

patch FSS (SP-FSS), the capacitance is large, but the inductance is very small (thus having high resonance frequency). While, in the case of square loop FSS (SL-FSS), both the inductance and capacitance show large values, and therefore the resonance frequency is the lowest. The two FSS combination makes it possible to design the low-reflectivity absorbers in a broader frequency span, while maintaining a small layer thickness. A planar electromagnetic absorber with a combination of two resistive FSS layers is illustrated in Fig. 5.54, where the SP-FSS is placed at the 1st layer (FSS1) and the SL-FSS at the 2nd layer (FSS2). The dimensions of two FSSs are given in Table 5.5.

Figure 5.55 shows the maximum absorption bandwidth optimized with surface resistance of FSSs ($R_{s,1}$, $R_{s,2}$) and thickness of substrates (t_1, t_2). At $R_{s,1} = 100\ \Omega/\text{sq}$, $R_{s,2} = 20\ \Omega/\text{sq}$, $t_1 = 2.0$ mm, $t_2 = 3.5$ mm, the reflection loss can be lowered to $-$ 10 dB or less in the frequency range of 6.3–40.0 GHz. In the higher-frequency region above 40 GHz, the unit cell period is larger than the wavelength ($p > \lambda$) and the grating lobe appears. Compared to the single-layer absorber of SL-FSS, the

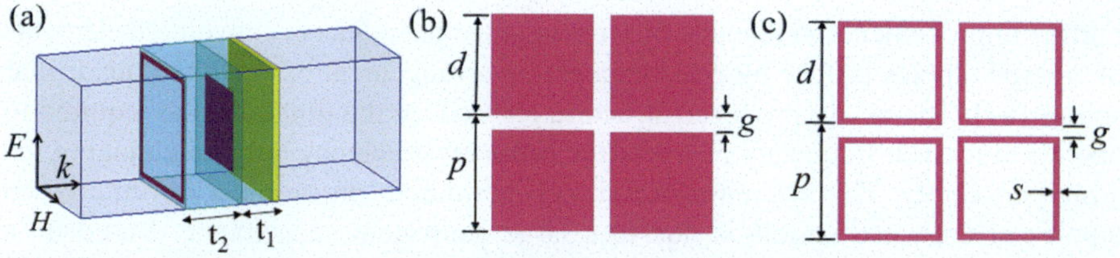

Fig. 5.54 **a** Schematic of the double-layered FSS absorbers with a combination of **b** square patch FSS, and **c** square loop FSS. Reproduced from Ref. [68], CC-BY license

Table 5.5 Dimensions of FSS (square loop and square patch)

FSS pattern	*p* (mm)	*d* (mm)	*s* (mm)	*g* (mm)
Square patch (FSS1)	7.5	4.5	–	3.0
Square loop (FSS2)	7.5	7.0	0.3	0.5

absorber thickness can be reduced. Compared to the SP-FSS absorber, the absorption band can be extended to lower frequencies. That is the advantage of the combined double-layered FSS absorber [68].

(3) High-Capacitive FSS

In ultrawide band FSS absorbers, the bandwidth is limited by the lowest frequency f_L, which is a theoretical value determined from a given reflectance and thickness [13]. The highest absorption frequency f_H is limited by the periodic structure of the absorber in relation to the grating lobes or the harmonic waves of fundamental resonance [68, 69]. In order to enlarge the absorption bandwidth by decreasing the lower bound and increasing the upper bound at the same time, FSS with large inductance and capacitance is required to obtain a low resonance frequency at as a small periodicity as possible.

In an effort to create a high-capacitive FSS, a new approach was made, in which the liner conductor was spirally convoluted or wrapped in a spiral configuration [77–79].

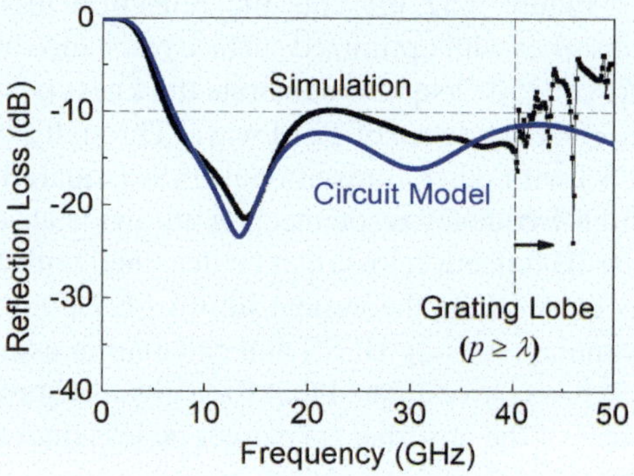

Fig. 5.55 Simulation and calculated results of the optimized reflection loss for the double-layered absorber consisting of a square patch FSS and square loop FSS. Reproduced from Ref. [68], CC-BY license

Further improvements were made by weaving together adjacent convoluted elements of crossed dipoles [80] or by creating a planar arrangement of entwined quadrifilar spirals [81]. The newly proposed structure can reduce the unit cell size required to operate at a given frequency or lower the resonant frequency without changing the array periodicity. This decrease in unit cell periodicity in the interwoven crossed dipoles or entwined spirals is due to a large equivalent capacitance. Based on a similar design concept, a high-capacitive FSS with a folded arm spiral conductors has been proposed [82, 83]. This can be combined with other existing FSS (square loops, square patches) with medium- and high-frequency resonances and makes it possible to design the ultrawide band absorbers with its advantage of a small array periodicity and a low-frequency resonance.

Figure 5.56a illustrates a multilayer EM absorber designed with a combination of three resistive FSS layers, whose geometries are shown in Fig. 5.56b. The operation of the FSS absorber is explained using the equivalent circuit presented in Fig. 5.56c. For the triple-layered FSS absorber, FSS1 (square patch) is placed on the grounded substrate with a thickness of t_1, FSS2 (square loop) is placed on the second substrate with a thickness of t_2, and FSS3 (spiral) is placed on the third substrate with a thickness of t_3, which is in the order of increased $\lambda/4$ thickness from the ground plane.

The spiral FSS (designated S-FSS) has a structure of double-folded strips of width ($w = 0.3$ mm) and gap ($s = 0.2$ mm) convoluted in a clockwise spiral configuration with three turns and terminated at the corner as a square patch with a size $a = 0.8$ mm. The spacing between adjacent folded strips is $g = 0.2$ mm throughout the unit cell. A unit cell periodicity could be lowered to $p = 6$ mm with a resonance frequency of 8.1 GHz [82]. With the same periodicity, the square loop FSS (SL-FSS) has a dimension $d = 5.6$ mm and $w = 0.3$ mm, and the square patch FSS (SP-FSS) has $d = 4.0$ mm.

Table 5.6 summarizes the L and C values of the FSS determined using equivalent circuit model under the assumption that the substrate is air. For the S-FSS, a large capacitance of $C = 288.46$ fF (more than three times that of SL-FSS with the same periodicity) is predicted, which is due to the large effective segment length and a small spacing [82, 83]. The resonance frequency f_0 calculated from the L and C values is given in Table 5.6 in good agreement with the simulation results.

Figure 5.57 presents the reflection loss determined in the triple-layered FSS absorber with optimized surface resistance of FSSs ($R_{s,1} = 100$ Ω/sq, $R_{s,2} = 20$ Ω/sq, $R_{s,3} = 250$ Ω/sq) and substrate thickness ($t_1 = 2.0$ mm, $t_2 = 3.0$ mm, $t_3 = 2.0$ mm) for normal incidence of TEM wave. Through a combination of the high-capacitive spiral FSS with other conventional FSSs (square loop, square patch) with a medium- and high-frequency resonance, an ultrawide absorption bandwidth (4.7–56.0 GHz for −10 dB reflection loss) can be designed with a small total thickness of 7.0 mm, which is close to the theoretical limit (~ 6.7 mm). A good agreement is verified between simulation using HFSS and calculation using circuit model.

It was investigated that for a given absorption frequency response, the total thickness of the absorber (t) cannot be less than the theoretical limit [13], which is given by

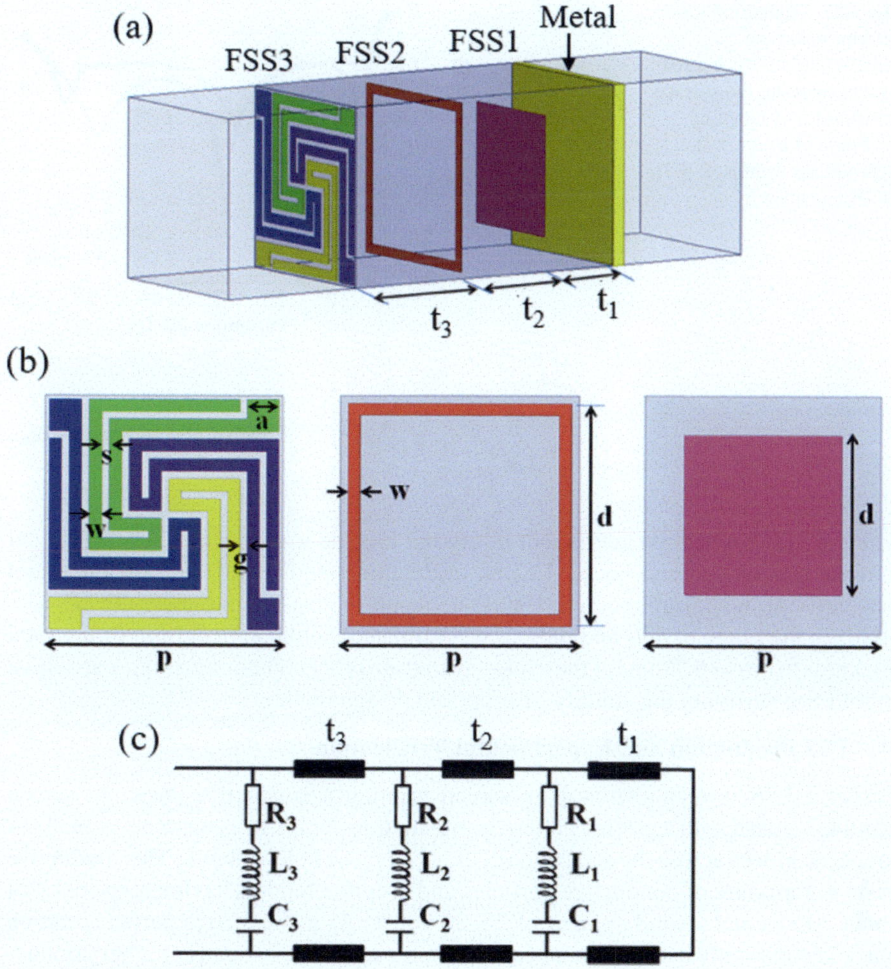

Fig. 5.56 Schematic representation of **a** triple-layered FSS absorbers with a combination of FSS1/FSS2/FSS3 (square patch/square loop/spiral), **b** their dimensions, and **c** equivalent circuit. Reproduced from Ref. [82], CC-BY license

Table 5.6 Circuit parameters (L, C), resonance frequencies (f_0), and quarter wavelengths ($\lambda/4$) at f_0 for three types of FSS

FSS pattern	L (nH)	C (fF)	f_0 (calc.) (GHz)	f_0 (sim.) (GHz)	$\lambda/4$ (mm)
Spiral	1.16	288.46	8.7	8.1	9.3
Square loop	2.77	90.62	12.4	12.4	6.0
Square patch	0.31	35.40	46.1	46.4	1.6

Reproduced from Ref. [82], CC-BY license

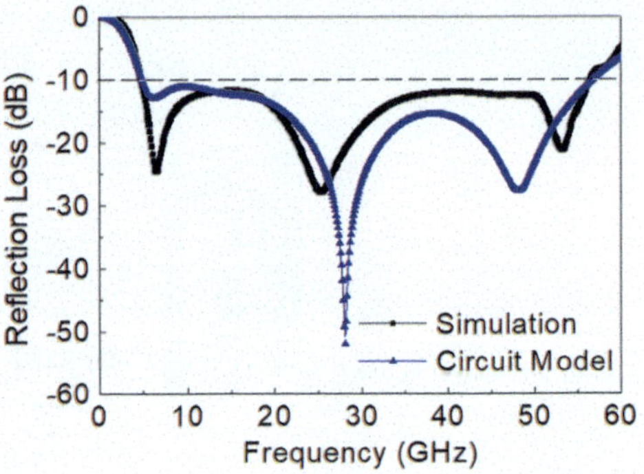

Fig. 5.57 Reflection loss determined in the triple-layered FSS absorber with optimizing the surface resistance and substrate thickness of FSS. Reproduced from Ref. [82], CC-BY license

$$t \geq \frac{1}{2\pi}\left|\int_{0}^{\infty} \ln|\Gamma(\lambda)|d\lambda\right| \tag{5.59}$$

where Γ is the reflection coefficient in dB and λ is the wavelength. This theoretical minimum thickness can be used as an evaluation standard for the efficiency and feasibility of the design method. Using the result of absorption bandwidth ($\Gamma = -10$ dB in the range of 4.7–56.0 GHz), we estimated the theoretical limit of the total thickness to be $t = 6.7$ mm for the triple-layered absorber containing a high-capacitive spiral FSS, verifying the validity of proposed design method.

(4) **FSS Production and Experimental Verification**

Resistive FSS is manufactured by screen printing a conductive paste. In screen printing, a mesh with a printed pattern is used to transfer ink to a substrate. A blade or squeegee moves across the screen to fill the open mesh hole with ink. The conductive paste is a mixture of conductive particles and polymer binders having a viscosity of about 200 Ps and a solid content of 25–30 vol.%. As a conductive particle, carbon black is commonly used. In some cases, metal particles are used for higher conductivity. As a printing substrate, fiber glass–epoxy resin composite (FR4) plates (with $\varepsilon_r = 4.1 - j0.03$) are used. FR4 is installed on the printing machine and the bottom is sucked in by a vacuum pump for the stability of printing. The conductive paste is pressed through a mask pattern that is fixed on the substrate. The printed pattern is dried at 120 °C for 10 min. The average thickness of the conductor film after drying is about 20 μm.

Figure 5.58 shows a prototype sample of multilayer FSS absorber consisting of two square patches of different size [84]. Sample size is 50 cm × 50 cm. FSS with an appropriate surface resistance and thickness is printed on FR4 substrate with a 0.2 mm thickness. A polyethylene foam ($\varepsilon_r = 1.05 - j0.02$) with an open cell structure is an appropriate material for the spacer of low dielectric constant and dielectric loss. The back side is terminated by Cu plate. The power reflection loss is measured by free-space method.

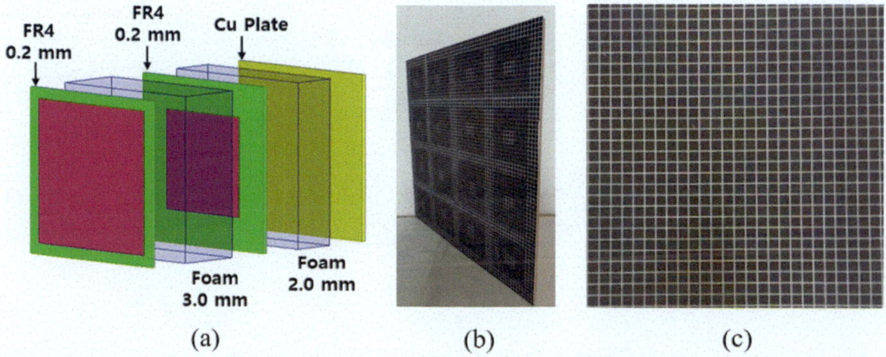

Fig. 5.58 A double-layered absorber of square patch FSSs: **a** schematic structure, **b** test sample, **c** screen-printed pattern. Reproduced from Ref. [84] with the permission of The Korean Institute of Metals and Materials

Fig. 5.59 Reflection loss measured in double-layered FSS absorber consisting of square patch FSSs with different size. Reproduced from Ref. [84] with the permission of The Korean Institute of Metals and Materials

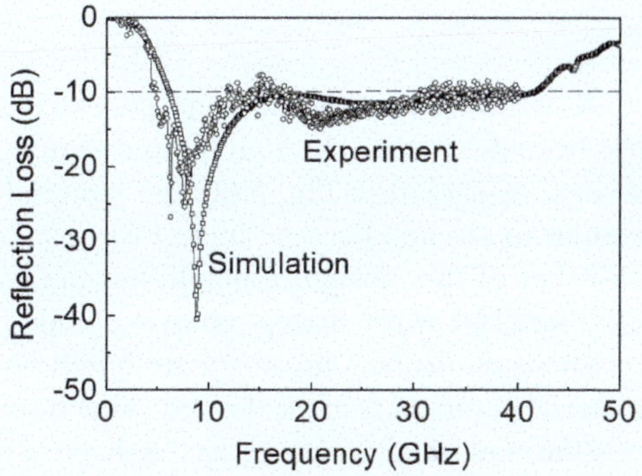

Figure 5.59 shows the experimental result of reflection loss, demonstrating a 10 dB absorption bandwidth of 4.7–40.0 GHz, which is in good agreement with the simulation result for the same multilayer structure of test sample [84]. It is noted that the total thickness is 5.4 mm equal to $0.08\lambda_L$ (λ_L: wavelength at the lowest operation frequency), which is quite thin as compared with the typical values at the level of $0.11–0.15\lambda_L$. The theoretical limit of the total thickness calculated using Eq. (5.59) is $t = 5.4$ mm, verifying the validity of design method with the same experimental result. Good experimental verifications can also be found in other structures of multilayered FSS absorbers [68, 76, 82].

5.4.5 Shape-Controlled Absorbers

The basic conditions required for an EM wave absorber are, firstly, the impedance of the material must be equal to the free space, so that the reflection of the EM

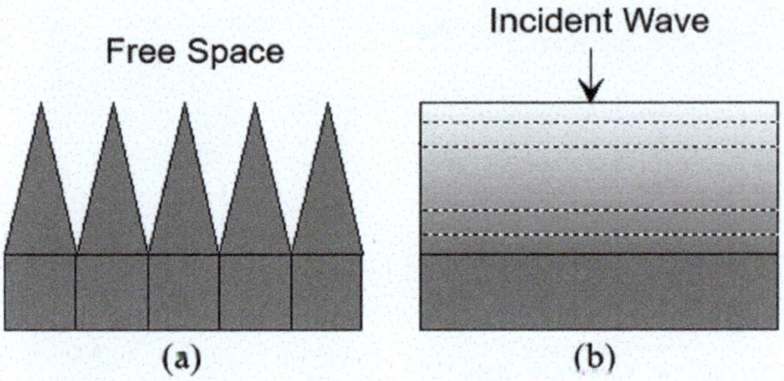

Fig. 5.60 Geometries of wide bandwidth absorbers: **a** pyramid and **b** multilayer

wave does not occur at the interface, and secondly, the material loss coefficient is large, so the penetrated EM wave must be attenuated rapidly. In order to achieve the first impedance matching condition, the following equation must be satisfied, but the synthesis of these materials is not easy,

$$\varepsilon_r = \mu_r \tag{5.60}$$

As a solution to this, the shape control or multilayering method of lossy materials has been proposed. Through multilayering or changing the shape into a pyramid or wedge, as shown in Fig. 5.60, the material properties of the surface can be made similar to air, and become more lossy as the waves propagate to the absorber. The absorber of this design principle has the problem of increasing thickness, but has excellent EM wave absorption over a wide frequency band.

Shape-controlled absorbers are based on a method of controlling the equivalent material parameters to the desired value by changing the size of the air cavity inserted into the material or by changing the shape of the absorber [85, 86]. Figure 5.61a shows a grid-type absorber with air cavities arranged periodically on a planar single-layer plate. The inverted grid absorber is shown in Fig. 5.61b, in which a columnar layer of absorbing material is stacked on a flat base plate. Modifying the columnar shape to a pyramidal shape produces the pyramidal absorber shown in Fig. 5.61c. This structure utilizes the concept of density-gradient absorbers to make the material properties near the surface similar to free space, and the waves suffer more losses when they enter the absorber. In these shape-controlled absorbers, transmission line theory can be applied for the design of wide bandwidth absorption based on the effective medium theory.

(1) **Grid-Type Absorber**

The grid-type absorber has a structure of air cavity periodically arranged in a ferrite plate or composite sheet of which unit cell structure is shown in Fig. 5.62a. If the period is small compared to the wavelength of incident wave, it can be treated as an effective medium and its material parameters can be controlled by the size of inserted air cavity. The effective material parameters can be obtained using the equivalent circuit model shown in Fig. 5.62b [87].

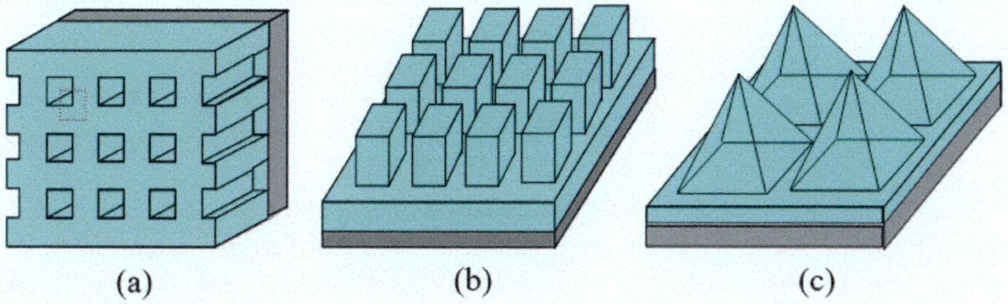

Fig. 5.61 Several types of shape-controlled EM wave absorbers: **a** grid-type, **b** reverse grid-type, and **c** pyramidal

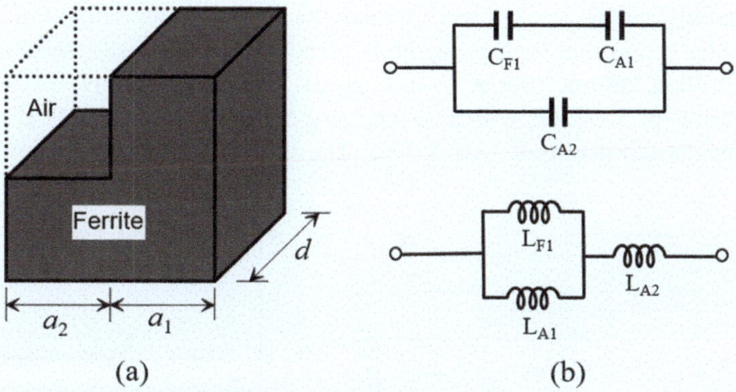

Fig. 5.62 Grid-type absorbers: **a** unit cell and **b** equivalent circuit

From the synthesized capacitance and inductance model for the periodic structure, the equivalent permeability μ_{eq} and permittivity ε_{eq} are given by [87, 88]

$$\mu_{eq} = K\mu_r + \frac{(1-K)\mu_r}{K + (1-K)\mu_r} \tag{5.61}$$

$$\varepsilon_{eq} = K\varepsilon_r + \frac{(1-K)\varepsilon_r}{K + (1-K)\varepsilon_r} \tag{5.62}$$

where $K = a_1/(a_1 + a_2)$ is the volume ratio of ferrite part, and μ_r and ε_r are the intrinsic relative permeability and permittivity of absorber material, respectively.

Using the equivalent material parameters, the characteristic impedance Z_c and propagation constant γ of the grid structure can be obtained by Eqs. (5.63) and (5.64), respectively. Substituting Z_c and γ into Eq. (5.65) yields the input impedance Z_{in} for the single-layer grid absorber and the reflection loss can be calculated:

$$Z_c = Z_0\sqrt{\frac{\mu_{eq}}{\varepsilon_{eq}}} \tag{5.63}$$

$$\gamma = j\frac{2\pi}{\lambda}\sqrt{\mu_{eq}\varepsilon_{eq}} \tag{5.64}$$

$$Z_{in} = Z_c \tanh \gamma d \tag{5.65}$$

As an example, Fig. 5.63 shows the variation of μ_{eq} and ε_{eq} as a function of K in the composite sheet of Ni–Zn ferrite particles [89]. For the bulk sheet ($K = 1$), both the real and imaginary part of complex permeability (μ_r', μ_r'') decrease with the frequency, as shown in Fig. 5.63a. The real part of complex permittivity is $\varepsilon_r' = 6.2$ and the imaginary part ε_r'' is negligibly small. By inserting the air cavity in the sheet, the decrease in the effective material parameters is estimated. Especially, the reduction in ε_{eq}' and μ_{eq}'' is evident. μ_{eq}' is slightly increased after air cavitation, because the bulk sample has a lower μ_r' value than the air as shown in Fig. 5.63c.

Figure 5.64 shows that the bandwidth is broadened by the insertion of air cavity with a controlled volume. For $K = 1/2$, 20 dB bandwidth can be extended to a frequency range of 5–9 GHz with a thickness of 5 mm. While, for bulk absorber ($K = 1$), the bandwidth is only 9.5–10.5 GHz. The insertion of air cavity in the ferrite

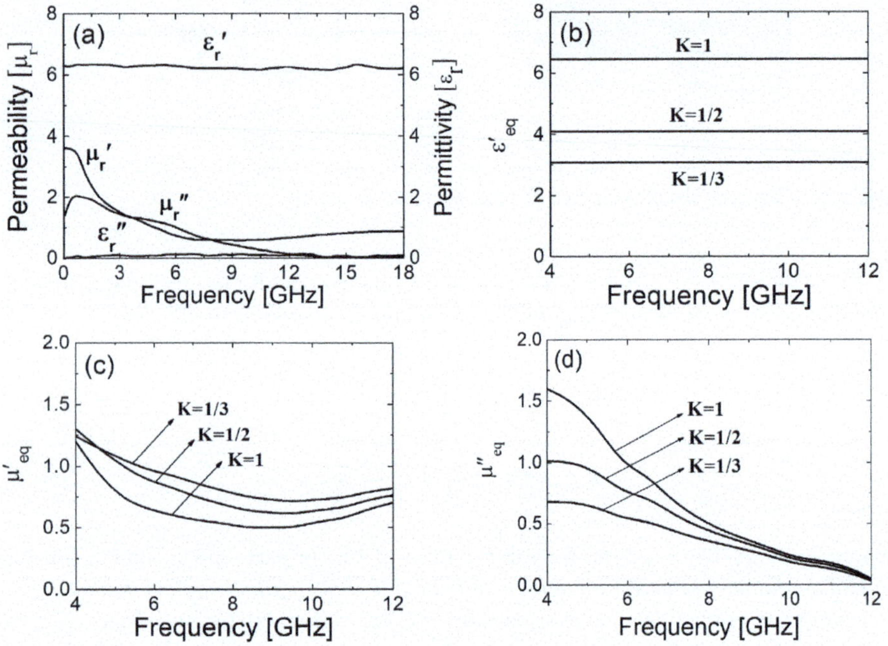

Fig. 5.63 Variations in effective material parameters by insertion of air cavity in the composite sheet of Ni–Zn ferrite particles: **a** intrinsic material parameters, **b** ε_{eq}', **c** μ_{eq}', **d** μ_{eq}''. Reproduced with permission from Ref. [89]

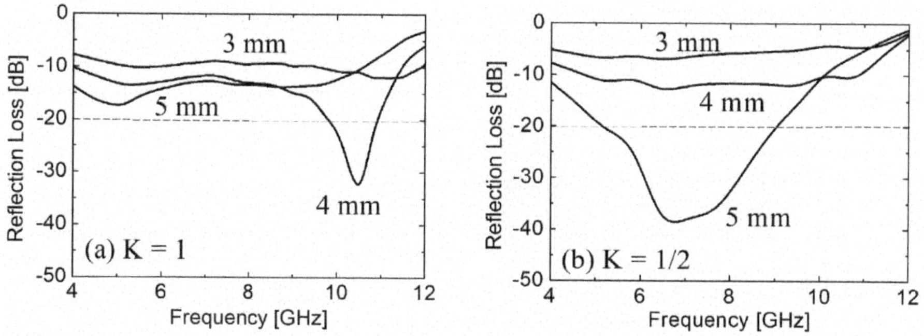

Fig. 5.64 Reflection loss determined in the composite sheet of Ni–Zn ferrite particles: **a** bulk ($K = 1$) and **b** grid-type ($K = 1/2$). Reproduced with permission from Ref. [89]

composites reduces the difference between the permittivity and the permeability and thereby gives a positive effect for the impedance matching over the wide frequency band.

(2) **Reverse Grid-Type Absorber**

A reverse grid-type absorber can be treated as the two-layer absorber with upper square-rod structure and bottom bulk layer, as shown in Fig. 5.65a, where a_1 and a_2 are width of area occupied by ferrite composite and air, respectively. The square-rod part can be represented by the equivalent circuit of synthesized inductance and capacitance model represented in Fig. 5.65b. With a control of the period $(a_1 + a_2)$ smaller than wavelength, the periodic structure can be treated as an effective medium and the equivalent permeability μ_{eq} and permittivity ε_{eq} are given as a function of ferrite volume rate (defined by $K = a_1/(a_1 + a_2)$) [87–90]:

$$\mu_{eq} = (1 - K) + \frac{K\mu_r}{(1 - K)\mu_r + K} \tag{5.66}$$

$$\varepsilon_{eq} = (1 - K) + \frac{K\varepsilon_r}{(1 - K)\varepsilon_r + K} \tag{5.67}$$

From these equivalent material parameters, the characteristic impedance Z_{c2} and propagation constant γ_2 for the upper columnar layer can be obtained. Transmission line approximation of the two-layer structure gives the input impedance at the surface of bottom and upper columnar layers ($Z_{in,1}$ and $Z_{in,2}$, respectively):

$$Z_{in,1} = Z_{c1} \tanh \gamma_1 d_1 \tag{5.68}$$

$$Z_{in,2} = Z_{c2} \frac{Z_{in,1} + Z_{c2} \tanh \gamma_2 d_2}{Z_{c2} + Z_{in,1} \tanh \gamma_1 d_1} \tag{5.69}$$

where d_1 and d_2 are the thickness of bottom and upper layers, respectively.

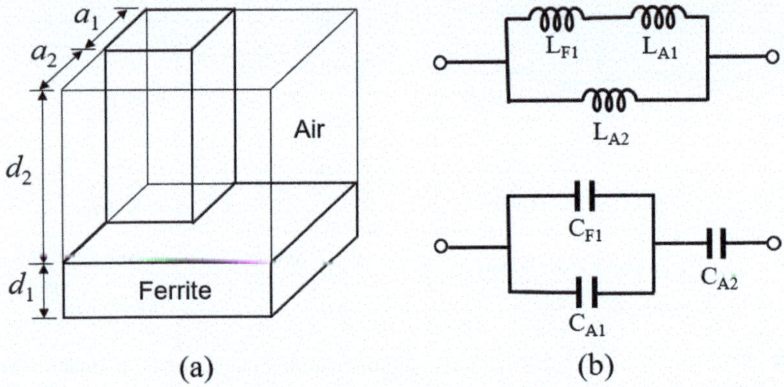

Fig. 5.65 Reverse grid-type absorber: **a** unit cell and **b** equivalent circuit

Figure 5.66 shows the reflection loss of a reverse grid-type absorber designed using a Zn_2Y ($Ba_2Zn_2Fe_{12}O_{22}$) ferrite composite as the base material. A significant increase in absorption bandwidth can be obtained with an optimized dimension of the size (represented by K) and height (d_2) of square rod in the upper layer. At $K = 1/2$, the widest bandwidth (3.8–9.0 GHz with respect to − 20 dB reflection loss) is predicted with a thickness of bottom layer ($d_1 = 4$ mm) and a height of square rod ($d_2 = 8$ mm), as shown in Fig. 5.66b. At the other dimensions of $K = 1/3$ and $K = 2/3$, the bandwidth is greatly reduced. A similar result was observed in other composites of NiZn ferrite and Co_2Z ferrite [90].

(3) **Pyramidal Absorber**

Figure 5.67 shows the schematic and equivalent multilayer structure of a pyramidal absorber [91]. It consists of base plate terminated by metal and pyramidal cones, as shown in Fig. 5.67a. If the period a_1 is shorter than the wavelength, the periodic cone structure can be replaced by a density gradient effective medium [88, 91], as shown in Fig. 5.67b, where d_1 is the thickness of bottom base layer and d_p is the height of

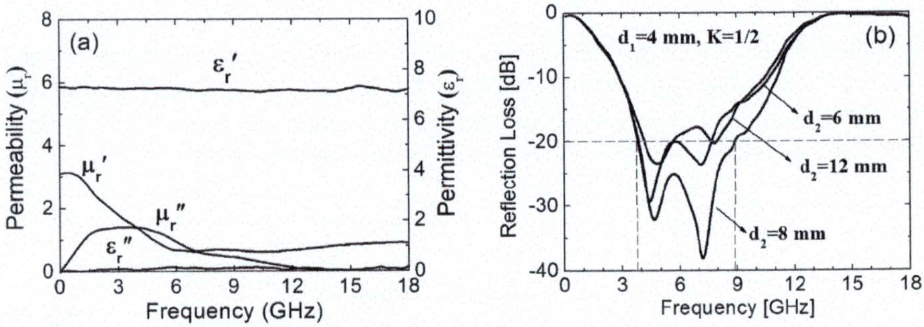

Fig. 5.66 **a** Material parameters and **b** reflection loss determined in the composite of Zn_2Y ferrite particles with a reverse grid-type structure

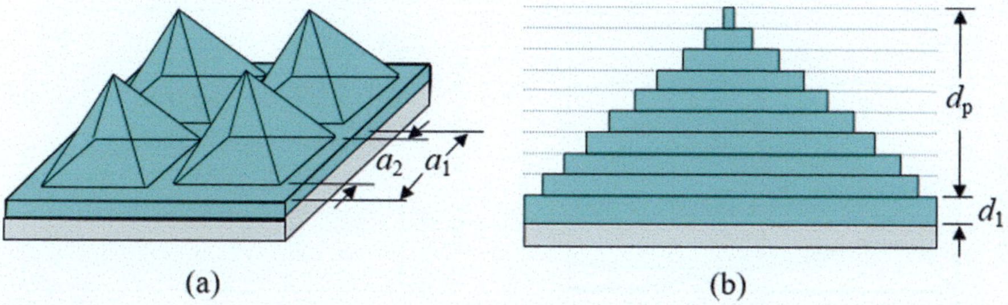

Fig. 5.67 Schematic representation of **a** pyramidal absorber and **b** equivalent multilayer approximation. Reproduced from Ref. [91] with the permission of Springer Nature

pyramidal cone. Each layer can be treated as a composite of absorbing material plus air space.

The equivalent permittivity $\varepsilon_{\mathrm{eq},i}$ and permeability $\mu_{\mathrm{eq},i}$ of the ith layer is then given by [87, 92]

$$\varepsilon_{\mathrm{eq},i} = (1 - K_i) + \frac{K_i \varepsilon_{\mathrm{r}}}{(1 - K_i)\varepsilon_{\mathrm{r}} + K_i} \tag{5.70}$$

$$\mu_{\mathrm{eq},i} = (1 - K_i) + \frac{K_i \mu_{\mathrm{r}}}{(1 - K_i)\mu_{\mathrm{r}} + K_i} \tag{5.71}$$

where K_i is the volume ratio occupied by the absorber material in the ith layer. The characteristic impedance $Z_{\mathrm{c},i}$ and the propagation constant γ_i of the ith layer are then given by

$$Z_{\mathrm{c},i} = Z_0 \sqrt{\frac{\mu_{\mathrm{eq},i}}{\varepsilon_{\mathrm{eq},i}}} \tag{5.72}$$

$$\gamma_i = j\frac{2\pi}{\lambda}\sqrt{\mu_{\mathrm{eq},i}\varepsilon_{\mathrm{eq},i}} \tag{5.73}$$

The input impedance for the 1st layer terminated by metal plate $Z_{\mathrm{in},1}$ is calculated by Eq. (5.68), and the input impedance at the surface of the ith layer $Z_{\mathrm{in},i}$ can be calculated by a recurrence formula, which is given by

$$Z_{\mathrm{in},i} = Z_{\mathrm{c},i} \frac{Z_{\mathrm{in},i-1} + Z_{\mathrm{c},i} \tanh \gamma_i d_i}{Z_{\mathrm{c},i} + Z_{\mathrm{in},i-1} \tanh \gamma_i d_i} \quad (i = 2 \sim N) \tag{5.74}$$

where d_i is the thickness of the ith layer. The wideband absorber is designed in such a way that the input impedance at the absorber surface $Z_{\mathrm{in},N}$ is matched to the free-space impedance.

Figure 5.68 shows a design example of a pyramidal absorber using a composite material of Co_2Z hexagonal ferrite ($Ba_3Co_2Fe_{24}O_{41}$) particles [91]. The Co_2Z ferrite

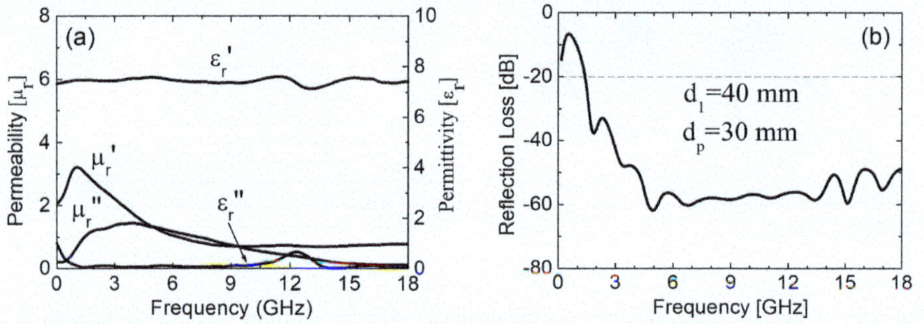

Fig. 5.68 **a** Material parameters of Co_2Z ferrite composite and **b** reflection loss determined in the pyramidal absorber using Co_2Z ferrite composite. Reproduced from Ref. [91] with the permission of Springer Nature

composite has a magnetic loss $\mu_r'' \geq 0.2$ even in the high-frequency range above 12 GHz, as shown in Fig. 5.68a. With a dimensional control of base layer (d_l = 40 mm) and pyramidal cone (d_p = 30 mm), the reflection loss can be lowered below − 20 dB in the frequency range above 1.5 GHz (Fig. 5.68b). The proposed absorbers have a thickness advantage over the classical pyramidal Ohmic absorbers and thus can be suitable for small or semi-anechoic chambers.

Pyramidal absorbers made of ferrite composites have a problem of poor absorption performance in the lower-frequency range below 2 GHz. As a complement, a modified structure is designed by inserting a sintered ferrite plate between the ground metal and the pyramidal ferrite composite, as shown in Fig. 5.69. Ferrite tile has the high absorption properties in the hundred MHz frequency region (as demonstrated in Fig. 5.8), and thus has the positive effect in improving the low-frequency absorbance in combination with pyramidal absorbers.

Figure 5.70 shows the EM wave absorption of this hybrid pyramidal absorber. When using the Co_2Z ferrite composites as the pyramid cone and the ferrite tiles of NiZn ferrite as the base plate, it is possible to design the ultrawide bandwidth absorption extending the lower limit as low as 150 MHz with respect to − 20 dB reflection loss with a total thickness of about 100 mm, as shown in Fig. 5.70 [89].

Fig. 5.69 A schematic representation of hybrid absorber composed of sintered ferrite (base plate) and ferrite composite (pyramidal cone)

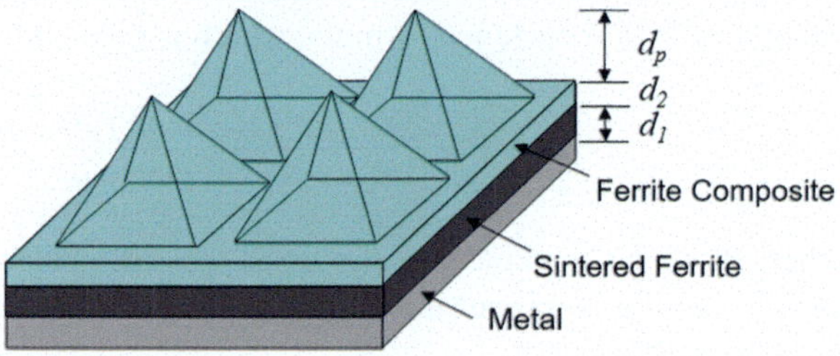

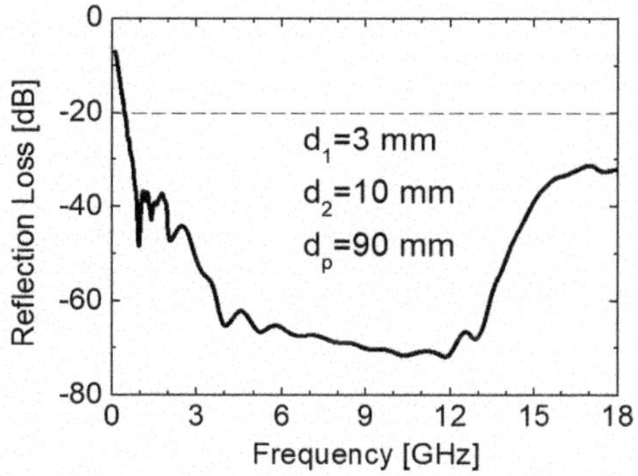

Fig. 5.70 Wideband absorption of hybrid pyramid absorber optimized with dimensions (thickness of sintered ferrite $d_1 = 3$ mm, base part $d_2 = 10$ mm, and pyramid cone $d_p = 90$ mm). Reproduced with permission from Ref. [89]

5.5 Oblique Incidence Absorbers

For oblique incidence of plane waves, two types of polarizations must be considered, as shown in Fig. 5.71: transverse electric (TE) and transverse magnetic (TM). For transverse electromagnetic waves (TEM), both electric and magnetic fields are parallel to the interface (Fig. 5.71a). TE polarization is shown in Fig. 5.71b, where the *E*-field is perpendicular to the plane of incidence, which is defined as a plane consisting of a unit normal vector to the boundary and a propagation vector *k* of incidence. When the *E*-field is on the plane of incidence (the *H*-field is perpendicular), it is called the TM polarization, as shown in Fig. 5.71c.

The characteristic impedance of TE and TM polarization in the layer *n* is given by the following equations (expressed as $Z_{c,n}(\text{TE})$ and $Z_{c,n}(\text{TM})$, respectively):

$$Z_{c,n}(\text{TE}) = \frac{E}{H\cos\theta_n} = \frac{Z_0\sqrt{\mu_{r,n}/\varepsilon_{r,n}}}{\cos\theta_n} \tag{5.75}$$

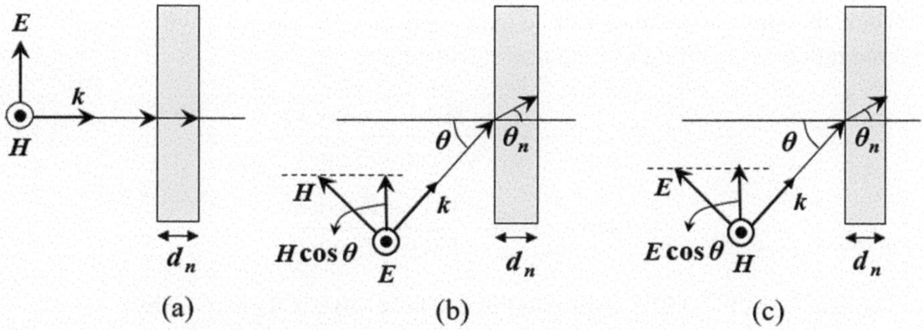

Fig. 5.71 Polarization of plane waves: **a** TEM, **b** TE, and **c** TM. Reprinted with permission from Ref. [70], copyright 2010 Elsevier

$$Z_{c,n}(\mathrm{TM}) = \frac{E\cos\theta_n}{H} = Z_0\sqrt{\frac{\mu_{r,n}}{\varepsilon_{r,n}}}\cos\theta_n \tag{5.76}$$

where Z_0 is the wave impedance of free space and θ_n is the refraction angle in layer n for the incidence angle θ. $\mu_{r,n}$ and $\varepsilon_{r,n}$ are the complex permeability and complex permittivity of layer n, respectively. Propagation constant along the plane normal in layer n is given by

$$\gamma_n = j\frac{2\pi}{\lambda}\sqrt{\varepsilon_{r,n}\mu_{r,n}}\cos\theta_n \tag{5.77}$$

where λ is the wavelength in free space.

For a microwave absorbing structure containing N layers backed by a metal, the input impedance at the surface of the 1st layer and the layer n (= 2 ~ N) are given by [70]

$$Z_{in,1} = Z_{c,1}\tanh\gamma_1 d_1 \tag{5.78}$$

$$Z_{in,n} = Z_{c,n}\frac{Z_{c,n-1} + Z_{c,n}\tanh\gamma_n d_n}{Z_{c,n} + Z_{in,n-1}\tanh\gamma_{n-1}d_{n-1}} \quad (n = 2 \sim N) \tag{5.79}$$

where d_n is the thickness of layer n. Substituting the characteristic impedance and propagation constant into Eqs. (5.78) and (5.79) yields the input impedance of layer n for the TE and TM polarizations. If this is calculated sequentially from the first layer to the surface layer, the input impedance of total N-layered structure $Z_{in,N}$ can be calculated. According to Snell's law, the angle of refraction θ_n can be written in terms of angle of incidence θ, which was explained in Sect. 3.3.2:

$$\cos\theta_n = \sqrt{1 - \frac{\sin^2\theta}{\varepsilon_{r,n}\mu_{r,n}}} \tag{5.80}$$

Thus, the input impedance can be expressed as a function of angle of incidence, and the reflection coefficient Γ can be calculated by

$$\Gamma(\mathrm{TE}) = \frac{Z_{in,N}(\mathrm{TE}) - Z_0/\cos\theta}{Z_{in,N}(\mathrm{TE}) + Z_0/\cos\theta} \tag{5.81}$$

$$\Gamma(\mathrm{TM}) = \frac{Z_{in,N}(\mathrm{TM}) - Z_0\cos\theta}{Z_{in,N}(\mathrm{TM}) + Z_0\cos\theta} \tag{5.82}$$

The reflection loss (RL), expressed in decibels (dB), is then given by

$$\mathrm{RL} = 20\log|\Gamma| \tag{5.83}$$

For example, the oblique incidence properties are analyzed in a single-layer absorber consisting of a carbonyl iron powder (CIP) composite, in which the filler content is about 35 vol.%. The μ_r and ε_r of the CIP composite is given in Fig. 5.72a [66]. Calculating the reflection loss for normal incidence, the minimum value is − 25 dB at the frequency of 6 GHz with a layer thickness of 2 mm, as shown in Fig. 5.72b. For this composite of optimized thickness and frequency, reflection loss is estimated for oblique incidence. Reflection coefficient increases with increasing the incident angles for both TE and TM polarization, as shown in Fig. 5.72c. In TE polarization, the minimum reflection loss is observed at an angle of incidence ($\theta \approx 23°$) which is understood to be the Brewster angle.

Illustrating the double-layered absorber as another example, the input impedance at the surface of layer 2 is given by for TE and TM polarizations:

$$Z_{in,2}(TE) = \frac{Z_0 \mu_{r,2}}{\sqrt{\varepsilon_{r,2}\mu_{r,2} - \sin^2\theta}}$$

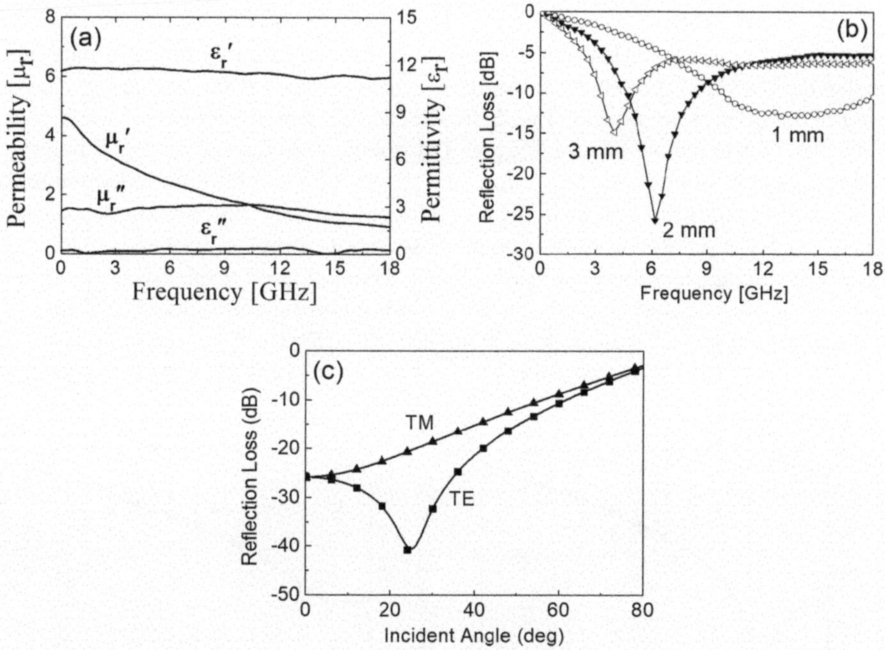

Fig. 5.72 Material parameters and reflection loss determined in a single-layer absorber of carbonyl iron powder (CIP) composites: **a** material parameters, **b** reflection loss for normal incidence (reproduced from Ref. [66] with the permission of Springer Nature), and **c** reflection loss for oblique incidence

$$\times \frac{\frac{\mu_{r,1}}{\sqrt{\varepsilon_{r,1}\mu_{r,1}-\sin^2\theta}}\tanh\gamma_1 d_1 + \frac{\mu_{r,2}}{\sqrt{\varepsilon_{r,2}\mu_{r,2}-\sin^2\theta}}\tanh\gamma_2 d_2}{\frac{\mu_{r,2}}{\sqrt{\varepsilon_{r,2}\mu_{r,2}-\sin^2\theta}} + \frac{\mu_{r,1}}{\sqrt{\varepsilon_{r,1}\mu_{r,1}-\sin^2\theta}}\tanh\gamma_1 d_1 \tanh\gamma_2 d_2} \tag{5.84}$$

$$Z_{in,2}(\mathrm{TM}) = \frac{Z_0\sqrt{\varepsilon_{r,2}\mu_{r,2} - \sin^2\theta}}{\varepsilon_{r,2}} \times \frac{\frac{\sqrt{\varepsilon_{r,1}\mu_{r,1}-\sin^2\theta}}{\varepsilon_{r,1}}\tanh\gamma_1 d_1 + \frac{\sqrt{\varepsilon_{r,2}\mu_{r,2}-\sin^2\theta}}{\varepsilon_{r,2}}\tanh\gamma_2 d_2}{\frac{\sqrt{\varepsilon_{r,2}\mu_{r,2}-\sin^2\theta}}{\varepsilon_{r,2}} + \frac{\sqrt{\varepsilon_{r,1}\mu_{r,1}-\sin^2\theta}}{\varepsilon_{r,1}}\tanh\gamma_1 d_1 \tanh\gamma_2 d_2} \tag{5.85}$$

where

$$\tanh\gamma_1 d_1 = \tanh\left(j\frac{2\pi d_1}{\lambda}\sqrt{\varepsilon_{r,1}\mu_{r,1} - \sin^2\theta}\right) \tag{5.86}$$

$$\tanh\gamma_2 d_2 = \tanh\left(j\frac{2\pi d_2}{\lambda}\sqrt{\varepsilon_{r,2}\mu_{r,2} - \sin^2\theta}\right) \tag{5.87}$$

For a double-layered absorber consisting of a composite of iron flake particles for the 1st layer and the other composite of carbon black for the 2nd layer, whose material parameters are given in Fig. 5.46, we can calculate the reflection coefficient using Eqs. (5.81) and (5.82) for TE and TM polarizations, respectively. When the thickness is controlled to be $d_1 = 1$ mm and $d_2 = 7$ mm, a minimum reflection loss of − 36 dB is predicted at 6 GHz for normal incidence (see Fig. 5.46d). At 6 GHz, both TE and TM polarization show an increase in reflection loss with increasing the angle of incidence, as shown in Fig. 5.73 [70]. The measured results show the same tendency. In this double-layered absorber, a minimum reflection loss (less than − 50 dB) is also observed at an angle of incidence ($\theta \approx 15°$) for TE polarization, which is understood to be the Brewster angle.

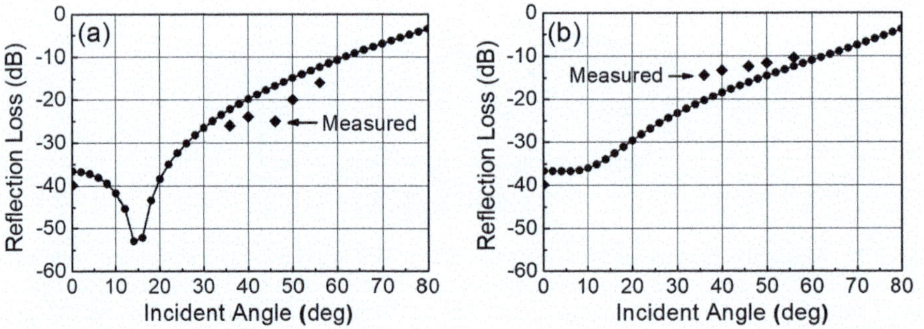

Fig. 5.73 Reflection loss determined in two-layered composite laminate of iron flake particles (1st layer) and carbon black (2nd layer) for oblique incidence: **a** TE and **b** TM. Reprinted with permission from Ref. [70], copyright 2010 Elsevier

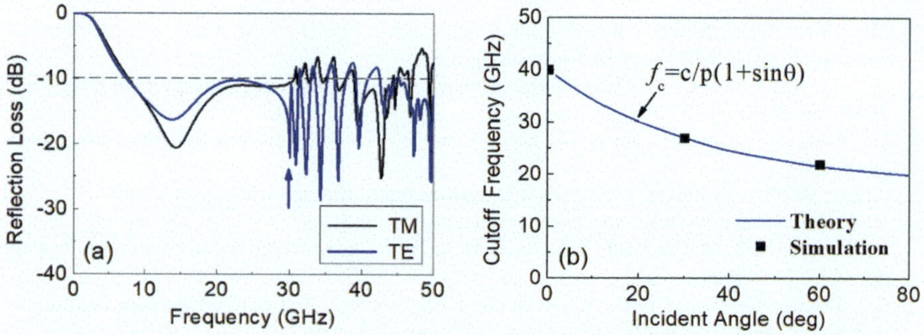

Fig. 5.74 **a** Simulated results of reflection loss showing the grating lobes at the incidence angles of $\theta = 20°$ in a double-layered FSS absorber (whose structure is illustrated in Fig. 5.54a) and **b** theoretical cutoff frequency with simulation results

In the periodic array structure such as an FSS absorber, the increase in incidence angle causes a problem of decrease in bandwidth by the appearance of grating lobes. Figure 5.74 shows the simulation result of oblique incidence properties for a double-layered FSS absorber, whose structure is composed of square patch FSS (1st layer) and square loop FSS (2nd layer) on a grounded dielectric substrate, as illustrated in Fig. 5.54a. Grating lobes or high-frequency harmonics are observed in high-frequency regions where the unit cell period is greater than the wavelength ($p \geq \lambda$) [69, 93]. In a periodic lattice structure, the grating lobe appears as many small reflection peaks above the cutoff frequency, as shown in Fig. 5.74a.

For normal incidence, the critical frequency is about 40 GHz, at which the wavelength is the same as the FSS period ($p = 7.5$ mm), as shown in Fig. 5.55. As the angle of incidence increases, the grating lobe occurs at a lower frequency for both TE and TM polarizations, as indicated by the arrows in Fig. 5.74a. For $\theta = 20°$, the critical frequency decreases to 30 GHz for both TE and TM polarization. The period of the unit cell must be controlled below the threshold value, which is given by approximately, $p \leq \lambda/(1 + \sin\theta)$ [94]. The cutoff frequency, given by $f_c = c/p(1 + \sin\theta)$ where c is the velocity of light, decreases to a value of $c/p = 20$ GHz as the incidence angle increases to $\theta = 90°$, which is shown in Fig. 5.74b. It is seen that the simulation result is in good agreement with the theoretical values.

References

1. CST Computer Simulation Technology. www.cst.com/
2. ANSYS HFSS. www.ansoft.com/products/hf/hfss/
3. Comsol Multiphysics. www.comsol.com/
4. B. David, *Davidson, Computational Electromagnetics for RF and Microwave Engineering*, 2nd edn. (Cambridge University Press, New York, 2010)
5. J. Van Hese, J. Sercu, D. Pissoort, H.-S. Lee, *State of the Art in EM Software for Microwave Engineers, Agilent Technologies Application Note 5990-3225EN* (2009)

6. M.D. Maas, Open-Source Electromagnetic Simulation: FDTD, FEM, MoM. https://www.matecdev.com/posts/differences-fdtd-fem-mom.html
7. D.L. Logan, *A First Course in the Finite Element Method*, 6th edn. (Cengage Learning, USA, 2016), pp. 1–30
8. G. Dhatt, G. Touzot, G. Cantin, *The Finite Element Method Displayed*, 1st edn. (Wiley, New York, 1984)
9. Ansoft HFSS User Guide. https://www.slideshare.net/hillnerpaiva/hfss-user-guide
10. RF Design House Website. http://www.rfdh.com
11. R.L. Fante, M.T. McCormack, T.D. Syst, M.A. Wilmington, IEEE Trans. Antennas Propag. **36**, 1443–1454 (1988)
12. C.J. David, *Radar and Laser Cross Section Engineering*, 2nd edn. (American Institute of Aeronautics and Astronautics Inc., USA, 2005), pp. 352–359
13. K.N. Rozanov, IEEE Trans. Antennas Propag. **48**, 1230–1234 (2000)
14. L.J. Du Toit, IEEE Antennas Propag. Mag. **36**, 17 (1994)
15. L.J. Du Toit, J.H. Cloete, IEEE Trans. Microw. Theory Tech. **44**, 2238 (1996)
16. B. Chambers, A. Tennant, IEE Proc.-Radar Sonar Navig. **143**, 23 (1996)
17. Y. Naito, *Radio Wave Absorbers* (Ohm, Japan, 1987), pp. 77–89 (in Japanese)
18. D. Han, Dissertation (MS thesis), Chungbuk National University, 1995
19. Y. Naito, K. Suetake, IEEE Trans. Microw. Theory Tech. **19**, 65 (1971)
20. H.M. Musal Jr., H.T. Hahn, IEEE Trans. Magn. **25**(5), 3851 (1989)
21. S.-S. Kim, D.H. Han, S.B. Jo, IEEE Trans. Magn. **30**, 4554 (1994)
22. S.S. Kim, S.B. Cho, K.K. Choi, K.I. Gueon, J.M. Kim, K.S. Churn, IEEE Trans. Magn. **27**, 5462 (1991)
23. O. Hashimoto, *Story of Radio Wave Absorbers* (Nikkan Kogyo Shimbun, Japan, 2001), pp. 43–52 (in Japanese)
24. E.F. Knott, J.F. Shaeffer, M.T. Tuley, *Radar Cross Section*, 2nd edn. (Artech House, Boston, 1993), pp. 240–271
25. F. Costa, A. Monorchio, G. Manara, IEEE Trans. Antennas Propag. **58**, 1551 (2010)
26. F. Costa, A. Monorchio, IEEE Trans. Antennas Propag. **60**, 4650 (2012)
27. H.-B., Zhang, P.-H. Zhou, H.-P. Lu, L.-J. Deng, IEEE Trans. Antennas Propag. **61**, 976 (2013)
28. L. Sun, H. Cheng, Y. Zhou, J. Wang, Opt. Express **20**, 4675 (2012)
29. A. Kazemzadeh, A. Karlsson, IEEE Trans. Antennas Propag. **57**, 2307 (2009)
30. T. Liu, S.-S. Kim, Opt. Commnun. **359**, 372 (2016)
31. B.A. Munk, *Frequency Selective Surfaces: Theory and Design* (Wiley, New Jersey, 2000), pp. 63–78
32. M. Kominami, H. Wakabayashi, S. Sawa, H. Nakashima, Electron. Commun. Jpn. (Part I Commun.) **77**, 85 (1994)
33. I. Bardi, R. Remski, D. Perry, Z. Cendes, IEEE Trans. Magn. **38**, 641 (2002)
34. R.J. Langley, E.A. Parker, Electron. Lett. **18**, 294 (1982)
35. F.M. Smits, Bell Syst. Tech. J. 711 (1958)
36. N. Marcuvitz, *Waveguide Handbook* (Lightening Source Ltd., Milton Keynes, UK, 2012), pp. 280–285
37. H.-B. Zhang, P.-H. Zhou, H.-P. Lu, L.-J. Deng, IEEE Trans. Antennas Propag. **61**, 976 (2013)
38. D. Singh, A. Kumar, S. Meena, V. Agarwala, Prog. Electromagn. Res. B **38**, 297 (2012)
39. K.R. Jha, G. Singh, R. Jyoti, Prog. Electromagn. Res. B **45**, 165 (2012)
40. X.F. Luo, P.T. Teo, A. Qing, C.K. Lee, Microw. Opt. Tech. Lett. **44**, 156 (2005)
41. B.C. Wadell, *Transmission Line Design Handbook* (Artech House, Boston, 1991)
42. J.B. Pendry, A.J. Holden, W.J. Stewart, I. Youngs, Phys. Rev. Lett. **76**, 4773 (1996)
43. J.B. Pendry, A.J. Holden, D.J. Robbins, W.J. Stewart, IEEE Trans. Microw. Theory Tech. **47**, 2075 (1999)
44. N.I. Landy, S. Sajuyigbe, J.J. Mock, D.R. Smith, W.J. Padilla, Phys. Rev. Lett. **100**, 207402 (2008)
45. F. Costa, S. Gonovesi, A. Monorchio, IEEE Trans. Antennas Propag. **61**, 1201 (2013)
46. H. Li, L.H. Yuan, B. Zhou, X.P. Shen, Q. Cheng, J. Appl. Phys. **110**, 014909 (2011)

47. J.F. Wang, S.B. Qu, Z.T. Fu, H. Ma, Y.M. Yang, X. Wu, Prog. Electromagn. Res. Lett. **7**, 15 (2009)
48. H. Tao, N.I. Landy, C.M. Bingham, X. Zhang, R.D. Averitt, W.J. Padilla, Opt. Express **16**, 7181 (2008)
49. X. Liu, T. Starr, A.F. Starr, W.J. Padilla, Phys. Rev. Lett. **104**, 207403 (2010)
50. Y. Avitzour, Y.A. Urzhumov, G. Shvets, Phys. Rev. B **79**, 045131 (2009)
51. K. Aydin, V.E. Ferry, R.M. Briggs, H.A. Atwater, Nat. Commun. **2**, 1528 (2011)
52. J.-H. Lim, T. Liu, S.-S. Kim, Appl. Phys. A **117**, 1401 (2014)
53. C.G. Hu, X. Li, Q. Feng, X. Chen, X.G. Luo, Opt. Express **18**, 6598 (2010)
54. A. Ourir, H.H. Ouslimani, Appl. Phys. Lett. **98**, 113505 (2011)
55. N.T. Tung, V.T.T. Thuy, J.W. Park, J.Y. Rhee, Y. Lee, J. Appl. Phys. **107**, 023530 (2010)
56. J. Zhou, E.N. Economon, T. Koschny, C.M. Soukoulis, Opt. Lett. **31**, 3620 (2006)
57. L. Huang, D.R. Chowdhury, S. Ramani, M.T. Reiten, S.N. Luo, A.J. Taylor, H.T. Chen, Opt. Lett. **37**, 154 (2012)
58. H. Luo, T. Wang, R.Z. Gong, Y. Nie, X. Wang, Chin. Phys. Lett. **28**, 034204 (2011)
59. H. Luo, Y.Z. Cheng, R.Z. Gong, Eur. Phys. J. B **81**, 387 (2011)
60. C. Gu, S. Qu, Z. Pei, H. Zhou, J. Wang, B. Lin, Z. Xu, P. Bai, W. Peng, Prog. Electromagn. Res. Lett. **17**, 171 (2010)
61. Q.Y. Wen, H.W. Zhang, Y.S. Xie, Q.H. Yang, Y.L. Liu, Appl. Phys. Lett. **95**, 241111 (2009)
62. H. Tao, C.M. Bingham, D. Pilon, K. Fan, A.C. Strikwerda, D. Shrekenhamer, W.J. Padilla, X. Zhang, R.D. Averitt, J. Phys. D Appl. Phys. **43**, 225102 (2010)
63. M.H. Li, H.L. Yang, X.W. Hou, Y. Tian, D.Y. Hou, Prog. Electromagn. Res. **108**, 37 (2010)
64. H. Wakatsuchi, S. Greedy, C. Christopoulos, J. Paul, Opt. Express **18**, 22187 (2010)
65. H. Wakatsuchi, J. Paul, S. Greedy, C. Christopoulos, IEEE Trans. Antennas Propag. **60**, 3670 (2012)
66. J.-H. Lim, Y.-H. Ryu, S.-S. Kim, Electron. Mater. Lett. **11**, 447 (2015)
67. J.-H. Lim, S.-S. Kim, AIP Adv. **7**, 125223 (2017)
68. T. Liu, S.-S. Kim, Sci. Rep. **8**, 13889 (2018)
69. A. Kazemzadeh, IEEE Trans. Antennas Propag. **59**, 135 (2011)
70. J.-W. Kim, S.-S. Kim, Mater. Des. **31**, 1547 (2010)
71. S. Padhy, A. De, R.R. Debata, R.S. Meena, IEEE Trans. Electromagn. Compat. **60**, 1734 (2018)
72. M.B. Amin, J.R. James, Radio Electron. Eng. **51**, 209 (1981)
73. M.N. Iqbal, F. Malek, S.H. Ronald, M. Shafiq, K.M. Juni, R. Chat, Prog. Electromagn. Res. **131**, 19 (2012)
74. E.F. Knott, J.F. Schaeffer, M.T. Tuley, *Radar Cross Section*, 2nd edn. (Artech House, USA, 1993), pp. 247–252
75. P. Saville, Defence R&D Canada—Atlantic, Technical Memorandum TM-2005-003, Jan 2005
76. T. Liu, S.-S. Kim, Microw. Opt. Tech. Lett. **60**, 2013 (2018)
77. E.A. Parker, A.N.A. El Sheikh, Proc. Microw. Antennas Propag. **138**, 19 (1991)
78. E.A. Parker, A.N.A. El Sheikh, A.C. Lima, Proc. Microw. Antennas Propag. **140**, 378 (1993)
79. C.R. Simovski, P. Maagt, S.A. Tretyakov, M. Paquay, A.A. Sochava, Electron. Lett. **40**, 92 (2004)
80. F. Huang, J.C. Batchelor, E.A. Parker, Electron. Lett. **42**, 788 (2006)
81. A. Vallecchi, A.G. Schuchinsky, IEEE Antennas Wirel. Propag. Lett. **9**, 994 (2010)
82. T. Liu, S.-S. Kim, Sci. Rep. **9**, 16494 (2019)
83. T. Liu, S.-S. Kim, Microw. Opt. Technol. Lett. **62**, 301 (2020)
84. T. Liu, S.-S. Kim, Korean J. Met. Mater. **57**, 741 (2020)
85. Y. Naito, H. Anzai, T. Mizumoto, M. Takahashi, Electron. Commun. Jpn. (Part I Commun.) **77**, 106 (1994)
86. Y. Naito, T. Mizumoto, *Ferrites: The 6th International Conference on Ferrites* (1992), pp. 1320–1325
87. D.I. Kim, M. Takahashi, H. Anzai, S.Y. Jun, IEEE Trans. Electromagn. Compat. **38**, 173 (1996)
88. E.F. Kuester, C.L. Holloway, IEEE Trans. Electromagn. Compat. **36**, 300 (1994)
89. M. J. Park, Dissertation (MS thesis), Chungbuk National University, 2000

90. S.-S. Kim, T. Liu, MATEC Web Conf. **67**, 06066 (2016)
91. M.J. Park, S.-S. Kim, Electron. Mater. Lett. **12**, 610 (2016)
92. M.J. Park, J. Choi, S.-S. Kim, IEEE Trans. Magn. **36**, 3272 (2000)
93. F. Costa, A. Monorchio, G. Manara, IEEE Trans. Antennas Propag. Mag. **54**, 35 (2012)
94. Q. Chen, J. Bai, L. Chen, Y. Fu, IEEE Trans. Antennas Propag. Lett. **14**, 80 (2015)

Chapter 6
Electromagnetic Wave Absorption Materials

In order to satisfy the impedance matching with free space, the loss parameters are necessary in the absorbent medium. There are three types of material loss: electrical conductivity σ, dielectric loss ε_r'', and magnetic loss μ_r''. In this materials part, various lossy materials are reviewed that can be used as EM wave absorbers, including conductive materials (carbon black, carbon nanotube, graphene, MXene), magnetic materials (spinel and hexagonal ferrites, carbonyl iron, high-permeability metals), dielectric materials (ferroelectrics, silicon carbide), core–shell materials (with various structures and materials combination), fiber-reinforced composite materials (used for radar absorbing structures), and recently developed metamaterials or metasurfaces. Literature reviews on their high-frequency electromagnetic properties and EM wave absorption performance are also presented.

6.1 Types of Losses

The input impedance (Z_{in}) at the surface of the single-layer absorber terminated by metal is given as a function of the complex permittivity ($\varepsilon_r = \varepsilon_r' - j\varepsilon_r''$), complex permeability ($\mu_r = \mu_r' - j\mu_r''$), and the thickness of absorbent material d, which can be expressed as the following Taylor series polynomials:

$$\begin{aligned}\frac{Z_{\text{in}}}{Z_0} &= \sqrt{\frac{\mu_r}{\varepsilon_r}}\tanh\left(j\frac{2\pi fd}{c}\sqrt{\mu_r\varepsilon_r}\right),\\ &= j\frac{2\pi fd\mu_r}{c} + j(\frac{2\pi fd}{c})^3\frac{\mu_r^2\varepsilon_r}{3} + j(\frac{2\pi fd}{c})^5\frac{2\mu_r^3\varepsilon_r^2}{15} + \cdots .\end{aligned} \quad (6.1)$$

When Z_{in} is equal to the free-space impedance ($Z_0 = 377\ \Omega$), a non-reflected absorber through impedance matching can be realized. If the constituent material is lossless ($\varepsilon_r'' = 0$, $\mu_r'' = 0$), Z_{in} becomes a pure imaginary and cannot satisfy the impedance

S. Kim, *Fundamentals and Principles of Electromagnetic Wave Absorbers*, Springer Series in Advanced Microelectronics 66, https://doi.org/10.1007/978-981-97-6477-8_6

matching. Lossy materials with non-zero ε_r'' and μ_r'' must be used to implement the function of EM wave absorber. This is analogous to the principle that the electrical energy cannot be absorbed in a lossless transmission line ($R = 0$, $G = 0$ in Fig. 2.13). Equating the distributed element circuit and the constitutive material parameters in field theory, R corresponds to μ_r'' and G corresponds to ε_r''. Therefore, in order to design a non-reflected absorber, the loss parameters (ε_r'', μ_r'') are necessary in the absorbent medium.

The complex permittivity, including electrical conductivity σ, is expressed by $\varepsilon_r = \varepsilon_r' - j(\varepsilon_r'' + \sigma/\omega)$. Therefore, there are three types of material loss: electrical conductivity σ, dielectric loss ε_r'', and magnetic loss μ_r''. The loss due to conductivity σ is called the conduction loss, which corresponds to the resistance of the circuit theory due to the scattering of the charge carriers. Dielectric loss ε_r'' occurs when the reaction of an electric dipole does not follow the time-varying changes of external electric field. It is due to a loss in nature caused by the orientation and relaxation of electric dipoles or ferroelectric domains. Magnetic loss μ_r'' occurs from the delayed reaction of magnetic domains or magnetic moments to the changes of the external magnetic field. The atomic or molecular nature of dielectric and magnetic loss has been explained in Chap. 4.

6.2 Conduction Loss Materials

6.2.1 Carbon Allotropes

Magnetic materials such as iron, nickel, cobalt, or their alloys are frequently used as EM wave absorbers, because they have strong EM absorption capabilities in various spectral ranges owing to their magnetic loss. However, these magnetic materials, in general, have a problem of heavy weight. Meanwhile, carbon-based particles have been proposed as the potential materials that are light and have excellent functions of EM wave absorption [1]. Carbon has several allotropes, including diamonds, graphite, carbon black, graphene, fullerene (a nano-sized sphere of graphene), and nanotubes [2]. These carbon materials are promising in the application of microwave shielding or absorption due to their inherent high electrical conductivity. Carbon black and graphite have been used as the most preferred absorbent materials until carbon nanotube (CNT) was discovered in 1990s. The development of synthetic and production technologies has led to the emergence of other types of carbon-based materials with good EM wave absorption properties [3–5]. The most promising candidates, such as 2D graphene [6–8] and semi-1D CNTs [9, 10], opened up a new field of electromagnetic wave absorption technology.

6.2.2 Carbon Black

Carbon black (CB) is a fine powder produced by incomplete combustion or pyrolysis of hydrocarbon. EM wave absorber is produced by dispersing CB, which has a high electrical conductivity in the range of $10^2 - 10^4$ S/m, in a polymer matrix and by processing of composite materials. Its density is as low as 2.1 g/cc, so it is advantageous for manufacturing the lightweight absorbers. Kwon et al. [11] investigated the X-band absorption properties with the change of CB concentration in silicon rubber matrix. It was reported that if the CB concentration in the composite exceeds the threshold (about 14 wt.%), the conductivity increases rapidly due to the percolation of conductive particles. In order to design the impedance-matched EM wave absorber, the CB concentration should be adjusted below the threshold. In the composite of 10 wt.% CB, the reflection loss less than -10 dB can be obtained in the frequency range of 9.6–13.5 GHz with a sample thickness of 2 mm. Graphite powder was used to increase the permittivity of ferrite composite, and thereby to produce the matched-wave-impedance absorbers [12]. The glass fiber/epoxy composite containing 6 vol.% CB produced a radar absorbing structures with a maximum absorption of 32 dB in X-band frequencies [13].

6.2.3 Carbon Nanotubes

Carbon nanotube (CNT) is one of carbon allotropes having a cylindrical nanostructure. CNT exhibits superior electrical and mechanical properties, having electrical conductivity similar to copper (10^6–10^7 S/m) along the tube axis, high thermal conductivity equal to the best diamond in nature, and strength 100 times greater than steel [14, 15]. Generally, CNTs are classified into single-walled carbon nanotubes (SWCNTs) and multi-walled carbon nanotubes (MWCNTs), depending on the number of tubular graphene layers. SWCNTs are composed of a layer of graphene with a diameter of 0.42 nm and a thickness of less than one millionth of diameter. MWCNTs consist of two or more layers of graphene sheets with a diameter of 3 to 30 nm [16, 17].

A polymer composites in which CNTs are dispersed is usually used as an EM wave absorber and the absorption properties of such composites was reviewed in Ref. [1]. Since CNTs are non-magnetic materials, possible mechanisms for electromagnetic absorption are the dielectric polarization, ohmic loss, and multiple scattering from the large surface of CNTs. Since the aspect ratio of CNTs is usually between 100 and 1000, the percolation threshold is very low within the range from 0.1 to 1 vol.% [18, 19]. When incorporating CNTs into the polymer matrix, a high conductivity level (~1 S/m) can be obtained with a much smaller amount (~0.35 vol.%) compared to CB particles (~20 vol.%). These properties are advantageous in cost-effectiveness and producing a lightweight absorber [20].

In order to achieve the thin thickness, minimum reflection coefficient, and broadband microwave absorption in the CNT composites, studies have been conducted on filler and its concentration along with various synthesis techniques [21]. The reflection coefficient evaluated in monolayer absorber reaches -18 dB at 8 GHz using 5 wt.% MWCNTs [22]. A composite with 4 wt.% MWCNTs dispersed in a polymer resin was prepared and a reflection coefficient of -17 dB was obtained at 6 GHz [23].

In nanostructured materials of CNTs, the loss mechanisms are dielectric relaxation and electrical conduction. First of all, the large specific surface area (~300 m^2/g) and the high aspect ratio of CNTs result in the interfacial polarization with a relaxation frequency in the microwave region and the high conduction loss due to the percolation of conductive fillers [24]. In addition, the multiple reflections inside tubes increase the loss. The composite material obtained by mixing Fe-filled CNTs with epoxy resins shows a reflection loss of less than -10 dB at 11.8–14.7 GHz, and the minimum value reaches -31.71 dB at 13.2 GHz [25]. The enhanced microwave absorption of the CNT/epoxy composite filled with Fe is due to the additional effect of magnetic loss and improved impedance matching. A similar result was observed in CNTs/$CoFe_2O_4$ nanocomposites with better microwave absorption and enlarged effective bandwidth [26].

6.2.4 *Graphene*

Graphene is a hexagonal structured carbon which has a single atomic thickness (0.345 nm). Carbon atoms are covalently bonded in the plane, which enables the graphene to have the superior mechanical properties (200 times stronger than steel) [27]. In addition to its structural advantages, a number of unique characteristics are observed, such as large specific surface area, low electrical resistivity (~10^{-6} Ωcm) with a remarkable charge carrier mobility [28, 29]. The microwave absorption of graphene is primarily due to its high electrical conductivity. Multiple reflections between adjacent graphene sheets and multiple scattering on the surface of wrinkled graphene produce an additional loss mechanism.

A study on graphene nanocomposites was conducted to solve the problem of impedance mismatch due to the high electrical conductivity and non-magnetic properties of graphene [30]. The results of existing research on EM wave absorption of graphene nanocomposites are summarized in Refs. [21, 31]. As an example, graphene composites mixed with polyurethane exhibited a minimum reflection loss of -12.56 dB at 10.43 GHz [32].

As a microwave shielding and absorption material, graphene is generally deposited on various substrates or dispersed in a polymer matrix in the form of composite material [33–35]. Such a composite material includes many interfaces that reflect and scatter the electromagnetic waves. In addition, because of the large specific surface area, the graphene-based composites generally have a very low percolation threshold (~0.1 vol.%) [36]. Therefore, even with a very small amount of graphene in polymer composites, a continuous conductive network may be formed, which increases the conduction loss of propagation energy.

For the purpose of synthesizing a graphene material in large amount, it was intended to obtain a nano-graphene dispersion directly from graphite. But it was difficult to peel one graphene due to π-π binding or van der Waals force between graphene layers. In order to solve this problem, a method of exfoliating graphene oxide (GO) has been proposed, that is a chemical method involving oxidation, reduction, and exfoliation processes, starting from the bulk graphite [37–39]. The reduced graphene oxide (rGO) dispersion from graphite through GO was made using the hydrazine (NH_2NH_2) as a reducing agent [40]. Synthesis process and mechanical properties of graphene and graphene-based nanocomposites have been reviewed in Ref. [41].

In decades, studies on EM absorption of rGO are rapidly increasing. Due to its high dielectric loss, large specific surface area, and residual defects and oxygen, rGO will be a promising absorption material [33]. Examining the EM absorption performance of rGO as a filler in nitrile butadiene rubber (NBR) matrix, the minimum reflection loss of the composite (3 mm thick) reaches −57 dB at 9.6 GHz with a 10 dB bandwidth of 7.5–12 GHz using a very small amount of rGO (~10 wt.%) [42]. Even when the incorporation of rGO is reduced to 1 wt.%, the minimum reflection loss is reaching −37.8 dB at 12.3 GHz with a composite of 2.5 mm thickness [35]. By improving the dispersion of graphene sheet in polymer matrix, the interface between graphene and matrix can be increased. Accordingly, the multiple reflections improve the microwave absorption of the composites.

Since rGO composite has a high dielectric constant, studies have been attempted to improve the performance by mixing it with magnetic materials. A composite of rGO nanosheets decorated with Fe_3O_4 shows a minimum reflection loss of −34.4 dB (at 14 GHz) and −37.5 dB (at 2.7 GHz) with a thickness of 1.6 mm and 6.5 mm, respectively [43]. A nanocomposite of Al sheet coated with rGO (Al@rGO) (2 mm thick) shows a minimum reflection loss of −46 dB at the frequency of 13.68 GHz, due to the enhanced interfacial polarization and impedance matching [44]. Another polymer composite of rGO decorated with ZnO nanocrystals exhibits the minimum reflection loss of −54.2 dB at 15.2 GHz with a wide 10 dB bandwidth of 6.7 GHz (11.4–18 GHz) at a thickness of 2.4 mm [45].

6.2.5 MXene

MXenes are new two-dimensional inorganic materials consisting of a-few-atoms-thick layers of transition metal carbides, nitrides, or carbonitrides, first discovered in 2011 [46]. The general formula is expressed as $M_{n+1}X_nT_x$, where M represents a transition metal (Ti, Zr, Hf, V, Nb, Ta, Cr, Sc, etc.), X represents C or N and n is usually 1 to 3, and T represents a surface group (O^{2-}, OH^-, F^-, NH_3, NH_4^+, etc.). Currently,

MXene is obtained mainly by etching the weakly bound A elements (e.g., Al atoms) in the MAX phase using a mixture with hydrofluoric acid, hydrochloric acid, or fluoride [47, 48]. For example, when M_3AX_2 (Ti_3AlC_2) is etched with HF solution at room temperature, the A (Al) atom is selectively removed, and the surface of carbide layer (Ti_3C_2) is bonded with O, OH, or F atoms [49]. Subsequently, the MXene powders are produced by delaminating the etched powders of layered structure [50].

The MXene monolayers exhibit the metallic conduction properties with high electron density at Fermi level [51]. This suggests that MXene can be used as an EM wave absorption filler in polymer composite materials [52]. The research results on microwave absorption of MXene-based nanocomposites are summarized in Ref. [21]. Dielectric permittivity and EM wave absorption of the polymer composites containing Ti_3C_2 nanosheets were investigated [54, 55]. When the content of Ti_3C_2 reaches 6.67 vol.%, the dielectric loss tangent ($\tan\delta$) exceeds 1 [54]. The high dielectric permittivity even in a small Ti_3C_2 amount is due to the high electrical conductivity of Ti_3C_2 and the multi-reflections between the stacked Ti_3C_2 sheets. Optimizing the EM wave absorption gives a result that when Ti_3C_2 content is 2.39 vol.%, the reflection loss is lower than -10 dB in the entire Ku-band (12.4–18 GHz). When Ti_3C_2 content increases to 3.69 vol.%, the absorption band moves to lower 7–10 GHz band. This is due to the increase in dielectric constant and dielectric loss.

6.3 Magnetic Loss Materials

All magnetic materials have the various types of magnetic loss and EM wave absorbers are prepared using it. Some of the practical materials for microwave applications are ferromagnetic metals (iron, cobalt, nickel, and their alloys, etc.) and ferrimagnetic oxides (ferrites). Ferrites, electrically insulators, are used in the form of a sintered bulk body or a composite material incorporating powders with rubber or resin matrix. Magnetic metals, electrically conductors, are processed into powders to form a composite with polymer matrix for application to EM wave absorbers.

6.3.1 Spinel Ferrites

Spinel ferrite is a ferrimagnetic oxide with a chemical formula MFe_2O_4 where M represents a divalent metal cation. In the spinel structure with a formula AB_2O_4, oxygen ions O^{-2} are cubic close-packed and A^{2+} (M^{2+}) metal ions and B^{3+} (Fe^{3+}) ions are located in interstitial vacancy sites, as shown in Fig. 6.1.

Unit cell of spinel ferrite has 32 oxygen ions and one eighth of 64 A sites (tetrahedral sites) and half of 32 B sites (octahedral sites) are occupied. When M^{2+} ions fill A sites and Fe^{3+} ions fill B sites, it is called the normal spinel. In contrast, when M^{2+} ions fill B sites and half of Fe^{3+} ions fill A sites and the other half of Fe^{3+} ions fill B sites, it is called the inverse spinel. The normal spinel is represented by $M^{2+}[Fe_2{}^{3+}]O_4$, and

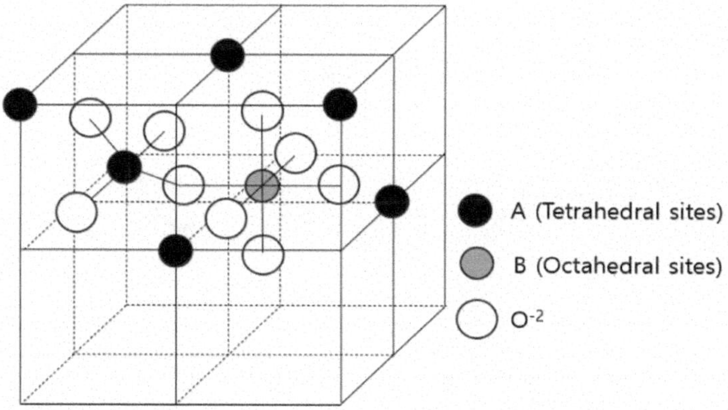

Fig. 6.1 Crystal structure of spinel ferrite **a**: tetrahedral sites, **b**: octahedral sites)

the inverse spinel is represented by $Fe^{2+}[M^{2+}Fe^{3+}]O_4$. NiZn ferrite with a chemical formula $Ni_{1-x}Zn_xFe_2O_4$ is a solid solution having a structure in which Zn ferrite having a normal spinel structure and Ni ferrite having an inverse spinel structure are mixed.

The frequency dependence of magnetic loss is expressed by complex permeability ($\mu_r = \mu_r' - j\mu_r''$). A typical example is the frequency dependence of μ_r' and μ_r'' of NiZn ferrite presented by Smit and Wijn [56] (see Fig. 4.15). μ_r' maintains a constant value and drops rapidly at a resonating frequency. μ_r'' shows a peak at the frequency at which μ_r' has dropped to about half of initial value. This μ_r''-peak frequency corresponds to the natural resonance frequency. The higher the initial permeability (higher Zn^{+2} content), the lower the natural resonance frequency. This means that the frequency range that can have high permeability is limited, and this relationship is called the Snoek's limit. For sintered NiZn ferrites, the frequency range of high permeability is less than 1 GHz. MnZn ferrite has a greater initial permeability than NiZn ferrite and is used as EM wave absorber at lower frequencies.

Figure 6.2a shows the complex permeability spectrum of sintered NiZn ferrites and polymer composites of ferrite powders [57]. In the case of sintered ferrite, μ_r' shows a value of about 1400 in low-frequency region and begins to decrease at about 1 MHz. The μ_r'' shows the maximum value of 800 at about 2 MHz. In the high-frequency region of 1 to 100 MHz, μ_r' decreases with frequency. As the ferrite content decreases in the polymer composite, the low-frequency permeability monotonically decreases. At the same time, the frequency at which μ_r' begins to decrease moves to a higher frequency as the ferrite content decreases. As a result, the permeability at high frequencies is greater in ferrite composites than that in sintered ferrite. The shift of resonance frequency can also be seen in μ_r'' spectrum. The μ_r''-maximum frequency increases as the ferrite content in composites decreases. When the ferrite content is 51.8 vol.%, it increases to about 400 MHz.

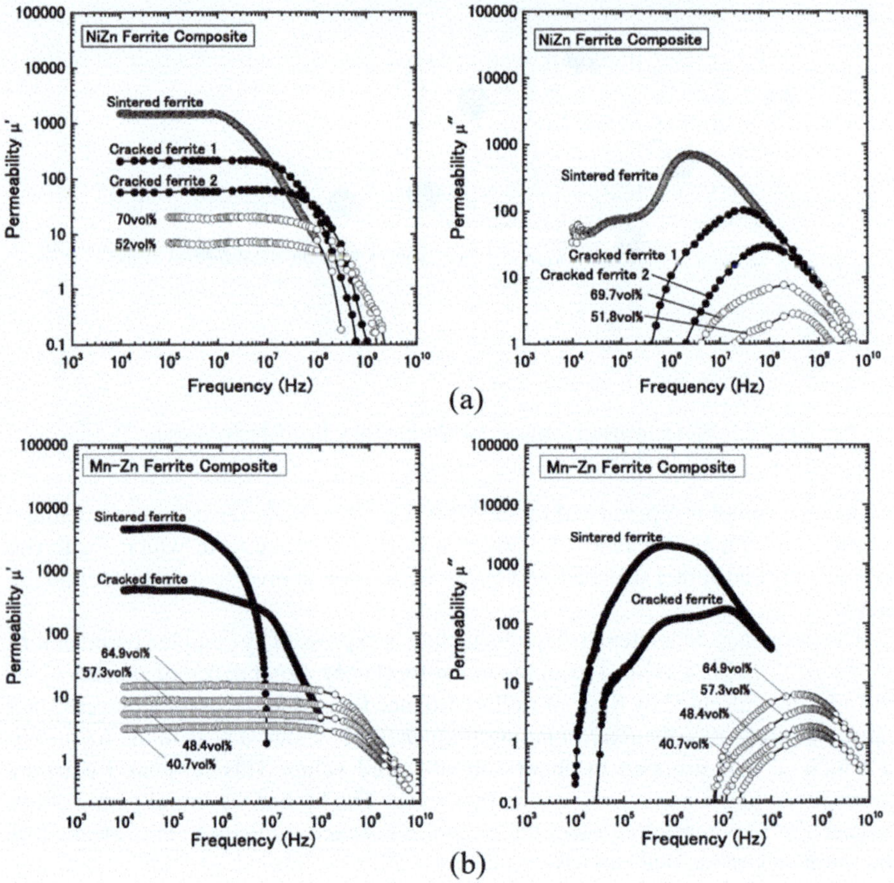

Fig. 6.2 Complex permeability of spinel ferrites (sintered, composite): **a** NiZn ferrite and **b** MnZn ferrite (reproduced from Ref. [57], with the permission of AIP Publishing)

Figure 6.2b shows the complex permeability spectrum of sintered MnZn ferrites and polymer composites of ferrite powders [57]. For sintered MnZn ferrite, the low-frequency permeability is about 4000 and the dispersion frequency is located at about 100 kHz, which is much lower than NiZn ferrite. μ_r' decreases sharply at higher frequencies above 200 kHz, and μ_r'' begins to increase and shows a peak around 1 MHz. In the polymer composites of MnZn ferrite powders, the low-frequency permeability decreases with a shift of the resonance frequency to a higher frequency, as the ferrite content decreases. The μ_r'' peak is observed at higher frequencies above 100 MHz and shows a relaxation type of magnetic loss similar to that of NiZn ferrite composites [57].

The above results are summarized as follows. The permeability μ_r' of sintered MnZn ferrite is about 3 times greater than that of sintered NiZn ferrite. As a result, μ_r' begins to decrease and a μ_r'' peak appears at a lower frequency. In the composites of ferrite powders, low-frequency value of μ_r' decreases and μ_r'' peak moves to a higher frequency as the ferrite content decreases. Therefore, sintered ferrites are used as EM wave absorbers in MHz band, and ferrite composites are used in GHz band. Sintered ferrite with a $Ni_{0.36}Zn_{0.64}Fe_2O_4$ composition exhibits a − 50 dB reflection loss at 280 MHz with a thickness of 5.4 mm [58]. While in the rubber composites of NiZn ferrite powder (44 vol.%), −50 dB reflection loss is observed at 10.5 GHz with a thickness of 3.5 mm [59].

Studies have been conducted to increase the initial permeability and Curie temperature by substituting CuO in NiZn ferrite [60]. The addition of CuO activates the sintering of ferrite and increases the density, leading to the increase in initial permeability. For this reason, NiCuZn ferrite is utilized in manufacturing the multilayer chip inductors [61]. Substituting the appropriate amount of CoO in NiZn ferrite, the saturation magnetization and permeability can be increased [62]. By adding 0.2 wt.% CoO to NiCuZn ferrite, it was possible to produce a high permeability and low-loss EM shield material at 13.56 MHz [63]. This provides an effective solution to increase the detection distance of radio frequency identification (RFID) tags in the near field communication (NFC) system equipped in smart phones. NFC is a type of RFID that uses a frequency of 13.56 MHz and transmits data at a distance of about 10 cm between reader and tag. However, the metal component in the battery-package creates an eddy current in responding to the transmitting magnetic flux from reader antenna, which interferes the communication between tag and reader. Therefore, a thin plate or sheet with an excellent magnetic shielding at 13.56 MHz is required between the battery and loop antenna. Here, the high-permeability and low-loss ferrite, such as shown in Fig. 6.3, is used.

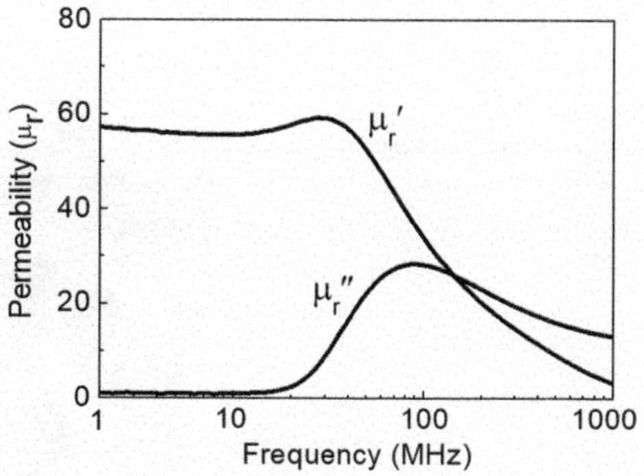

Fig. 6.3 Complex permeability spectrum of NiCuZn ferrites

6.3.2 Hexagonal Ferrites

(1) Types and Microstructures

Hexagonal ferrite is an oxide magnetic material having a complex hexagonal crystal structure, produced by the reaction of barium or strontium oxides (BaO, SrO) and iron oxides (α-Fe_2O_3) at high temperatures. It has a grain structure of hexagonal plates (Fig. 6.4), where the *c* axis is perpendicular to the plate and the *a-b* plane is parallel to it. These substances were discovered by Philips researchers in mid-1950s. Ferroxdure is the first material in this family, known as M-type barium ferrite with a chemical formula $BaFe_{12}O_{19}$ [64]. This compound has a high magneto-crystalline anisotropy along the *c* axis and is used as a permanent magnet material.

Conventional spinel ferrite has a cubic crystalline structure and because of its small crystalline anisotropy field, the operating frequency cannot exceed the Snoek's limit [56]. However, in the case of hexagonal ferrites, it is possible to escape the Sneok's limit due to their large uniaxial anisotropy along *c* axis or easy magnetization on *a-b* plane, which enables it to maintain the initial permeability up to a few GHz [56, 65–67]. This series of materials can be classified by the case where the magnetizing axis coincides with *c* axis (uniaxial anisotropy) and another case where the easy axis is parallel to *a-b* plane (planar anisotropy) [68, 69]. Ferroxdure is a representative material having the uniaxial anisotropy and ferroxplana is magnetic materials having the planar anisotropy.

Many detailed reports have been published on the crystal structure and composition of hexagonal ferrites [56, 68–70]. The hexagonal ferrites, ternary compounds of MeO (Me: divalent metal ion), BaO, and Fe_2O_3, are classified as M-type ($BaFe_{12}O_{19}$), U-type ($Ba_4Me_2Fe_{36}O_{60}$), W-type ($BaMe_2Fe_{16}O_{27}$), X-type ($Ba_2Me_2Fe_{28}O_{46}$), Y-type ($Ba_2Me_2Fe_{12}O_{22}$), and Z-type ($Ba_3Me_2Fe_{24}O_{41}$), depending on the composition, as shown in Fig. 6.5 [56, 68–70]. Of these, Y-type and some of Z-type have a planar magnetization on *a-b* plane and the other M-, U-, W-, X-, Z-type hexagonal

Fig. 6.4 The grain structure of sintered hexagonal ferrites

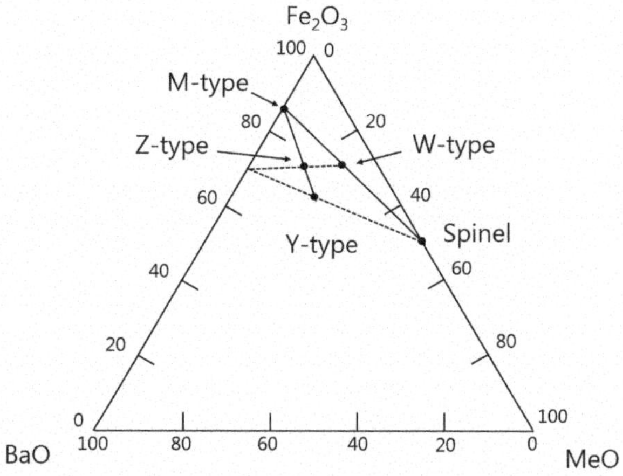

Fig. 6.5 Ternary composition of hexagonal ferrites (reproduced from Ref. [56])

ferrites have uniaxial anisotropy along *c* axis. The substitution of Co^{2+} ions to any composition of the hexagonal ferrites tends to change the crystal anisotropy from *c* axis to *a-b* plane [71–73].

In general, hexagonal ferrites with strong uniaxial anisotropy have difficulty in oscillating their magnetic moments in response to the applied magnetic field, and magnetic resonance cannot be expected up to a frequency of several GHz. However, in materials with *a-b* plane anisotropy, magnetization has no preference for a specific direction in the *a-b* plane, and the easy rotation of magnetization leads to large magnetic loss in the GHz frequency band [64–69]. A typical example is the Co_2Z hexagonal ferrite, which has a natural resonance frequency of about 1.5 GHz [56]. Compared to $NiFe_2O_4$ spinel ferrite with almost the same initial permeability, the natural magnetic resonance appears at much higher frequency band. This is due to the greater anisotropy field of Co_2Z hexagonal ferrite.

Figure 6.6 shows the crystal structures of M-, W-, Y-, Z-type hexagonal ferrites [56, 74]. The M-type hexagonal ferrite has a magnetoplumbite structure whose unit cell has an R block ($BaFe_6O_{11}$ composition) and an S block (Fe_6O_8 composition) along *c* axis. It is repeated in sequence R^*S^* | RSR^*S^*| RS, where * means an 180° rotation symmetry around *c* axis. The crystal structure of W-, Y-, and Z-type hexagonal ferrites consists of an appropriate combination of S block, R block, and T block ($Ba_2Fe_8O_{14}$). The W-type constructs a unit cell with two S blocks and one R block, and the build-up sequence along *c* axis is $SSRS^*S^*R^*$. The Y-type takes a structure 3(ST) in which the ST block combination is repeated three times. The Z-type takes a combination structure of Y + M (hence ST + SR), and the build-up sequence along *c* axis is $STSRS^*T^*S^*R^*$.

(2) **Magnetic Anisotropy**

The general formula for magneto-crystalline anisotropy energy in hexagonal crystals is given by [75]

$$E_a = K_1 \sin^2\theta + K_2 \sin^4\theta + K_3 \sin^6\theta \cos(6\phi), \tag{6.2}$$

where, K_1, K_2, and K_3 are the anisotropy constants, θ is the angle between magnetization vector M and c axis, and ϕ is the angle between M and a axis in a-b plane, as shown in Fig. 6.7. Given K_1, K_2, and K_3, the magnetization direction is determined with an E_a minimum in Eq. (6.2). The condition that the c axis becomes the easy direction ($\theta = 0°$) is $K_1 > 0$, and for easy a-b plane ($\theta = 90°$), $K_1 + 2K_2 < 0$ [75].

The anisotropy constants (K_1, K_2), saturation magnetization (M_s), and anisotropy field (H_θ^A) of hexagonal ferrites of various compositions is presented in Ref. [56]. Most M- and W-types have c axis anisotropy with $K_1 > 0$. The representative compound of in-plane anisotropy is Y-type ferrite with $K_1 + 2K_2 < 0$. In addition, when Co^{+2} ion is substituted, it shows a tendency to be a-b plane anisotropy, and the representative compound is Co_2Z. The Zn_2Z compound ($Ba_3Zn_2Fe_{24}O_{41}$) exhibits c axis anisotropy, but when Co^{+2} ion is substituted by more than 0.5 mol, it turns into a-b plane anisotropy [56].

The magneto-crystalline anisotropy energy in hexagonal ferrites having a-b plane anisotropy is determined by K_3 in Eq. (6.2). The anisotropy constant K_3

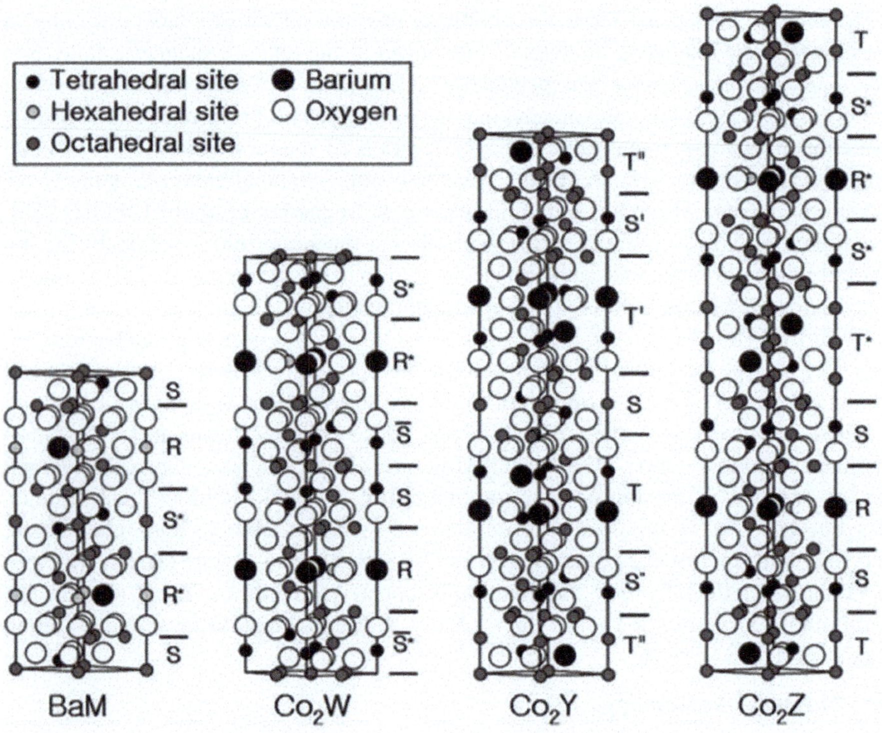

Fig. 6.6 The crystal structure of M-, W-, Y-, Z-type hexagonal ferrites (reproduced from Ref. [74] with the permission of The Korean Ceramic Society)

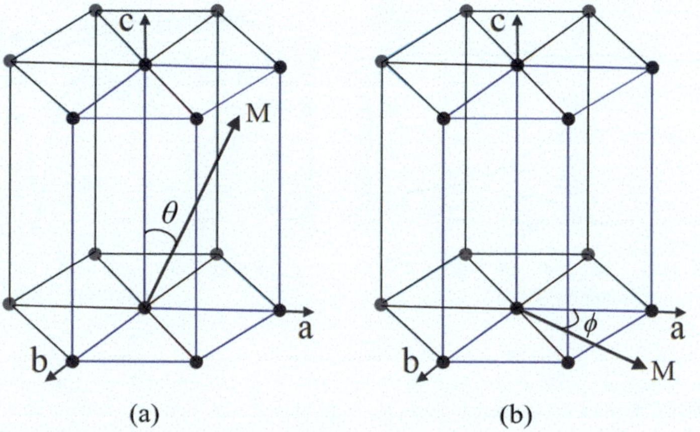

Fig. 6.7 The direction of spontaneous magnetization in hexagonal ferrites: **a** *c* axis anisotropy, **b** *a-b* plane anisotropy

and anisotropy field H_{ϕ}^{A} of hexagonal ferrites with *a-b* plane anisotropy is also presented in Ref. [56]. H_{ϕ}^{A} is small, approximately 10^3 lower than H_{θ}^{A}. In planar anisotropic ferrites, the more Co^{+2} ion substitutions are, the greater H_{ϕ}^{A} becomes. On the other hand, Zn^{+2} substitution decreases H_{ϕ}^{A}. For $Zn_{1.5}Fe_{0.5}Y$ hexagonal ferrite, H_{ϕ}^{A} is reduced to 1 Oe or less [56].

(3) **M-type Hexagonal Ferrites**

M-type compounds have the simplest crystal structure of hexagonal ferrites and are widely used as permanent magnet material because they have a large magneto-crystalline anisotropy along the *c* axis. Due to this strong uniaxial anisotropy ($K_1 = 3.2 \times 10^6$ erg/cm^3, $H_a = 16.8$ kOe) [67, 74], the natural resonance frequency of BaM ($BaFe_{12}O_{19}$) reaches about $f_r = 45$ GHz [64, 73, 76]. Therefore, to reduce the resonance frequency to the range of 1 – 18 GHz, the anisotropy field H_a must be reduced by substituting the appropriate elements, or the anisotropy must be changed from *c* axis to *a-b* plane.

Early in 1959 by Casimir, Smit, Wijn, et al. [56, 67], it was proposed that the substitution of Ti-Co into Fe^{3+} sites changed the magneto-crystalline anisotropy in M-type hexaferrites. Similar researches were conducted that the combined substitution of non-magnetic or weak magnetic + 4 ions (Ti^{4+}, Ir^{4+}, Ru^{4+}, etc.) and + 2 ions (Co^{2+}, Zn^{2+}, Mn^{2+}, etc.) in pure M-type material changes the magnetic anisotropy from *c* axis to *a-b* plane [64, 68–71, 77]. With a compound of $BaFe_{12-2x}Co_xTi_xO_{19}$, the resonance frequency decreases from 50 GHz for x = 0 (BaM) to less than 1 GHz for x = 1, which is due to the change of magneto-crystalline anisotropy from *c* axis to *a-b* plane [78]. Further substitutions above x = 1.3 strengthen the in-plane anisotropy and f_r increases again above 1 GHz [72]. In the case of Ru-Co substitution, coercivity decreases from 500 to 250 Oe with almost the same saturation magnetization as the substitution rate increases from x = 0.2 to x = 0.4 [79]. As a result, the initial

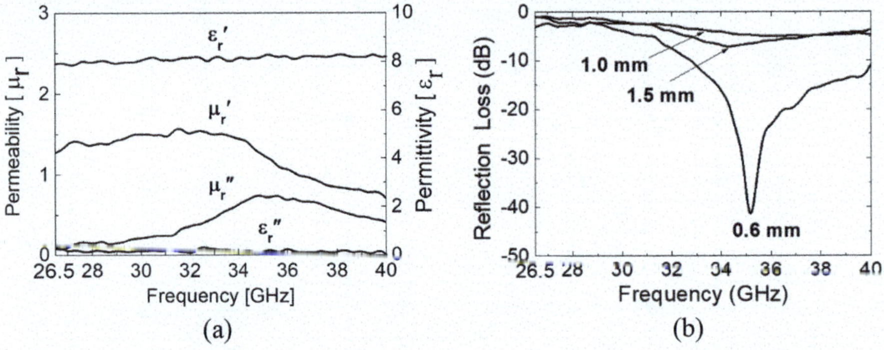

Fig. 6.8 **a** Material parameters and **b** reflection loss determined in the composites of $BaFe_{12-2x}Ti_xCo_xO_{19}$ (x = 0.4) ferrite particles (reproduced with permission from Ref. [82])

permeability increases significantly from 3 to 7, and f_r decreases sharply from 14 to 1 GHz.

High-frequency magnetic properties and EM wave absorption were investigated in M-type hexagonal ferrites substituted with Co-Ti and Mn-Ti [80–82]. For example, a polymer composite incorporating $BaFe_{12-2x}Ti_xCo_xO_{19}$ powders with x = 0.4 exhibits high magnetic loss and wide absorption bandwidth in Ka-band frequencies (26.5–40 GHz), as shown in Fig. 6.8 [82]. Increasing Ti-Co substitution to x = 1.2 shifts the loss peak to 4.5 GHz, and the EM absorption is maximized in this frequency band [70]. This is due to the change in crystalline anisotropy from *c* axis to *a-b* plane. Composites of $BaFe_{12-2x}Mn_xTi_xO_{19}$ ferrites with x = 1.5 and 2.0 showed the wide 20 dB absorption bandwidth of 14.1 − 20GHz and 7.4 − 12 GHz with a thickness of 1.6 mm and 2.7 mm, respectively [80, 81].

(4) **W-type Hexagonal Ferrites**

W-type hexaferrites, denoted as Me_2W with a composition $BaMe_2Fe_{16}O_{27}$ (Me: divalent metal ion), have *c* axis anisotropy, the highest saturation magnetization, and a relatively high Curie temperature [56]. Fe_2W, Ni_2W and Zn_2W, due to their uniaxial anisotropy, induces the natural resonance at high frequency (f_r = 36 GHz for Ni_2W ferrite) [83]. On the other hand, the substitution of Co^{2+} ions in these compounds reduces the anisotropy constant K_1, so that the magneto-crystalline anisotropy gradually changes to *a-b* plane [84–87]. Thus, as the Co^{2+} substitution increases, the natural resonance frequency and EM absorption band gradually shifts to lower frequencies.

When replacing Co^{2+} ions with non-magnetic Zn^{2+} ions in Co_2W hexagonal ferrite, the saturation magnetization M_s increases and the coercivity H_c decreases [87–89]. Zn^{2+} ions preferentially occupy the tetrahedral sublattices to increase M_s, and the weakening of exchange interaction by non-magnetic Zn^{2+} ions result in the decrease of H_c. Figure 6.9 shows the change in complex permeability of composite specimens with an increase in Zn^{2+} substitution x in $BaCo_{2-x}Zn_xFe_{16}O_{27}$ ferrites [90]. As the Zn^{2+} substitution increases, the resonance peak shifts to low frequencies. This is due to the increase in M_s and the decrease in H_c. The broad magnetic

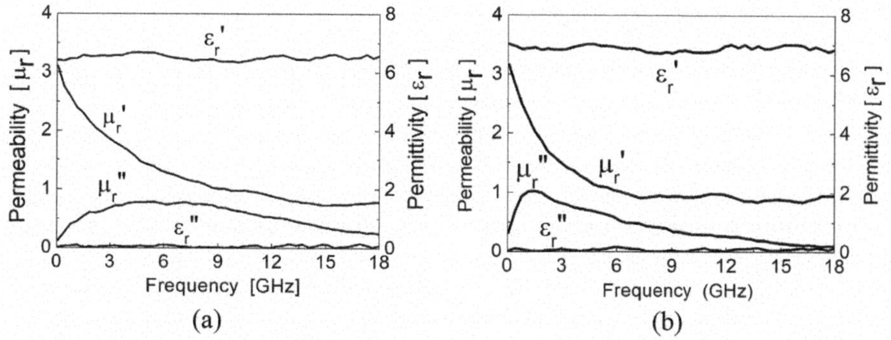

Fig. 6.9 Material parameters of polymer composites containing $BaCo_{2-x}Zn_xFe_{16}O_{27}$ particles: **a** x = 0.5 and **b** x = 1.3 (reproduced with permission from Ref. [90])

loss spectrum observed in Fig. 6.9 is due to the magnetic relaxation, which is the dominant loss mechanism in the composites of ferrite fine particles [90].

Figure 6.10 shows the reflection loss determined in the polymer composites of $BaCo_{2-x}Zn_xFe_{16}O_{27}$ particles. Improvement in EM absorption is evident by substitution of Zn^{2+} ions. The specimen substituted with x = 1.3 shows a reflection loss of −30 dB at 7 GHz at a thickness of 4 mm [90].

Several results have been reported to improve the EM absorption using Co_2W hexagonal ferrites doped with various metal oxides [91–94]. Doping of CaO, CuO, and MnO_2 increases the initial permeability and accordingly f_r moves to low-frequency band (2–3 GHz). V_2O_5 doping increases both real and imaginary parts of complex permeability, resulting in an improvement in absorption bandwidth [91]. 1 wt.% V_2O_5 doping enlarged 10 dB absorption bandwidth to 3.5–13.8 GHz. In addition, it was reported that the substitution of Ba ions with rare earth elements is effective to increase magnetic and dielectric loss in GHz frequencies [95–97].

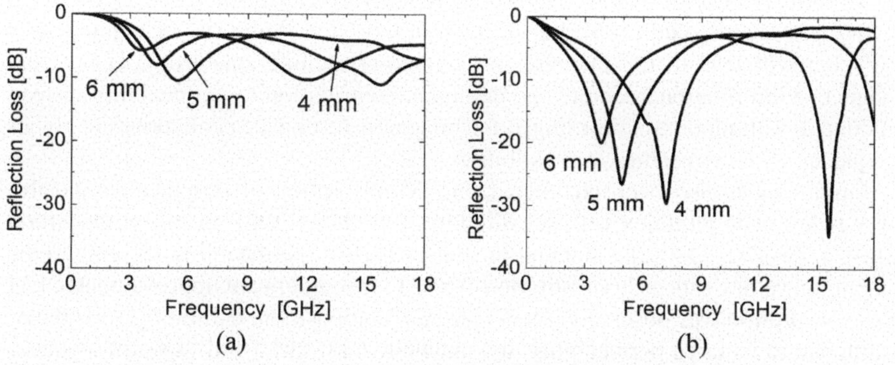

Fig. 6.10 Reflection loss determined in polymer composites containing $BaCo_{2-x}Zn_xFe_{16}O_{27}$ particles: **a** x = 0 and **b** x = 1.3 (reproduced with permission from Ref. [90])

(5) **Y-type Hexagonal Ferrites**

Y-type hexagonal ferrites, denoted by Me_2Y with a chemical formula $Ba_2Me_2Fe_{12}O_{22}$, have an *a-b* plane anisotropy, which is due to their special crystal structure of 3(ST) blocks. There is no R block in the Y-type structure. It was explained that the strong uniaxial anisotropy of M-, W-, and Z-type ferrites is due to the Fe^{3+} ions in the trigonal bipyramidal sites in R blocks [56, 98]. On the while, the Y-type ferrite exhibits in-plane anisotropy owing to the absence of the R blocks. In hexagonal compounds with R blocks, Co^{2+} has strong contribution to the planar anisotropy due to its negative effect to the first order magneto-crystalline anisotropy constant K_1 [99]. However, in Y-type ferrites, Co^{2+} substitution increases the in-plane anisotropy [100–102].

Static and high-frequency magnetic properties of Y-type hexagonal ferrites were reviewed in Ref. [103]. Compared to other types of hexagonal ferrites, Me_2Y (Me: divalent metal ion) ferrite has a relatively small M_s, low curie temperature T_c, and very small in-plane anisotropy field. The complex permeability and microwave absorption of $Ni_{2-x}Zn_xY$ ferrites were investigated with increasing the Zn substitution x [100]. In both sintered ceramics and polymer composite of large particles, two types of resonance, the domain wall and the spin rotational resonance, were observed. The first resonance frequency due to domain wall motion is nearly invariant with x. However, the second resonance due to spin resonance decreases sensitively with x from 6.4 GHz for Ni_2Y to 3.0 GHz for Zn_2Y. The anisotropy field of Ni_2Y and Zn_2Y at room temperature is 14 kOe and 9 kOe, respectively [56]. Thus, the spin resonance frequency decreases with increasing Zn content. For the ferrite composites, excellent microwave absorbing properties could be found in GHz frequencies. For the composite of Ni_2Y particles, it was predicted that the matching frequency and matching thickness are 8.7 GHz and 3.1 mm, respectively, and the 10 dB bandwidth is 6–11 GHz [104]. For Zn_2Y composite, impedance matching was found at two frequencies of 3.0 GHz and 11.8 GHz with a matching thickness of 5.9 mm and 2.7 mm, respectively [100].

The effect of Co substitution on the magnetization mechanism of Zn_2Y hexagonal ferrite was investigated [105]. Substitution of Co^{2+} ion in the site of nonmagnetic Zn^{2+} ion will lead to the increase in magneto-crystalline anisotropy. Hence, the magnetization rotation becomes the dominant contribution to permeability instead of domain wall motion. Accordingly, ferrimagnetic resonance peak moves to higher frequency with increasing Co substitution.

Studies have also been conducted on EM absorption of polymer composites of Y-type ferrite particles with substitution of Cu [103, 106]. In this composition ($Ba_2Cu_xZn_{2-x}Fe_{12}O_{22}$), the real and imaginary parts of permeability increase to the maximum values with a Cu substitution of $x = 1$. The significant improvement in EM wave absorption was observed at this optimum composition, exhibiting wide bandwidth due to its large permeability and magnetic loss. The 10 dB absorption bandwidth includes 2.6 – 9.5 GHz, particularly with an improvement in lower frequency region, resulting from the lower resonance frequency (about 1.6 GHz) of the ferrite particles [106].

(6) Z-type Hexagonal Ferrites

The Z-type ferrites ($Ba_3Me_2Fe_{24}O_{41}$) have a uniaxial anisotropy parallel to *c* axis, except for Co_2Z, which has a planar anisotropy in *a-b* plane. Thus, for the Co_2Z hexagonal ferrite ($Ba_3Co_2Fe_{24}O_{41}$), the out-of-plane anisotropy field is very large $H_\theta^A = 12$ kOe, and the in-plane anisotropy field is very small $H_\phi^A = 0.12$ kOe [107]. Easy magnetization in *a-b* plane makes it possible to have a relatively large permeability and be used as the microwave absorbing materials in GHz frequencies.

Polycrystalline sintered specimen of Co_2Z shows a higher dispersion with resonance frequency of $f_r = 1.4$ GHz and a high value of initial permeability $\mu_i \approx 12$ [56, 100, 108]. For the application to EM wave absorbers, polymer composites are usually used. Figure 6.11 shows the material parameters and reflection loss determined in rubber composites containing Co_2Z ferrite powders [109, 110]. A broad magnetic loss spectrum with $f_r = 4.0$ GHz is observed (Fig. 6.11a). Due to this high magnetic loss, a minimum reflection of −30 dB is observed at 4.4 GHz with a thickness of 4 mm (Fig. 6.11b). At a thickness of 3 mm, a broad absorption bandwidth is predicted (4–11 GHz for −10 dB reflection loss).

In order to improve the high-frequency properties of Co_2Z ferrite, it was conducted to replace Co^{2+} with other divalent ions. For example, Zn^{2+} substitution for Co^{2+} sites can increase the initial permeability to $\mu_i = 16$, which is due to the decrease in in-plane anisotropy [56]. High value of magnetic loss was observed in the composites of $Ba_3Co_xZn_{2-x}Fe_{24}O_{41}$ particles with Zn^{2+} substitution of x = 1.2 [111]. High permeability of $\mu_i = 21$ was observed in Fe_2Z ferrite [112]. Cu^{2+} substitution is effective in lowering the sintering temperature of Z-type hexagonal ferrites [113, 114]. Co-Ti substitution for Fe sites is effective to increase f_r with almost the same initial permeability, and thereby the enlargement of absorption bandwidth was possible [115]. For the $Ba_3Co_{2+x}Ti_xFe_{24-2x}O_{41}$ composite with x = 1.0, the 10 dB bandwidth covers 4.2–14.0 GHz at the optimum thickness of 2.5 mm.

The doping of oxides is effective in increasing the permeability and magnetic loss of Co_2Z ferrite. The most interesting dopant is TiO_2 [104], which enables it

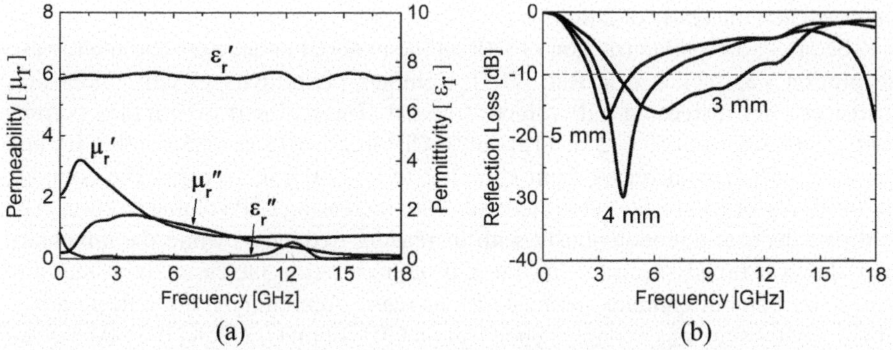

Fig. 6.11 **a** Material parameters and **b** reflection loss determined in the rubber composites of Co_2Z hexagonal ferrite ($Ba_3Co_2Fe_{24}O_{41}$) powders (reproduced with permission from Ref. [110])

to increase both μ_r' and μ_r''. The 10 dB bandwidth is 3–12.5 GHz for the polymer composite with a thickness of 3 mm. TiO_2-doped Co_2Z ferrite is an excellent EM absorption material with wide bandwidth and thin thickness. Doping effect of other oxides in Co_2Z ferrite composites was reviewed in Ref. [103].

6.3.3 *Carbonyl Iron Powder*

Carbonyl iron powder (CIP), developed by BASF in 1925 [116], is a fine and pure iron powder with unique morphological properties. CIP is a high-purity iron (97.5% for grade S, 99.5% for grade R) prepared by chemical decomposition of high-purity iron pentacarbonyl ($Fe(CO)_5$). The carbon content of the powder is 0.075 to 0.8% depending on the grade. The powder characteristics are as follows: (a) very fine spherical shape, (b) powder size of 10 μm or less, (c) unique onion husk structure with cubic crystal lattice, (d) uniform particle size distribution, (e) high-purity powder with an iron content of 99.5% or more, and (f) electrical resistivity of approximately 10^{-5} Ωcm.

With its unique properties, CIP is a good EM wave absorbing material above 1 GHz with broadband absorption characteristics. It can easily be incorporated in most polymer matrices to various form of composites. EM properties of CIP are dependent on chemical composition, shape, and particle size and distribution. Figure 6.12a shows the material parameters of a rubber composite specimen with a CIP content of 30 vol.% [117]. The dielectric constant has a constant value of about $\varepsilon_r' \approx 11$, and the dielectric loss shows a value close to zero. The real part of permeability shows a value that continuously decreases from $\mu_r' = 3.8$ (at 0.05 GHz) to $\mu_r' = 1.1$ (at 18 GHz) with increasing frequency, and the imaginary part shows a nearly constant value of $\mu_r'' \approx 1.2$ above 6 GHz. The magnetic loss is higher than that of conventional ferrites (with $\mu_r'' \approx 0.5$) with the same natural resonance frequency, which enables it to design the thinner-thickness absorber. Figure 6.12b shows the reflection loss of the CIP composite specimen [117]. At 8.4 GHz, a reflection loss of −40 dB can be observed at a thickness of 2 mm.

The effect of volume fraction of CIP on the material parameters and microwave absorption was reviewed in Ref. [103]. Complex permittivity of CIP composites increases with increasing CIP volume fraction. The real part of complex permittivity increases from $\varepsilon_r' = 5$ at 11.6 vol.% CIP to $\varepsilon_r' = 18$ at 42.6 vol.% CIP, and the imaginary part increases from $\varepsilon_r'' = 0.2$ to $\varepsilon_r'' = 0.8$ in average. The complex permeability of composites also increases with increasing CIP volume fraction. The real part decreases monotonously with increasing frequency, while the imaginary part shows a broad spectrum. As the CIP concentration increases, both real and imaginary parts of complex permeability increase. Accordingly, the reflection loss of CIP composites was strongly dependent on CIP volume fraction [103]. At a given CIP volume fraction, the minimum reflection loss less than − 30 dB is shown at a specified frequency and thickness, which is summarized in Table 6.1.

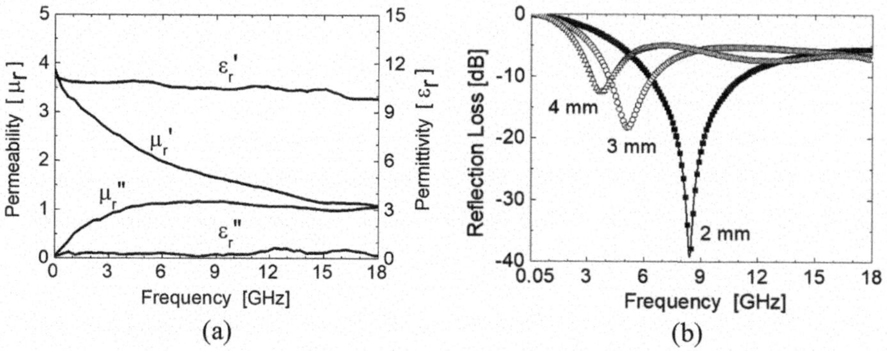

Fig. 6.12 **a** Material parameters and **b** reflection loss determined in the rubber composites of the carbonyl iron particles (30 vol.%) (reproduced from Ref. [117] with the permission of The Korean Institute of Metals and Materials)

Table 6.1 Reflection loss, matching frequency, and matching thickness of CIP composites with increasing CIP volume fraction (data from Ref. [103])

CIP vol.%	Reflection loss(dB)	Matching frequency(GHz)	Matching thickness (mm)
23	−37	17.2	1.3
28	−50	10.5	1.6
34	−40	6.5	2.0
43	−53	4.8	2.0

The frequency of the minimum reflection loss shifts to lower frequencies as the CIP volume fraction increases. Compared to ferrite composites, it exhibits better EM absorption at a thinner thickness. For example, for 28% CIP composites, the reflection loss is −50 dB at 10.5 GHz with a thickness of 1.6 mm. While for X-band absorber of ferrite composite, the thickness is usually greater than 3 mm. The enhanced property is due to the high permeability and magnetic loss, and high dielectric constant of the CIP composites.

6.3.4 High-Permeability Metals

Soft magnetic metals used in EM wave absorbers include pure iron, silicon steel (Fe–Si system), permalloy (Fe–Ni system), molypermalloy powder (MPP), Sendust (Fe–Si–Al system), amorphous (Fe-Si-B system), and nanocrystalline ribbon (Fe–Cu–Nb–Si–B system). The compositions and properties of typical soft magnetic materials are summarized in Table 6.2 [118].

Permalloy is a binary alloy of nickel and iron and the trade name for a high-permeability alloy. Among nickel-based alloys, MPP with a chemical formula

Table 6.2 Compositions and properties of high-permeability soft magnetic materials (reproduced with permission from Ref. [118])

Materials	Composition	μ_i	B_s (T)	H_c (Oe)	ρ (Ωcm)
NiZn ferrite	$Ni_\delta Zn_{1-\delta}$ Fe_2O_4	15~1500	0.3~0.5	0.3~0.5	10^{11}~10^{13}
MnZn ferrite	$Mn_\delta Zn_{1-\delta}Fe_2O_4$	750~10K	0.3~0.5	0.04~0.25	10^6~10^8
Permaloy	$Ni_{80}Fe_{20}$	5K	0.7	0.3	6×10^{-6}
High flux	$Ni_{50}Fe_{50}$	14~160	1.5	1	3×10^{-6}
MPP	$Ni_{80}Fe_{14.35}Mo_{4.8}$ $Mn_{0.5}Si_{0.35}$	12K~200K	0.66~0.82	0.02~0.04	6×10^{-6}
Sendust	$Fe_{85}Si_9Al_6$	26~125	1	0.5	8.5×10^{-6}
FeSiCr	$Fe_{87}Si_9Cr_2$	20~70	1.2~1.4	1	1.2×10^{-5}
Metgals2605 (Amorphos)	$Fe_{78}Si_9B_{13}$	5K	1.56	0.4~0.6	1.3×10^{-5}
Finemet	$Fe_{73.5}Si_{13.5}B_9$ Nb_3Cu_1	15K	1.23	0.01~0.03	1.2×10^{-5}
Iron	Fe	4~100	2.2	5~9	10^{-6}

$Ni_{80}Fe_{14.35}Mo_{4.8}Mn_{0.5}Si_{0.35}$, is the most commonly used material, first developed at Bell Lab in 1920s. It is an excellent soft magnetic material with a very small magnetostriction constant and magneto-crystalline anisotropy constant. On the other hand, due to the low electrical resistance (~6×10^{-6} Ωcm), eddy current losses increase rapidly as the frequency increases, limiting the used frequency to the low-frequency band from 50 Hz to 20 kHz. The most recently developed high flux material of nickel-based alloys ($Ni_{50}Fe_{50}$) has a high saturation magnetization and is used for high-current power supplies, although the core loss is somewhat higher than MPP and the stability of permeability is poor. The Sendust is an alloy of $Fe_{85}Si_9Al_6$ composition with nearly zero value of magneto-crystalline anisotropy and magnetostriction constant. It was invented as an alternative to the permalloy in inductor applications. While it has the advantage of relatively low price, it is brittle and not easy in metal processing. Fe–Si–Cr-based alloys are high-permeability soft magnetic metals with high corrosion resistance due to the added Cr. The amorphous Fe–B–Si or Fe–B–Si–C alloy is manufactured in the form of thin stripes by rapid quenching method. Unlike the permalloy or Sendust, the electrical resistance is high and there is no magnetic anisotropy due to their amorphous structure. Finally, iron-based materials include a pure iron core made of sponge iron powder and a carbonyl iron core with a unique microstructure of onion husks. The ferrous iron core is inferior to the permalloy or Sendust in soft magnetic properties, but it is the cheapest powder cores and used as a noise filter and choke coil in the switching power supply. Carbonyl iron cores have a higher quality factor than iron cores, but have a higher raw material price.

The high-frequency behavior of metallic magnetic materials was discussed in Ref. [103] based on Snoek's law with the following equation:

$$f_r \mu_s = \frac{\gamma}{3\pi \mu_0} M_s, \tag{6.3}$$

where f_r: resonance frequency, μ_s: static permeability, μ_0: permeability of vacuum, γ: gyromagnetic ratio, and M_s: saturation magnetization. The right-hand side of Eq. (6.3) is defined as the Snoek constant which is proportional to M_s. Iron-based soft magnetic metals have the advantage of greater permeability at the GHz frequency because their saturation magnetization is more than twice as large as that of ferrites [56, 118]. Furthermore, such magnetic metals can be easily deformed into thin flake particles with a planar anisotropic structure and a thin thickness less than the skin depth. This means that it is possible to improve simultaneously the initial permeability and the resonance frequency using the flake metal particles. Many reports have been published, demonstrating the improved high-frequency magnetic properties of composite materials made of flaked particles of carbonyl iron [119–123] and other magnetic alloys [124–129]. Moreover, when the flake particles are stacked in insulating matrix, the composite exhibits a very high dielectric constant due to a large spatial charge polarization, which is effective in reducing the thickness of composites when used as EM wave absorbers [124, 130].

High-energy ball milling, such as an attrition milling, is the most useful for producing the metal flake particles. When milling the commercial carbonyl iron powders with a size distribution of 1 to 5 μm, spherical particles can be forged into flakes 100 to 500 nm thick [120]. It is of significance that the thickness of the milled powder is smaller than the skin depth of pure carbonyl iron (~1 μm at 1–5 GHz). Since the electrical resistivity is low in other magnetic metal powders, it is necessary to control the powder shape to thin flakes to reduce the eddy current loss. Milling the metal powders in an attritor produces the thin flake particles with about 1 μm thickness, as shown in Fig. 6.13a. When mixed with polymer resin and rolled into a thick film, a microstructure with aligned flake particles is produced, as shown in Fig. 6.13b. This particle alignment in insulating resin matrix also contributes to reducing the eddy current loss.

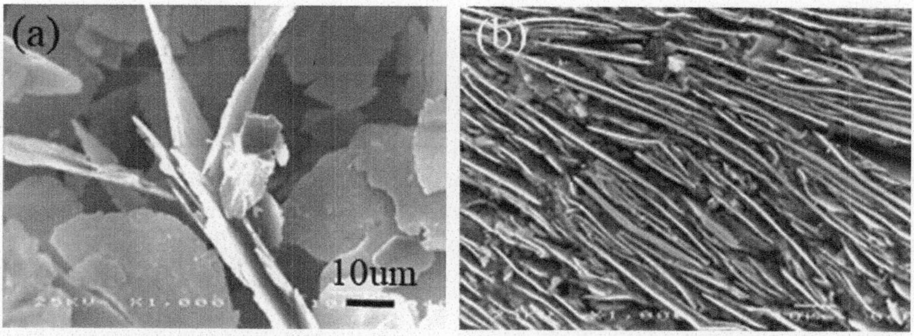

Fig. 6.13 SEM images of **a** attrition-milled iron alloy flake particles and **b** cross-section of rubber composites containing the aligned flake particles

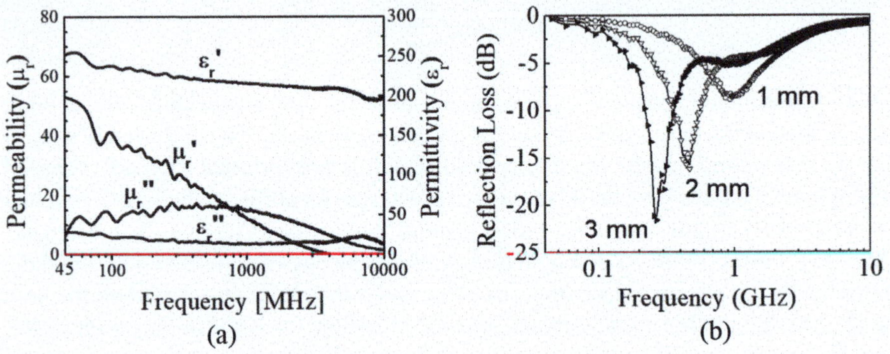

Fig. 6.14 **a** Material parameters and **b** reflection loss determined in the rubber composite of Sendust flake particles

With increase in magnetic permeability by particle forging process, fairly large magnetic loss by ferromagnetic resonance can be obtained at high-frequency band. A typical example is the rubber composite of Sendust particles [131]. As shown in Fig. 6.14a, the real part μ_r' of complex permeability decreases sharply with increasing frequency from 52 at 45 MHz to 0 at 3.5 GHz, and the imaginary part μ_r'' has a broad spectrum with a peak value of 15 at 1 GHz. The real part ε_r' of dielectric permittivity shows high values greater than 200, and the imaginary part ε_r'' also exhibits a high value of 10~15, which is due to the enhanced space charge polarization arising from the aligned thin flake particles.

Figure 6.14b shows the reflection loss determined in the Sendust particles-rubber composites with increasing thickness. When the thickness is 3 mm, the reflection loss drops to −23 dB at a very low frequency of 0.25 GHz. Compared to ferrite composites of the same thickness, it is seen that the absorption frequency band shifts to a fairly low frequency, which is due to the high magnetic permeability and dielectric permittivity of the composites containing the magnetic flake particles. As the low-frequency EM absorbers below 1 GHz, the metal powder composites are very promising.

The impedance matching for zero-reflection absorbers is given by Eq. (5.9). Figure 6.15 shows the trajectory of frequency × thickness ($f \cdot d$) satisfying impedance matching in the complex permeability plane of $\mu_r' - \mu_r''$. In the case of high-permeability metal composites, the dielectric loss is not so small ($\tan\delta_\varepsilon \approx 0.1$). Thus, the map of $\tan\delta_\varepsilon = 0.1$ is used. If we take the frequency band of interest 0.8–2 GHz (corresponding to the operating frequency of mobile phone) and the allowed absorber thickness 1 mm, the range of complex permeability for impedance matching in the range of $f \cdot d = 800 \sim 2000$ (MHz · mm) can be considered in two zones, A and B in Fig. 6.15.

Zone A is a region of low dielectric constant ($\varepsilon_r' < 30$) to which sintered ferrites are belong. The range of complex permeability is $\mu_r' = 3 \sim 6$ and $\mu_r'' = 20 \sim 60$, which

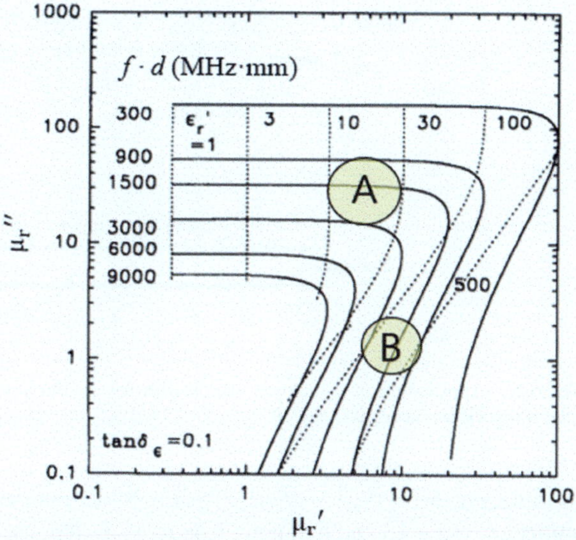

Fig. 6.15 Trajectory of $f \cdot$ d in $\mu_r' - \mu_r''$ plane that satisfies impedance matching in the magnetic composite of $\tan\delta_\varepsilon = 0.1$

requires a very large magnetic loss. However, in most ferrites, the permeability in GHz band has a low value of 5 or less [56]. Therefore, it is difficult to obtain a high magnetic loss $\mu_r'' > 20$. As a thin EM absorber for GHz band, the use of ferrites is restricted.

Zone B is characterized by a very large dielectric constant ($\varepsilon_r' > 100$), where the range of complex permeability is $\mu_r' = 7 \sim 10$ and $\mu_r'' = 1 \sim 3$. As the dielectric loss ($\tan\delta_\varepsilon$) increases, impedance matching requires a greater permeability because the trajectory of $f \cdot d$ shifts to high-permeability region. Polymer composites of soft magnetic metal particles are suitable for zone B, which requires a very large dielectric constant and high permeability rather than magnetic loss. In the case of composites of iron-based soft magnetic metal particles, high permeability can be obtained with their high saturation magnetization, and a high dielectric constant can be obtained from the space charge polarization enhanced by high electrical conductivity and increased surface area of metal flake particles aligned in polymer matrix.

In order to prevent the electromagnetic interference by conduction noise in printed circuit board (PCB), a method of mounting a magnetic thick film with excellent noise attenuation properties on the top of the microstrip line has been proposed, as illustrated in Fig. 6.16 [130, 132–134]. It is a method of attenuating the conduction noise by inserting or attaching the high-loss magnetic films or sheets on the signal line around which magnetic field is strong. Polymer composite sheets of the soft magnetic metal powders are suitable for this application.

For example, Fig. 6.17 shows the material parameters and conduction noise absorption properties of a rubber composite sheet containing Fe–Si–Cr flake particles. The reflection coefficient (S_{11}) and transmission coefficient (S_{21}) are measured using the microstrip line, and the power absorption (*PA*) was calculated using the

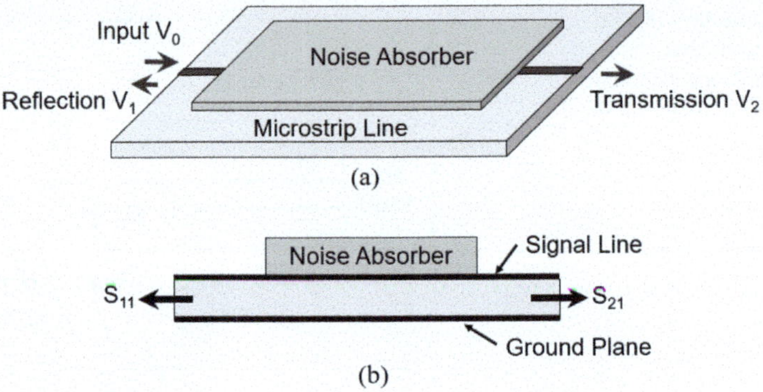

Fig. 6.16 Application of a magnetic thick film for conduction noise attenuation in a microstrip line: **a** top view and **b** side view

formula, $PA = 1 - |S_{11}|^2 - |S_{21}|^2$. The Fe–Si–Cr magnetic sheet exhibits high permeability $\mu_r' = 3 \sim 13$ and magnetic loss $\mu_r'' \approx 7$, and also high dielectric constant $\varepsilon_r' = 190 \sim 240$ and dielectric loss $\varepsilon_r'' \approx 20$, as shown in Fig. 6.17a. S_{11} shows an average value less than $-$ 10 dB after mounting the magnetic sheet (Fig. 6.17b). Most of the signal is transmitted through the microstrip line. S_{21} decreases sharply to about -40 dB as the frequency increases (Fig. 6.17c). Since the magnetic field is strong around the signal line, the magnetic loss of composite sheet makes a dominant contribution to the S_{21} reduction. Figure 6.17d shows the power absorption which increases with frequency and exhibits a high absorption rate greater than 80% above 1.5 GHz.

Similar studies have been conducted using Fe-based or Co-based granular or amorphous thin films for application to the conduction noise absorbers in more miniaturized electronic devices [135–139]. Ferrite thin films of spinel structure prepared by spin spray coating or physical vapor deposition have also been proposed as high-frequency noise suppressors with the advantages of high electrical resistance and low-temperature processing [140, 141].

6.4 Dielectric Loss Materials

6.4.1 Ferroelectric Materials

In the resonant-type EM wave absorbers, the power absorption is maximized when the thickness is equal to 1/4 of the wavelength (λ), and this matching thickness is inversely proportional to the square root of permeability and dielectric constant of constituent material. Therefore, in order to reduce the thickness of absorber, the choice of a high-permeability or high-permittivity material is important. In the case of spinel ferrite magnetic materials, the permeability decreases sharply beyond the Snoek's limit frequency, resulting in a permeability of only about 10 at 100 MHz

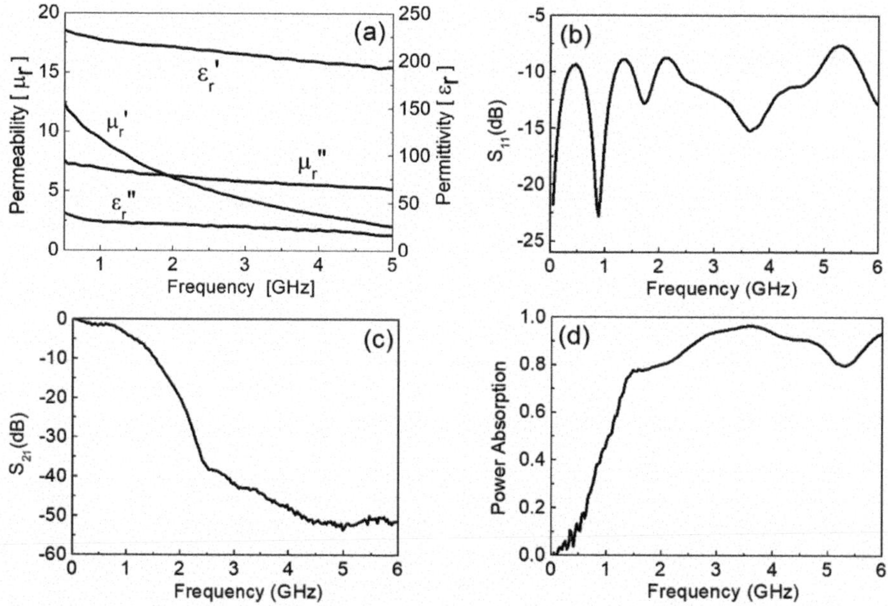

Fig. 6.17 EM properties of rubber composite sheet of Fe–Si–Cr flake particles: **a** material parameters, **b** reflection coefficient (S_{11}), **c** transmission coefficient (S_{21}), and **d** power absorption determined in the microstrip line

[143]. Even for hexagonal ferrites, which are used as the absorbers for high-frequency bands, the permeability at 300 MHz is only about 12 [143, 144].

On the other hand, ferroelectric materials ($BaTiO_3$, PZT, etc.) have a dielectric constant of greater than 100 and a large dielectric loss caused by domain polarization in the microwave band and draw attention for a potential material of thin EM wave absorbers [145]. Relaxor ferroelectrics are another candidate materials for high dielectric constant and high dielectric loss at microwave frequencies. $Pb(Mg_{1/3}Nb_{2/3})O_3$ is a typical material of the relaxor ferroelectrics, but it has paraelectric properties at room temperature because the phase transition temperature is only −15 °C [146]. Here, by substituting $PbTiO_3$ or $Pb(Zn_{1/3}Nb_{2/3})O_3$, ferroelectric properties can be obtained at room temperature, and high dielectric loss by domain polarization can be expected at high-frequency band [147].

Figure 6.18 shows the frequency dispersion of complex permittivity ($\varepsilon_r = \varepsilon_r' - j\varepsilon_r''$) of $BaTiO_3$ (BT), $0.9Pb(Mg_{1/3}Nb_{2/3})O_3$-$0.1PbTiO_3$ (0.9PMN-0.1PT), $0.8Pb(Mg_{1/3}Nb_{2/3})O_3$-$0.2Pb(Zn_{1/3}Nb_{2/3})O_3$ (0.8PMN-0.2PZN) ferroelectrics [148]. In the case of BT (Fig. 6.18a), ε_r' has a value of about 50 in the frequency range of 0.5 − 4 GHz, but decreases sharply and maintains a constant value of about 30 at frequencies above 6 GHz. Dielectric loss ε_r'' has a maximum value (about 25) around 6 GHz and exhibits a relatively broad frequency dispersion. Compared to BT, greater

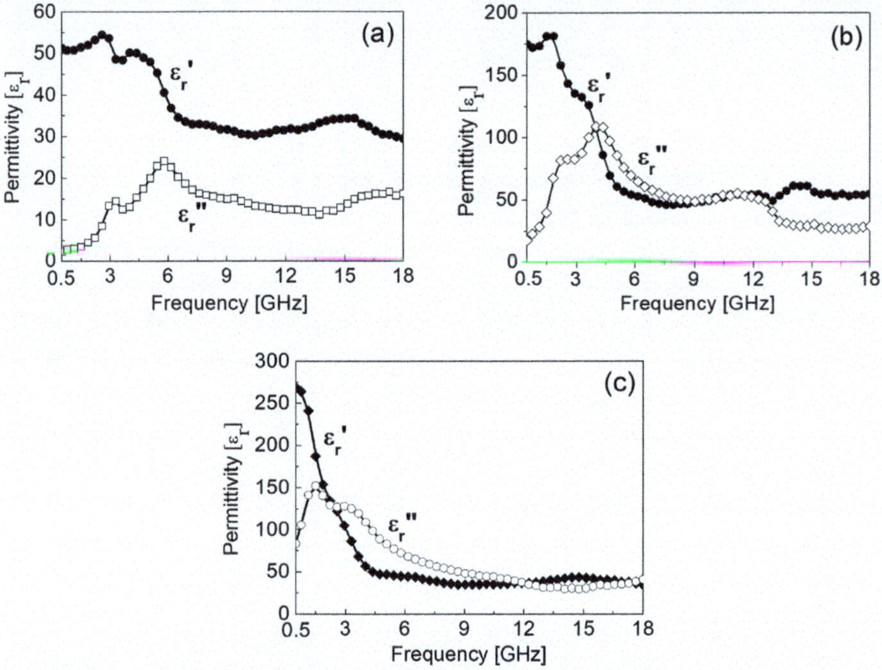

Fig. 6.18 Complex permittivity of ferroelectric ceramics: **a** $BaTiO_3$, **b** $0.9Pb(Mg_{1/3}Nb_{2/3})O_3$-$0.1PbTiO_3$, **c** $0.8Pb(Mg_{1/3}Nb_{2/3})O_3$-$0.2Pb(Zn_{1/3}Nb_{2/3})O_3$ (reprinted with permission from Ref. [148], copyright 2003 Springer Nature)

dielectric constants ($\varepsilon_r' = 180$ at $0.5 - 2$ GHz) and dielectric loss ($\varepsilon_r'' = 110$ at 4 GHz) are observed for 0.9PMN-0.1PT (Fig. 6.18b). For 0.8PMN-0.2PZN ceramics, a high value of $\varepsilon_r' = 270$ is observed at 0.5 GHz and ε_r'' shows a maximum value (about 150) at 2 GHz (Fig. 6.18c).

The domain polarization was proposed as the dominant polarization and dielectric loss mechanism for the ferroelectric ceramics in microwave frequencies [145, 149, 150]. In general, the dielectric constant is below 5 for electronic polarization, and no greater than 20 for ionic polarization. Accordingly, the high values of dielectric constant and dielectric loss presented in Fig. 6.18 can be explained by domain wall motion, which was explained earlier in Sect. 4.1. Because the relaxation frequency of domain wall oscillations is located in the microwave band [151], ferroelectric ceramics exhibit high dielectric constants and maximum losses at microwave frequencies above 1 GHz. Because the relaxation time of the domain wall movement is inversely proportional to the speed of movement, the larger the dielectric constant, the more likely it is that the maximum loss frequency will occur at lower frequencies, as shown in Fig. 6.18.

Figure 6.19 shows the reflection loss calculated from the complex permittivity of BT, 0.9PMN-0.1PT, and 0.8PMN-0.2PZN ceramics terminated by metal [152]. In the case of BT specimen, a reflection loss of −38 dB is predicted at about 2 GHz with a thickness of 4.5 mm (Fig. 6.19a). As the thickness decreases, the maximum absorption decreases. For the specimen with a thickness of 1.5 mm, it is seen that the measured reflection loss is in good agreement with the calculated value. For 0.9PMN-0.1PT specimen (Fig. 6.19b), the reflection loss increases to about −2 dB. For 0.8PMN-0.2PZN specimen (Fig. 6.19c), the reflection loss increases further to −1.5 dB (only 30% power absorption). The significant reduction in EM wave absorption in the 0.9PMN-0.1PT and 0.8PMN-0.2PZN ferroelectrics is due to the fact that the dielectric constant is too high to meet the impedance matching, resulting in large wave reflections from the surface.

It has been reported that the polymer composites containing the ferroelectric particles exhibits a good microwave absorbing properties in X-band frequencies (8.0–12.4 GHz). A maximum reflection loss of −15 dB (97% power absorption) at 10 GHz has been achieved with $BaTiO_3$-polyaniline composites [153]. Excellent microwave absorption performance in X- and Ku-bands was reported in the polymer composites of $BaTiO_3$ particles [154, 155].

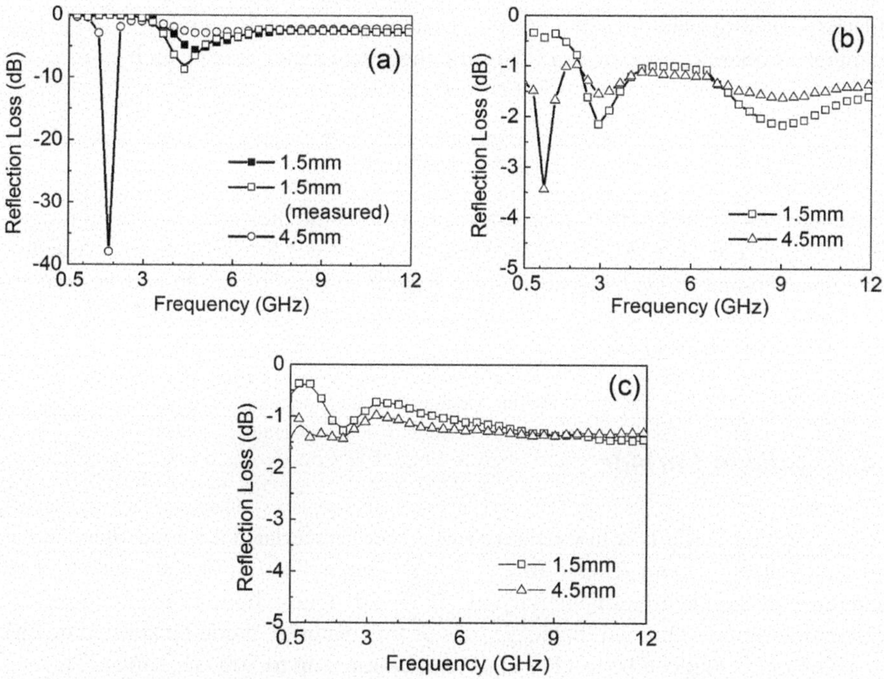

Fig. 6.19 Reflection loss determined in ferroelectric ceramics: **a** $BaTiO_3$, **b** $0.9Pb(Mg_{1/3}Nb_{2/3})O_3$-$0.1PbTiO_3$, **c** $0.8Pb(Mg_{1/3}Nb_{2/3})O_3$-$0.2Pb(Zn_{1/3}Nb_{2/3})O_3$ (reproduced from Ref. [152] with the permission of Materials Research Society of Korea)

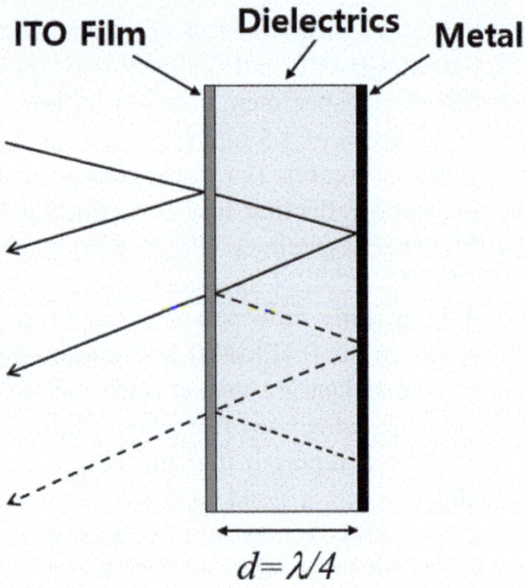

Fig. 6.20 Schematic of EM wave absorbers consisting of ITO resistive film over a dielectric spacer backed by metal

The high-permittivity materials can be used as a spacer in resistive screen absorbers and attain the effect of reducing the thickness of absorber. Figure 6.20 illustrates an absorber consisting of ITO resistive film over a dielectric spacer backed by metal. A quarter wavelength (λ/4) for a dielectric material is given by.

$$\lambda = \frac{\lambda_0}{[\frac{1}{2}\varepsilon_r'\sqrt{1+\tan^2\delta}+1]^{1/2}}, \tag{6.4}$$

where λ_0 is the wavelength in free space and tanδ is the dielectric loss tangent. For 0.9PMN-0.1PT and 0.8PMN-0.2PZN dielectrics, $\lambda/4 = 2.5$ mm at a target frequency of 2 GHz. Coating an ITO thin film on the dielectric spacer of $\lambda/4$ thickness reduces the reflection loss at the appropriate surface resistance (~360 Ω/sq) of ITO thin film, as shown in Fig. 6.21 [148].

6.4.2 *Silicon Carbide*

Silicon carbide (SiC) is a low-density (~3.2 g/cc) material with good mechanical properties at both room temperature and high temperature. It is a semiconductive material with an electrical resistivity of $10^2 - 10^4$ Ωcm. Taking advantage of the high dielectric constant and dielectric loss of this material, it can be applied to EM wave absorbers that can be used at high temperatures up to 700 °C [156, 157].

SiC exists in many crystalline forms [158]. Hexagonal silicon carbide (α-SiC) is the most common polymorph, and is formed at temperatures higher than 1700 °C. The cubic silicon carbide (β-SiC), with a zinc blende crystal structure, is formed at temperatures below 1700 °C [159]. Strong covalent bonds in β-SiC crystal forms a

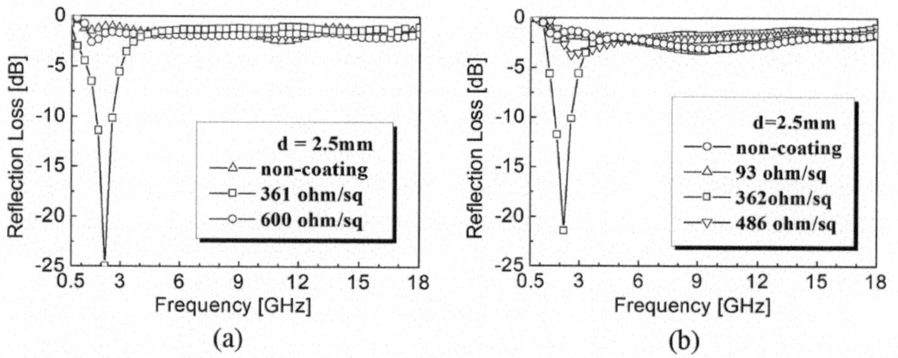

Fig. 6.21 A reduction of reflection loss through ITO coating on $\lambda/4$-thick dielectric spacer with compositions: **a** 0.9PMN-0.1PT, **b** 0.8PMN-0.2PZN (reprinted with permission from Ref. [148], copyright 2003 Springer Nature)

very hard and chemically stable material. SiC maintains corrosion resistance even in acids, alkalis, and melt salts up to 800 °C. In the air, SiC forms a protective silicon oxide film at 1200 °C and can be used up to 1600 °C. Low thermal expansion and high thermal conductivity give this material the excellent thermal shock resistance.

However, the low fracture toughness of monolithic SiC ceramics limits their application. On the other hand, the continuous SiC fiber/SiC ceramic matrix (SiCf/SiC) composites exhibit the pseudo-ductile fracture behavior with high fracture energy [160–163]. SiCf/SiC composites are very promising as the high-temperature EM wave absorbers because they maintain both mechanical and electrical performance at high temperatures [164]. In order to obtain the composite materials with good microwave absorption properties, the composition and microstructure of SiC materials (both fibers and matrices) must be controlled to ensure the electrical conductivity and dielectric properties under the specified application conditions (temperature, frequency, environment, etc.).

Dielectric loss in ceramics generally consists of ion movement loss (ion jump and relaxation), ionic polarization loss, and electronic polarization loss. In the case of polycrystalline and polyphase materials, losses due to space charge polarization should be taken into account. When the frequency is less than about 10^{10} Hz, the dielectric loss tangent ($\tan\delta$) is expressed as the sum of conduction loss ε''_C, ion shift loss ε''_R, and space charge polarization loss ε''_I:

$$\tan\delta = \frac{\varepsilon''_C + \varepsilon''_R + \varepsilon''_I}{\varepsilon'}. \tag{6.5}$$

The loss mechanisms from SiC powders or fibers include conduction loss and dielectric loss. The loss caused by electrical conduction is given by

$$\tan\delta_c = \frac{\varepsilon''_C}{\varepsilon'} = \frac{\sigma}{2\pi f \varepsilon_0 \varepsilon'_r}, \tag{6.6}$$

where σ: electrical conductivity, f: frequency, ε_0: permittivity of free space (= 8.854 $\times$ 10^{-12} F/m), ε_r': dielectric constant. The electrical conductivity of pure SiC is $\sigma \approx 10^{-3}$ S/cm, so the conduction loss is approximately $\tan\delta_c = 0.1$ at 10 GHz [165]. The electrical conduction of SiC has little effect on total loss. Other mechanism of dielectric loss includes the dielectric relaxation of defect dipoles and space charge polarization [166]. The most likely defect in β-SiC is the carbon vacancies, which break the balance of charge distribution and produces the electric dipoles [166]. This results in orientation polarization of the dipoles under an applied electric field, which induces dipole relaxation loss following the Debye relaxation process.

In addition, because the carbon vacancies change the band structure and conductivity of the crystals, both dipole relaxation and electrical conductivity affect the dielectric properties of β-SiC at high temperatures.

Dielectric permittivity measured in sintered SiC ceramics is in good agreement with Debye relaxation theory [167]. The frequency that makes up ε_r'' peak corresponds to the relaxation frequency f_r, which has a reciprocal relationship with the relaxation time τ, expressed as $f_r = 1/2\pi\tau$, as explained earlier in Sect. 4.3.1. The ε_r'' peak measured in SiC ceramics indicates $f_r = 0.7$ GHz, which yields $\tau = 2.55 \times 10^{-10}$ second and very consistent with the theoretical curve fitting [167]. As the temperature increases, f_r moves to higher frequencies because τ decreases. Thus, the dielectric constant and dielectric loss in microwave frequency band increase as the temperature is elevated [166, 168].

Dielectric loss at high temperatures is due to the dielectric relaxation and the increased electrical conductivity of SiC crystal. Therefore, SiC powders exhibit good EM absorption properties, and the effect is more pronounced as the temperature increases. In order to further improve the inherent dielectric loss and electrical conductivity of SiC crystals, surface treatment, doping, and compounding with other materials (e.g., the addition of magnetic materials) have been proposed [169].

SiC nanocrystals are processed into several forms, including powders, wires, and fibers [170]. EM wave absorption properties have been studied in the composites where these materials are mixed with polymer resins. Chiu et al. [171] examined the EM absorption in the epoxy composite (2 mm thick) containing 35 wt.% β-SiC nanowires and achieved the minimum reflection loss of −31.7 dB at 8.3 GHz. It was also reported that in 30 wt.% β-SiC nanowire/silicon composite (3 mm thick), a minimum reflection loss of −17.4 dB was obtained at 11.2 GHz with 10 dB absorption bandwidth of 2.5 GHz [172].

6.5 Core–Shell Materials

6.5.1 Core–Shell Structures

Core–shell nanomaterials have an assembled structure composed of two or more materials with a core and outer shell(s) layer [173]. Core–shell structure integrates the properties of both internal and external materials and can provide synergistic or new properties owing to the interactions between the core and the shell(s). Figure 6.22 shows the various forms of core–shell structures. The most common form is a solo sphere core covered with a continuous shell (Fig. 6.22a) or multishells (Fig. 6.22b). Another form has the yolk-shell structure where a small core exists within a hollow shell (Fig. 6.22c). The solid core can be replaced by the hollow core which is useful to produce the light-weight absorbers (Fig. 6.22d). In some cases, the core–shell structures may take the shape of rod, tube, or hexagons (Fig. 6.22e and f).

The core–shell heterostructures can provide several positive effects in their application to EM wave absorbers, which was reviewed in Ref. [174]. First, the most important factor is related with the impedance matching, which can be obtained from a proper combination of electromagnetic properties of core and shell layers. The next is the structural factor of multiple interfaces which is effective in reducing and attenuating the power of incoming EM waves through multiple scattering and reflection. In addition, interfacial polarization can be responsible for EM wave absorption. The more the layers and interfaces, the greater the interfacial polarization loss, particularly for the structure combination of dielectric core and conductive shells.

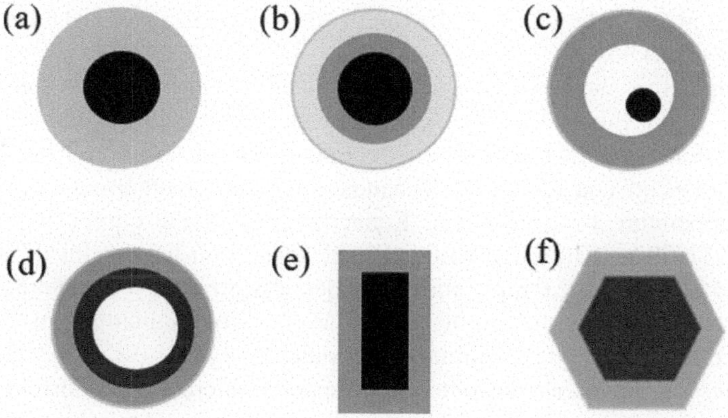

Fig. 6.22 Schematic representation of various core–shell nanoparticles: **a** core–shell, **b** core-multishells, **c** yolk-shell, **d** hollow core–shell, **e** rod core–shell, and **f** hexagonal core–shell

6.5.2 *Magnetic Core-Conductive Shell*

The design of core–shell nanomaterials can be modified by altering the constituent materials or by changing the size and thickness of the core and shell. Many combinations of core and shell materials have been examined to obtain superior EM wave absorbers with low reflection loss, wide bandwidth, and small thickness [174]. From the perspective of impedance matching, the combination of magnetic core and conductive shell is the most efficient way to take advantage of their intrinsic magnetic permeability and dielectric permittivity.

As a magnetic core material, Fe_3O_4 is attracting attention due to its low cost and inherent magnetic properties, which offers great application potential as an efficient EM wave absorber. As a shell material, carbon has the inherent advantages of light weight, good chemical stability, and high electrical conductivity, resulting in interfacial polarization and conduction losses. Liu et al. [175] have synthesized Fe_3O_4/C core–shell nanospindles by a wet chemical method and subsequent high-temperature carbonization, and obtained the minimum reflection loss (RL) of − 38.8 dB at 11 GHz in the flexible Fe_3O_4@C@PVDF nanocomposites. The low RL is attributed to the improved impedance matching by combined effect of magnetic Fe_3O_4 nanospindles and encapsulated conductive carbon shell. Chen et al. [176] synthesized Fe_3O_4/carbon core/shell nanorods by annealing treatment of α-Fe_2O_3/ carbon (core/shell) nanorods which was prepared by a hydrothermal method. They attained a minimum RL of − 28 dB at 15 GHz in the polymer composite of a 2 mm thickness. A more enhanced microwave absorption was observed in the three-dimensional (3D) Fe_3O_4@C core–shell nanocomposites with foam-like structure [177]. The synergy between the core and the shell produced a minimum RL value of −55.5 dB at 8.3 GHz with a 10 dB bandwidth of 4.0 GHz (6.8–10.8 GHz) at 2.5 mm absorber thickness.

As a conductive shell material, transition metal dichalcogenides (TMDC) have been proposed for their advantages of enhanced dielectric losses and strong electric polarization, which is useful for efficient microwave absorption. Pan et al. [178] prepared the Fe core coated with MoS_2 nanosheets via a solvothermal route followed by high-temperature heat treatment. Reflection loss of the composite with wax reached − 37 dB at 13.4 GHz with a 10 dB absorption bandwidth of 4.73 GHz (11.13 − 15.86 GHz) at 2.0 mm composite thickness. Li et al. [179] synthesized the flower-like core@shell structure of Fe_3O_4@MoS_2 through a simple hydrothermal method, and demonstrated the minimum RL value of − 50.75 dB at 8.20 GHz at 2.9 mm thickness of polymer composite. The improved microwave absorption properties have been interpreted to be attributed to the double loss mechanisms, synergistic effect, and special morphology.

The primitive Fe_3O_4 core can be replaced by other ferrites or magnetic metals. Liu et al. [180] reported $BaFe_{12}O_{19}$ as the core with carbon as the shell. For the composite of optimum filler content, the RL value reached − 87 dB at 10 GHz with a 10 dB bandwidth of 3.8 GHz at 3 mm layer thickness. Other ferrites with various compositions, such as $ZnFe_2O_4$ and $CoFe_2O_4$, have been investigated as core materials with conductive shells of carbon or MoS_2, demonstrating improved microwave absorption (low reflection loss, wide bandwidth) [181–183]. The enhanced microwave absorption was interpreted to be due to (a) facilitated impedance matching and interfacial polarization between the magnetic core and the conductive shell, (b) the dual loss mechanisms (magnetic loss and dielectric loss) and their synergy effect, (c) the microstructure enhancing EM wave scattering, and (d) the defect dipole capable of producing dielectric relaxation losses [183]. CoNi and FeCo were reported as the core materials with various forms of shell, such as C nanocapsules, graphitic carbon co-doped with B and N, nitrogen-doped CNT, and demonstrated the superior EM wave absorbing properties [184–187].

6.5.3 Other Core–Shell Combinations

A promising core–shell nanostructure was reported where the carbon nanosphere acts as a core and the MoS_2 sheet acts as a shell [188]. In this research, carbon sphere coated by MoS_2 nanosheets (CS@MoS_2) were synthesized via a simple hydrothermal method. CS@MoS_2-wax composite with 30 wt.% filler achieved a minimum RL of −52.6 dB and a 10 dB absorption bandwidth of 4.9 GHz (13.1–18 GHz) at 1.4 mm thickness. Kuchi et al. [189] proposed honeycomb core–shell nickel@copper silicate (Ni@$CuSiO_3$) nanostructures to promote impedance matching. $CuSiO_3$ exhibits good dielectric and anti-oxidation properties. In particular, the dielectric constant can be easily adjusted by controlling the shell thickness of the Ni@$CuSiO_3$ nanosphere. The polymer composite exhibited a minimum RL of − 39.5 dB at 9.6 GHz with a 10 dB absorption bandwidth of 4.8 GHz (7.8–12.6 GHz) at 2.5 mm thickness.

Similarly, core–shell nanostructures consisting of NiS_2 cores and MoS_2 shells (NiS_2@MoS_2) have been reported [190]. The metal sulfides were used as the core to prevent the particle agglomeration and utilizing their dielectric loss. In the composites with a filler loading of 20 wt.%, the minimum RL reached − 41.05 dB at 12.08 GHz, and the 10 dB absorption bandwidth is 4.4 GHz with a thickness of 2.2 mm. This is mainly attributed to the improved impedance matching, dipole and interface polarization, and multiple reflections. Zhu et al. [191] reported the flower-like CoS_2@MoS_2 core–shell microspheres coated with reduced graphene oxide (rGO) synthesized by a facile hydrothermal method. The introduction of rGO has significantly improved the dipole and interfacial polarization due to the abundant residual defects of rGO and the large interface between CoS_2, MoS_2, and graphene, which has significantly improved the dielectric properties of core–shell CoS_2@MoS_2 composites. It is noted that the composite sample with 20 wt.% loading of CoS_2@MoS_2/rGO microspheres exhibited the minimum RL of − 58 dB with a broad 10 dB bandwidth of 6.24 GHz (11.76 − 18 GHz) at 2.4 mm thickness.

A large absorption bandwidth was reported in the composites of core–shell particles of SiC whiskers@carbon heterostructure synthesized via chemical vapor deposition followed by a hydrothermal and carbonization process using the glucose as a carbon precursor [192]. In contrast to the bare SiC whiskers, the EM absorption has been greatly improved after coating the carbon shell. 10 dB absorption bandwidth is 6.3 GHz (8.2 − 14.5 GHz) at 3.5 mm thickness, and it covers the entire Ku-band with a smaller layer thickness of 2.7 mm. Wei et al. [193] reported excellent microwave absorbing properties of core–shell particles comprised of hollow silicon carbide spheres and nano-Ni layer. The hollow SiC particles were prepared by sol–gel method and the magnetic nickel nanoparticle layer was coated by electroless plating. The polymer composites of this core–shell particles exhibited excellent absorbing performance with the minimum RL reaching − 22.43 dB at 8.06 GHz and 10 dB absorption bandwidth of 8.2 GHz (6.2–15.0 GHz) at 3 mm thickness. It was interpreted that the excellent EM absorption performance is attributed to (a) the reflection and scattering effects at interfaces, (b) magnetic resonance and eddy current loss of Ni nanoparticles, (c) dielectric polarization occurred not only in the SiC particles but also at various interfaces, and (d) the interaction of magnetic Ni shell and dielectric SiC inner core.

6.5.4 Yolk-Shell Materials

Yolk-shell structure is a special class of core–shell configurations possessing a hollow shell and interior core with empty void. Due to their unique properties, such as a low density, large surface area, and synergy effect of interior core and outer shell, they can be used as the highly potential EM wave absorbing materials. Liu et al. [194] fabricated CoNi@air@TiO_2 yolk-shell microspheres through a facile process and investigated their microwave absorption properties in the sample of epoxy composites. Optimizing the core sizes, void volumes, and shell thicknesses, a minimum RL of − 33.4 dB at 7 GHz could be obtained with a composite thickness of 2 mm. Moreover, it was noted that the absorption bandwidth was enlarged up to 7.8 GHz with respect to − 10 dB reflection loss. Highly dispersed Fe_3O_4@ZrO_2 yolk-shell structure was successfully prepared via a polymer surfactant (hydroxypropyl cellulose) assisted sol–gel method [195]. The material has great potential as a high-temperature microwave absorber. Even at a temperature of 500 °C, the material still retained more than 90% of its room temperature microwave absorption.

The uniform yolk-shell C@C microspheres was fabricated via a 'coating-coating-etching' route, and the enhanced microwave absorption properties were demonstrated [196]. In the paraffin wax composites, the minimum RL reached − 39.4 dB at 16.5 GHz with 10 dB absorption bandwidth ranging from 12.6 to 18.0 GHz with an absorber thickness of 2.0 mm. The presence of voids between the core and the shell is favorable for impedance matching, and multiple reflections inside the shell can increase dielectric losses at the matched characteristic impedance. Zhang et al. [197] proposed an efficient strategy to design and fabricate the multi-core yolk-shell Co@C nanospheres by high-temperature carbonization. The minimal RL value of samples carbonized at 750 °C reached − 40 dB and the 10 dB absorption band reached 9 GHz with a controlled layer thickness in the range of 2~9 mm. A double-cavity yolk-shell structure with magnetic hollow Fe_3O_4 (H-Fe_3O_4) core and carbon shell was proposed as the high-performance microwave absorbers [198]. They fabricated yolk-shell H-Fe_3O_4@C microspheres via a 'coating–coating–etching' route using the phenolic resin as the carbon source. When H-Fe_3O_4@C microspheres are carbonized at 650 °C and compounding with polymer matrix, the optimized RL reached − 58.44 dB and the 10 dB bandwidth is 6.0 GHz (12.0–18.0 GHz) with a small layer thickness of 1.9 mm.

6.5.5 Hollow Core–Shell Materials

As a low-density core material, hollow microspheres are good candidates. In order to reduce the weight compared to conventional ferrite or metal powder absorbers, a method has been proposed to form the magnetic or conductive films on the surface of hollow microspheres with a very low density [199, 200]. For this purpose, various metal or ferrite films were deposited on the surface of the hollow microspheres, including Cu [201, 202], Ag [203, 204], Co [205], Co–Fe [206], Ni–Fe [207], Ni–P [208], Ni–Co–P [209], and ferrites [210, 211].

Hollow microspheres are classified into glass microspheres and ceramic microspheres. Main component of glass microspheres is SiO_2, and Q-Cel 520 hollow microspheres of PQ Corporation is a commercial product. The shape is sphere with an average size of 70 μm and the density is very low about 0.2 g/cm^3 [206]. It is made of glass material and the disadvantage is its weak strength. The ceramic microspheres, known as the cenospheres, are produced as a coal-fired by-product in power plants. It consists mainly of silica and alumina, and the price is cheap and the strength is high. The size is about 5~500 μm and the bulk density is 0.25~0.7 g/cm^3 [212].

The efficient way of coating the metal films on the surface of hollow microspheres is electroless plating, which is a solution process of plating metal ions onto the catalytically activated surface of substrates using a reducing agent with a low electrochemical potential. For the stability of plating solution, several auxiliary components are added, such as pH adjusters, pH buffering agents, and complexing agents. In the process of plating on powders, the specific surface area of substrates is very large at the level of 0.1 to 10 m^2/g for the average particle size of 0.01 to 1 mm. A sufficient amount of metal salt must be provided that can cover the large specific surface area, and the plating speed must be reduced for a uniform plating.

Based on EM absorption theory, impedance matching at the absorber surface requires an appropriate combination of magnetic permeability and dielectric permittivity of medium. From this point, magnetic metals are recommended as the deposited shell materials. Figure 6.23 shows the SEM morphologies of hollow microspheres coated with Ni–Fe films by electroless plating [207]. A uniform coating with smooth surface morphology and a networking grain structure is observed.

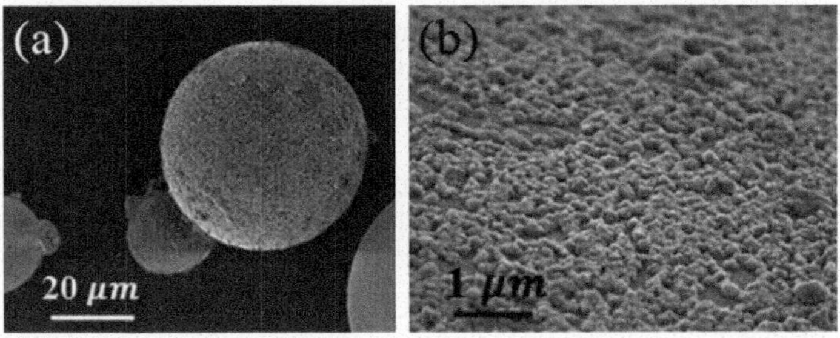

Fig. 6.23 SEM morphology of **a** surface and **b** grain structure of Ni–Fe coated microspheres prepared by electroless plating (reprinted with permission from Ref. [207], copyright 2016 Elsevier)

In those plated metallic films, material parameters at high frequencies are very sensitive to the composition and thickness of films due to their strong dependence of electrical conductivity and magnetic permeability. In general, the complex permittivity of composites containing the metallized particles sensitively increases with the electrical conductivity of the coated film due to the enhanced space charge polarization in polymer matrix. High permeability can be obtained in the magnetic thin films with low magnetic anisotropies and high saturation magnetizations.

A typical example can be found in Co-Fe films where the substitution of Fe has an effect of increasing magnetic permeability and magnetic loss [206]. Figure 6.24 shows the complex permeability of a rubber composite containing the hollow microspheres plated with Co-Fe films. The real part μ_r' increases with Fe content in the low-frequency region below 9 GHz (Fig. 6.24a). The magnetic loss μ_r'' increases with increasing Fe content and the dispersion peak moves to a lower frequency (Fig. 6.24b). According to the bulk magnetism of Co-Fe alloy [213], the magneto-crystalline anisotropy is sensitively dependent on Fe content. Easy axis is changed from *c* axis to *a-b* plane at 1.0 at.% Fe and then to cubic anisotropy of FCC at 5.2 at.% Fe. Fe substitution in Co lattice reduces the magneto-crystalline anisotropy and thereby increases the permeability and magnetic loss in microwave frequency range.

Figure 6.25 shows the reflection loss of rubber composites containing the hollow microspheres plated with Co-7 at.% Fe films. At 1.5 mm thickness, reflection loss less than − 10 dB can be obtained in X-band frequencies (8 − 12.4 GHz). Increasing the thickness to 2.0 mm shifts the absorption band to a lower C-band frequencies (4 − 8 GHz), and the maximum absorption (22 dB) is obtained at 7 GHz. In addition to the advantage of lightweight (~0.8 g/cc) compared to conventional ferrite absorbers (~5.7 g/cc), a substantial decrease in absorber thickness can be achieved, which is attributed to the high permeability and magnetic loss of the Co-Fe plating film [206].

The effect of film thickness on the high-frequency electromagnetic and microwave absorbing properties was investigated in the Co-coated hollow microspheres [205]. The thickness of Co film can be increased up to 3 μm by repeating the batch of plating process. Accordingly, the density increases from the initial density of hollow microspheres (~0.2 g/cc) to 0.84 g/cc, but still very small compared with that of ferrite or metal fillers. Figure 6.26 shows the material parameters of rubber composites

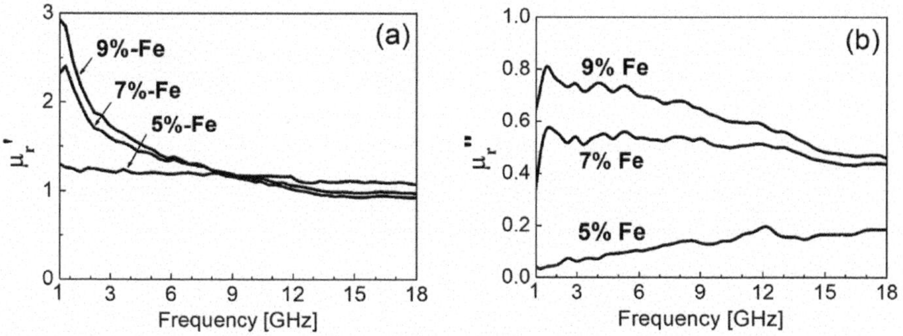

Fig. 6.24 Complex permeability of polymer composites containing Co-Fe plated hollow microspheres: **a** μ_r' and **b** μ_r'' (reprinted with permission from Ref. [206], copyright 2004 Elsevier)

Fig. 6.25 Reflection loss determined in the rubber composites containing the hollow microspheres plated with Co-7 at.% Fe films

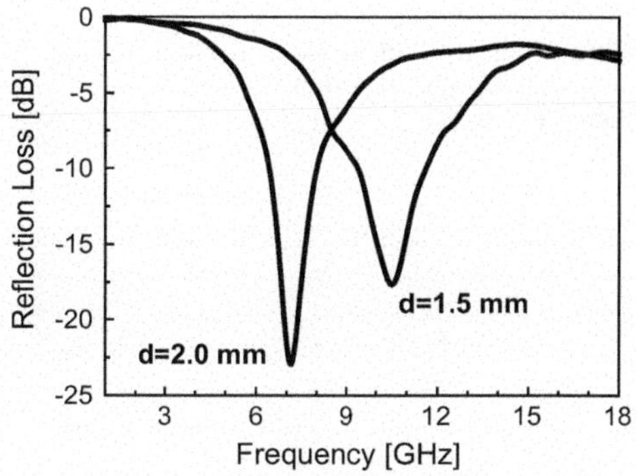

containing the Co-plated hollow microspheres. With increase in plating batches, the increase in magnetic loss (μ_r'') and dielectric constant (ε_r') is distinct. Since Co is a ferromagnetic metal, the magnetic loss and dielectric constant increase with film thickness.

Figure 6.27 shows the EM absorption properties of rubber composite containing the Co-plated hollow microspheres. When the Co is plated twice (film thickness is about 2 μm), a reflection loss of − 27 dB is observed at 9.1 GHz with a composite thickness of 2.0 mm. When plated three times (film thickness ≈ 3 μm), reflection loss of − 25 dB is observed at 5.5 GHz at 2.5 mm thickness. It is noted that the density of the absorbent filler can be reduced to a level of 0.66~0.84 g/cc with a high EM absorption of greater than 20 dB at X-band and C-band frequency bands.

For the thin film materials on hollow microspheres, other magnetic alloys [214–217] and ferrite compounds [218–220] were investigated. For example, using the polymer composite containing Ni-coated hollow spheres, RL = − 35 dB could be obtained at 5 GHz with a 3.0 mm layer thickness [214]. Plating the Fe–Ni–P nanoparticles on hollow glass microspheres, RL = − 50 dB at 7.7 GHz could

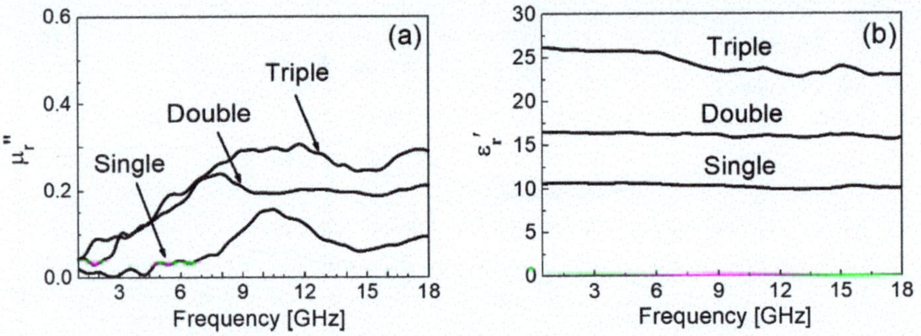

Fig. 6.26 Material parameters of rubber composites containing Co-plated hollow microspheres with increase of plating batches: **a** μ_r'' and **b** ε_r'

Fig. 6.27 Reflection loss determined in the composites containing Co-plated hollow microspheres

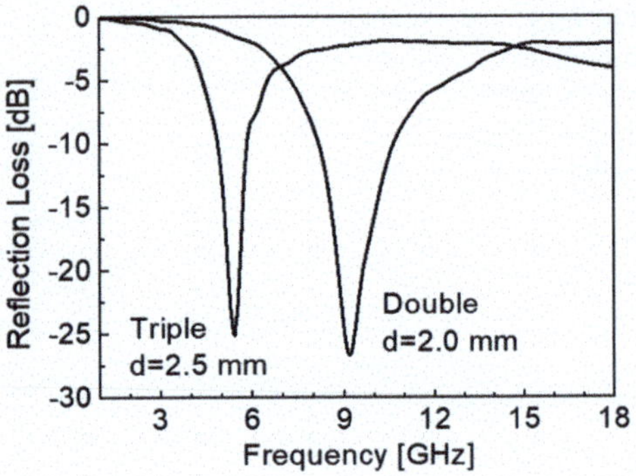

be achieved in the composite of 5 mm thickness [216]. $CoFe_2O_4$ coating by co-precipitation method [219] and M-type Ba ferrite coating by sol–gel technique [220] were proposed as the ferrite thin films on hollow microspheres. Combining the Fe_3O_4-coated hollow glass spheres and reduced graphene oxide (rGO) in epoxy resin matrix produced the lightweight composites with better impedance matching and enhanced electromagnetic wave attenuation [221].

6.6 Radar Absorbing Structures

6.6.1 Fiber Reinforced Composites

EM wave absorbing materials are classified into radar absorbing (coating) materials (RAM) and radar absorbing structures (RAS). The RAM has a function to suppress the reflected waves from the surface of metal structure, in the form of paint or polymer films composed of absorbent materials. As the absorption materials, carbonyl iron particles, ferrite grains, or conductive carbon black are widely used. However, in the case of RAM, mechanical properties are weak, need to be repaired frequently, and there is a problem to increase the volume and weight of device or structure. Furthermore, the coating thickness should be precisely controlled to be 1/4 wavelength during the painting for the optimal absorption performance. A solution to these problems is to replace the structure of object with a composite material with lightweight and radar absorption capability, that is RAS [222, 223].

In order to impart EM absorption capacity to composites, a loss mechanism is required. In the case of conventional fiber-reinforced composites, a loss mechanism expected is dielectric loss. However, for various fibers (glass, quartz, and Kevlar-49, etc.), the value of dielectric constant is 4 to 6 at 10 GHz and dielectric loss tangent is 0.01 or less [222]. In the case of matrix resin (epoxy, phenolic, polyester, etc.), the dielectric constant has a small value near 3 and the dielectric loss tangent is negligibly small [222]. Therefore, it is difficult to expect excellent EM wave absorption in the commercial composite materials.

Carbon fibers, having a high electrical conductivity, can be a strong potential material for EM wave absorption using their conduction loss. Currently, carbon black powder is used as EM absorption fillers in polymer composites. However, it has been reported that carbon fiber composites exhibit the much higher EM wave reflections. The reflection loss is as high as − 0.05 dB at 12 GHz for a carbon fiber-epoxy composite material [222], indicating that it can be treated as an EM wave reflector almost the same as metals. A composite slab made of four layers of carbon fiber prepregs provides more than 99.9% of electromagnetic shielding [224]. Therefore, it is difficult to configure EM wave absorbers with only carbon fiber composite materials.

These results suggest that EM wave absorption of conventional composites is not good, leading to multilayer design to solve the problem. A composite material impregnated with lossy fillers is used as the EM wave absorbing layer and a carbon fiber composite is used as the EM wave reflector. EM wave absorbers constructed in this way can also possess excellent mechanical properties. The multilayer design of RAS can be illustrated in Fig. 6.28a and an equivalent circuit for this structure is shown in Fig. 6.28b. The backside of reflector ends in free space with wave impedance $Z_0 = 377\ \Omega$. Input impedances at the surface of reflector and absorbing layer are represented by Z_{in1} and Z_{in2}, respectively. The reflection loss of the double-layered RAS is determined by the difference between Z_{in2} and Z_0 [225].

A requirement for the rear reflecting layer is a low impedance. Carbon fiber composite with high electrical conductivity is suitable for this low-impedance substrate [225–227]. When the first layer behaves as a perfect reflector, an impedance matching can be achieved by giving the loss characteristics in the second layer. The lossy fillers can be magnetic ferrite particles or conductive carbon black [228, 229]. Impedance matching in the double-layered RAS is determined by appropriate combination of fibers, matrix, and absorption fillers, which constitute the absorbing layer.

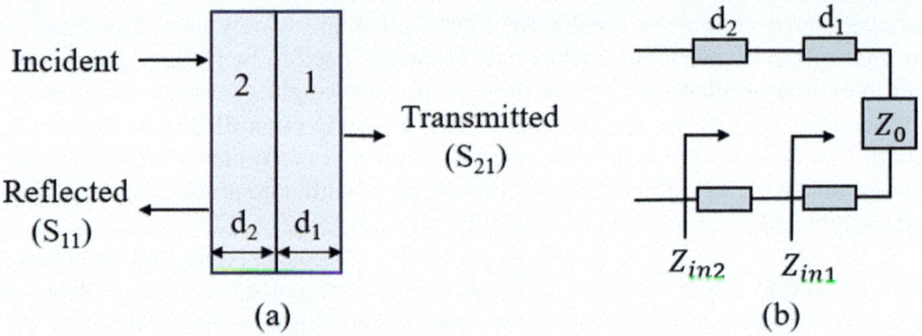

Fig. 6.28 **a** Illustration of the multilayer RAS and **b** equivalent circuit

Figure 6.29 shows the reflection coefficient (S_{11}) and the transmission coefficient (S_{21}) measured in the composite of double-layered RAS [229]. For the monolayer carbon fiber-epoxy composite, S_{11} is nearly 0 and S_{21} is very low (−40 dB) (Fig. 6.29a). The observed perfect reflection is due to the high electrical conductivity of carbon fibers. Figure 6.29b shows S_{11} and S_{21} measured in double-layered RAS consisting of glass fiber composites (thickness = 2.5 mm) with 3 wt.% of carbon black and the reflector of carbon fiber composites [229]. S_{21} maintains a low value (−40 dB), which means the perfect reflecting property of carbon fiber composites. S_{11} decreases significantly to a minimum value of − 40 dB at 11.75 GHz with a broad absorption bandwidth. It is a typical design result of multilayer RAS satisfying the impedance matching. The matching frequency can be adjusted by varying the filler content in the absorbing layer.

Lossy fillers used in the absorbing layer can be carbon black, carbon nanotubes, and ferrite powders [228, 230–234]. In single-layer RAS (2.9 mm thickness) composed of conductive carbon black mixed into E-glass/polyester composites, the minimum reflection loss of − 55 dB was observed in X-band frequencies [230].

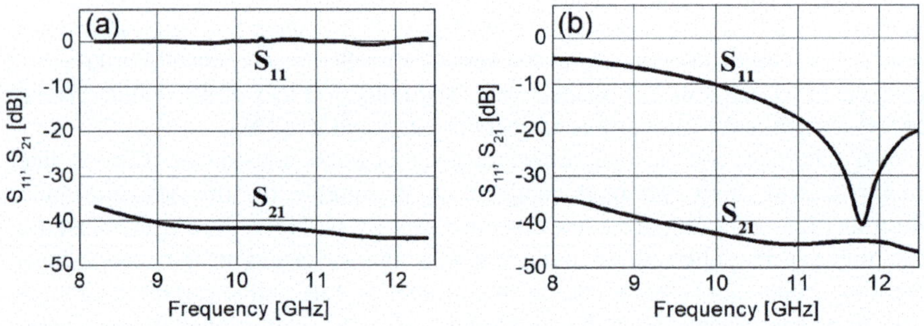

Fig. 6.29 Reflection coefficient (S_{11}) and transmission coefficient (S_{21}) measured in **a** monolayer carbon fiber composite and **b** double-layered RAS consisting of glass fiber composite containing 3 wt.% carbon black and reflector of carbon fiber composite (reproduced from Ref. [229] with the permission of Trans Tech Publications)

The fiber glass/epoxy composites containing multiwalled carbon nanotubes showed a power absorption of 90% in full X-band frequencies [231]. Multilayer RAS with different carbon black contents showed a 10 dB absorption bandwidth of 2.4 GHz in X-band frequency band [233]. When ferrite powder/epoxy composites were attached to carbon fiber composites, reflection loss of – 30 dB was achieved at 8.5 GHz [228].

6.6.2 Silicon Carbide Fiber Composites

From the point of view of mechanical properties, the addition of lossy fillers in a fiber-reinforced composite has a negative effect on the mechanical strength. The material that can overcome this problem is silicon carbide (SiC) fibers. As described in Sect. 6.4, SiC is a semiconductive material that has an electrical resistivity of $10^2 - 10^4$ Ωcm, a low density (~3.2 g/cc), and excellent mechanical properties at both room and high temperatures. Therefore, it is easy to control its dielectric constant and dielectric loss ($\varepsilon_r' = 3 \sim 20$, $\varepsilon_r'' = 0 \sim 10$), and RAS can be manufactured without any need of additional absorptive filler materials. Furthermore, the SiC fibers draw attracting attention as the high-temperature RAS, because they maintain the stable mechanical and electrical properties at high temperatures.

Commercial products of SiC fibers are manufactured by melting and spinning the polymer precursors, and the pyrolysis of green fiber makes it be converted to SiC ceramics. These fibers have a diameter of about 10 μm. Nicalon fibers from Nippon Carbon (Japan) and Tyranno fibers from Ube Industries (Japan) are some of the most common fibers that have been commercialized. These fibers are divided into 1st, 2nd, and 3rd generations according to the technological advancements, and their production and compositional details are summarized in Ref. [235].

1st-generation fibers choose PCS (polycarbosilane with a chemical formula, $[SiCH_3H\text{-}CH_2]_n$) as the starting precursors. PCS is melt-spun and produced into weak fibers. Heating in air at about 200 °C creates a precursor cross-linking. At this time, a significant amount of oxygen is introduced into fibers. Vacuum heat treatment at about 1200 °C produces the small-diameter SiC fibers. These include Nicalon 200 fibers from Nippon Carbon and Tyranno LOX-M fibers made with Ube's polytitano carbosilane (PTC) precursors. 1st-generation fibers are inferior to bulk SiC, due to the non-stoichiometric composition of excess carbon and amorphous oxygen-rich phase [236].

In the production of 2nd-generation fibers, electron irradiation methods are used during cross-linking process to reduce the oxygen content [237]. Although the radiation cross-linking process is expensive, it has succeeded in reducing the oxygen content of fibers to a level of 0.5 wt.% (Hi-Nicalon fiber from Nippon Carbon). Ube Industries did not use the expensive irradiation process, but instead applied a process of changing the metal added to PCS polymer. Titanium was replaced by zirconium to reduce the oxygen content (Tyranno ZMI fiber).

A fiber whose composition has been adjusted close to the stoichiometry is the 3rd-generation SiC fibers to achieve the high-temperature stability up to 1400 °C

[238]. To reduce the excess carbon in Hi-Nicalon fiber, Nippon Carbon lowered the C/Si ratio from 1.39 to 1.05 by heating to about 1500 °C in hydrogen (Hi-Nicalon Type S fiber). Ube used aluminum to improve the chemical properties of precursor polymers. Aluminum improves the cross-linking of precursor fibers and also plays a role as the sintering agent. These fibers are designated the Tyranno SA fibers and have an oxygen-free composition.

The microstructure of SiC fibers consists of SiC nanocrystals (β-SiC), amorphous oxides (SiC_xO_y), excess carbon, and thin carbon layer on fiber surface. The electrical conductivity σ of fibers and dielectric permittivity ε_r of composites are greatly dependent on these microstructural factors. In particular, there is a marked difference in electrical conductivity depending on the grade of fibers. This is because the electrical conductivity of constituent elements or phases is very different, as shown in Table 6.3. The SiO_xC_y oxide phase can be treated almost as an insulator ($\sigma = 10^{-14}$ to 10^{-12} S/cm), SiC nanocrystals are semiconductors ($\sigma = 10^{-4}$ to 10^{-2} S/cm), and free carbon within fibers or on fiber surface exhibits a very high conductivity ($\sigma =$ 1 to 10^2 S/cm) [239].

The EM wave absorption properties of SiC fibers at room temperature were investigated in the composites with polymer resin as a matrix [239–244]. The dielectric constant and dielectric loss of composites that mixed Tyranno ZMI fibers with resin were $\varepsilon_r' = 11.3$, $\varepsilon_r'' = 10.5$, respectively [240]. The main loss mechanism for the fiber is the conduction loss, which is due to the pyrolytic carbon deposited on fiber surface. The 2.0 mm thick ZMI fiber/resin composite showed an 80% power absorption in X-band frequencies. The EM wave absorption of SiC fibers close to the stoichiometric composition (C/Si = 1.125) and with a low oxygen content (about 0.51%) was investigated [241]. The minimum reflection loss of composites is – 30 dB at 8.2 GHz when the fiber volume fraction is 22.3% and the sample thickness is 4.7 mm. For the composite of SiC fiber fabric (fiber content = 40 vol.%) impregnated with epoxy resin, a conductive network is formed by the free carbon on fiber surface, exhibiting a high dielectric constant and dielectric loss [242]. As a result, the reflection loss of the fiber fabric composite is very high at −3.5 dB level.

The relationship between the electrical conductivity of SiC fibers and the EM wave absorption has been studied in epoxy composites (SiCf/epoxy) in X-band frequencies [245]. SiCf/epoxy composites with high conductive fibers of $\sigma > 10^{-1}$ S/cm exhibit a high EM wave reflectivity. Conversely, when $\sigma < 10^{-3}$ S/cm, SiCf/epoxy composite is transparent for EM wave. For the fibers of $10^{-3} < \sigma < 10^{-1}$ S/cm, an excellent EM wave absorption can be achieved in SiCf/epoxy composites with an optimum fiber content and layer thickness. As the electrical conductivity of SiC fiber increases, the optimum fiber content and layer thickness decrease [239].

Table 6.3 Electrical conductivity of constituent phases of SiC fibers

Phase	Electrical conductivity (S/cm)
SiO_xC_y oxide phase	10^{-14}–10^{-12}
SiC and amorphous carbon	10^{-4}–10^{-2}
Nanocrystalline carbon	1–10^2

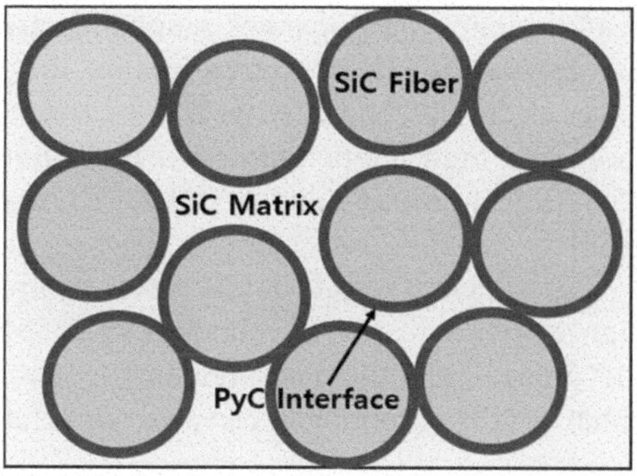

Fig. 6.30 Schematic representation of typical microstructure of SiCf/SiC composites

RAS for high-temperature applications uses SiC fiber/SiC matrix (SiCf/SiC) composites. The microstructure of a typical SiCf/SiC composite is composed of SiC fibers, SiC matrix, and the interfacial layer between these two phases [246], as schematically represented in Fig. 6.30. Many researchers have studied the requirements and properties of SiC fibers, interface control, and matrix manufacturing methods for high-temperature RAS [161, 162, 246–249].

The manufacturing of SiCf/SiC composites begins with the fabrication of SiC fiber preforms with various fiber arrangements from one dimensional to multidimensional structures. These preforms have a unit cell with a variety of shapes and sizes, which is filled with SiC matrix to increase the density. The densification process includes polymer infiltration and pyrolysis (PIP) and chemical vapor infiltration (CVI). In the PIP process, polycarbosilane (PCS) preceramic polymer is infiltrated into SiC fiber preforms and then converted to SiC ceramics through pyrolysis at about 1000 °C. This process is limited by high shrinkage and low SiC conversion rates. CVI is a well-established process for a uniform coating, but has the disadvantage of being expensive and long time to complete the manufacturing process.

The interface between the fiber and the matrix plays an important role in dielectric and microwave absorption properties of SiCf/SiC composites. It is to be noted that the most commonly observed interfacial phase is the pyrolytic carbon (PyC) which has a negative effect on microwave absorption of SiCf/SiC composites. This is because the PyC interface, which has a high electrical conductivity and is percolated above a critical loading of fibers, strongly reflects EM waves [250–253]. These PyC interfaces are observed in both CVI and PIP processes.

In order to optimize dielectric and microwave absorption properties at room and high temperatures, it is important to inhibit the formation of PyC interfaces. For this purpose, thermal cross-linking has been proposed in the PIP process [254]. In the case of the pyrolyzed SiC matrix, it was found that excess carbon was present when it was not subjected to the cross-linking process. However, a large amount of oxygen is introduced during the cross-linking process, which helps to form the amorphous silica in the SiC matrix during high-temperature pyrolysis. At the same time, the injected oxygen is efficient in removing excess carbon or intermediate carbon that decomposes during the pyrolysis process of PCS precursors. The produced silica is an insulator that separates semiconductive SiC particles, weakening the conductive network. Thus, after cross-linking, the composite material exhibits significantly lower electrical conductivity, and microwave absorption is greatly improved at both room and high temperatures. At room temperature, the minimum RL reaches −15.5 dB at 11.4 GHz, and at 300 °C, RL reaches the lowest −41.71 dB at 10.9 GHz [254].

Experimental results of EM wave absorption of SiCf/SiC composites at room and high temperatures can be found in Ref. [255]. SiCf/SiC RAS was manufactured using PIP process, and the measurements by free-space method show a reflection loss of about −8 dB at room and elevated temperatures (700 °C).

6.7 Metamaterials

6.7.1 Negative Index Materials

Electromagnetic metamaterials are composed of sub-wavelength unit elements that are smaller than the wavelength expressed by electrical permittivity ε and magnetic permeability μ. The early interest in metamaterials began with their special properties that could not be obtained from natural materials. The biggest attraction of metamaterials is that they can implement a negative refractive index (NRI) [256, 257]. The optical properties of transparent materials are explained by the refractive index N expressed as a function of ε and μ, that is

$$N = \pm\sqrt{\varepsilon\mu} \tag{6.7}$$

A conventional material has a positive refractive index because of positive ε and μ. In the case of metamaterials, however, ε and μ are simultaneously negative and the refractive index can be negative. Accordingly, metamaterials are also called the negative index materials (NIM), which were first suggested theoretically by Veselago in 1967 [258]. Metamaterials with negative N have the following unique properties. According to Snell's law, incident and refracted rays are on the same side with respect to the surface normal to interface. The pointing vector is opposite to the phase velocity, which means that the wave vector is opposite to the direction of energy transfer. Therefore, when a plane wave propagates within a metamaterial, the electric field, magnetic field, and wave vector follow a left-handed rule. In this sense, it is called the left-handed materials (LHM) to distinguish it from conventional right-handed materials (RHM).

The idea of NIM remained a theoretical hypothesis for 30 years after it was first proposed. This is because there are no materials that naturally have a negative refractive index. In 1999, Pendry proposed a practical way to make a NIM with metallic thin wires (TW) for negative permittivity and split ring resonators (SRR) for negative permeability [259, 260]. In 2000, NIM was artificially fabricated by

Smith et al. [261]. The periodic combination of SRR and TW exhibited negative permittivity and negative permeability simultaneously in microwave frequency band [261, 262]. These results were reproduced by similar experiments [263, 264] and improved by numerical simulations [265]. The TW structure exhibits a complex permittivity spectrum of Drude model, and negative dielectric constant is observed below the plasma frequency ω_{pe}. In the SRR structure consisting of two concentric conductor rings, *LC* resonance occurs and exhibits a complex permeability spectrum of Lorentzian model. Therefore, a negative permeability is observed between the resonance frequency ω_{rm} and the magnetic plasma frequency ω_{pm}, and a negative refractive index can be obtained in this frequency region [266].

The negative refractive index of metamaterials enables applications in a variety of new fields. It was proposed that a metamaterial having a negative refractive index can be applied to the manufacture of super lenses (or perfect lenses) [267, 268]. Soon afterward, an optical super lens beyond the diffraction limit was proposed by Zhang et al. [269, 270]. They showed for the first time a negative refractive index phenomenon in visible light region with a metamaterial manufactured by processing a laminated film of silver and MgF_2 into a nano-scale grid structure. Metamaterials can be used as the basis for clocking in microwave region [271, 272] and visible light region [273–276]. In addition, it can be applied to various circuit devices including antennas [277] and magnetic resonance devices [278] in microwave band.

6.7.2 Metamaterial Absorbers

From the point of view of effective medium, the electromagnetic properties of metamaterials are determined by effective magnetic permeability ($\mu = \mu' - j\mu''$) and effective dielectric permittivity ($\varepsilon = \varepsilon' - j\varepsilon''$). Research on metamaterials has mainly been limited to lossless materials ($\mu'' = 0, \varepsilon'' = 0$). For example, in the case of a material with a negative refractive index, the real parts of effective magnetic permeability and effective dielectric permittivity are designed to have negative values ($\mu' < 0, \varepsilon' < 0$). By using the loss component of metamaterials, EM waves can be absorbed, and a perfect absorber (100% absorption) can be realized. Landy et al. [279] proposed a metamaterial composed of electric ring resonator (ERR) and cut wire (CW), and first suggested the possibility of realizing a perfect absorber. Since then, metamaterial absorbers with various structures (patch, split cut wire, cross resonator, etc.) have been proposed [280–283]. Figure 6.31 shows an example of the composite structure of a metamaterial absorber. In Fig. 6.31, (a) is the split cut wire (SCW) resonator exhibiting electrical response, and (b) are the split ring resonator (SRR) responding to the magnetic field. By combining the resonators with a ground conductor on the back side of a dielectric substrate, a metamaterial absorber can be constructed, as shown in (c) and (d).

The basic structure of the metamaterial absorber is the same as that of FSS absorber illustrated in Fig. 5.16a. It is composed of a metal conductor on front side and a ground plane on back side with a dielectric substrate between them. However, in the case of

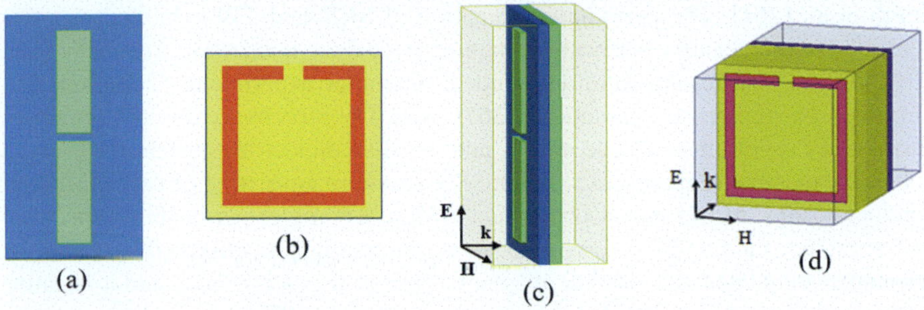

Fig. 6.31 Unit cell structure of metamaterial absorber: **a** split cut wire resonator showing electrical response, **b** split loop resonator showing magnetic response, **c** and **d** the structure of an absorber combining a resonator and a ground conductor

an FSS absorber, the thickness of the dielectric substrate is approximately $\lambda/4$, and the coupling between the front and rear conductors is very weak. However, in the case of metamaterials, the thickness of substrate is so small that the electrical coupling between the front and rear conductors dominantly contributes the resonating behavior of whole structure in addition to the inductance and capacitance of metallic resonators themselves. In particular, when the thickness of substrate becomes smaller than $0.3p$ (p: unit cell size), the capacitance between two conductors increases exponentially, which mainly contributes to the low reflection loss and the thin thickness of the metamaterial absorbers [284].

The conventional EM wave absorbers are divided into a resonant absorber and a broadband absorber according to the absorption mechanism. The most common method used so far for the design of resonant absorbers is a quarter-wavelength-thick ($\lambda/4$) lossy material (i.e., Dallenbach absorber) or a Salisbury screen [285, 286]. There are several problems that are difficult to solve in the conventional absorbers. The first problem is the thickness of absorber. In the case of a resonant absorber, a thickness of at least $\lambda/4$ is required, and if a multilayer structure and a pyramid absorber with broadband characteristics are designed, the thickness is significantly increased. It is also an unavoidable disadvantage that it exhibits a narrowband characteristic limited to a frequency corresponding to a quarter wavelength. It is also essential to secure the lossy materials that can induce impedance matching to the free space. The operation of these absorbers is usually confined to microwave frequency band, mostly below 30 GHz. Since the magnetic permeability drops to 0 when it exceeds the microwave band in usual magnetic materials, it is impossible to construct the absorber in the high-frequency band of terahertz (THz) or higher.

However, in the case of a metamaterial absorber, the effective material parameters (permeability and permittivity) are determined by the inductance L and capacitance C of metallic resonators and coupling between them. Accordingly, there are less restrictions on the thickness of substrate and the dimensional control of the artificial structure makes it possible to extend EM absorption up to the visible light band.

Therefore, the design flexibility is very wide and it is possible to realize the EM wave absorption that have not been obtained so far.

Metamaterial absorbers are classified into three different frequencies: microwave, terahertz (THz), and optical frequencies, that was reviewed in Ref. [287]. The first experimental research on metamaterial absorbers was conducted in the microwave frequency band [279]. In this study, the metamaterial design using split ring resonator and a cut wire achieved a 96% narrowband absorption at 11.48 GHz by simulation. Subsequent studies were conducted to increase the absorption bandwidth in metamaterial absorbers. Specifically, multiband or broadband absorption characteristics were developed using multiple resonant unit cells with different size or orientation through in-plane arrangement [288–295].

Similar studies were quickly reproduced in THz region. A terahertz (THz) wave is an electromagnetic wave that exists in the frequency band of 10^{11} to 10^{13} Hz that spans between microwave and infrared. Making the size of resonator smaller than the THz wavelength ($\lambda = 3$ mm to 300 μm for 0.1 to 10 THz) and utilizing the magnetic resonance with rear ground plane, it was possible to design a THz absorber [296–301]. Through the three-layered design (upper ERR/polyimide film/lower split wire), the simulation result showed a power absorption of 98% at 1.12 THz [302]. Metamaterial based absorbers with broad bandwidth at millimeter wave frequencies (77 GHz) were designed using two split ring resonators with dimensions slightly increased in millimeter wave region [303]. In the mid-infrared (MIR) region, an absorption rate of 97% was obtained at a wavelength of 6 μm by slightly reducing the size of unit cell [304]. The metamaterial absorber at near-infrared (NIR) frequency used a unit cell composed of disk-shaped conductor at the top of a grounded dielectric layer. In this study, absorption rate of 88~99% was demonstrated at approximately 1.6 μm wavelength [305, 306].

The optical metamaterial absorbers are a type of perfect absorbers operating in optical frequencies [307–309]. The optimized metamaterial absorbers in the optical regime have potential application in the field of electromagnetic cloaking and solar cells [307, 310–313]. The main constraint for designing perfect absorbers in the optical region is the small size of the unit cell which is on the order of several nanometers and requires using the practical nano-fabrication techniques. Due to the manufacturing difficulties, experiments in the visible region were not performed as much as in the microwave or THz frequencies.

6.7.3 *Simulation and Fabrication*

The power absorption of the metamaterial absorbers is given by $A = 1 - \left(|S_{11}|^2 + |S_{21}|^2\right)$, where S_{11}, S_{21} are the reflection coefficient and transmission coefficient, respectively. Simulation of the S parameters is conducted with varying the geometry (size of unit cells, dimensions of resonators, thickness of substrate), permittivity of substrate, and polarization and angle of incidence, using a commercial software based on 3D FEM or FDTD. From these results, it is possible to optimize

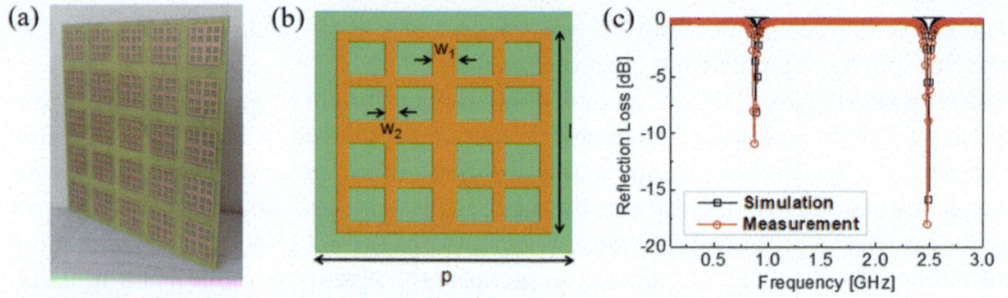

Fig. 6.32 **a** A test sample of metamaterial absorber with grid pattern fabricated by PCB process, **b** geometry of unit cell, and **c** reflection loss (reproduced from Ref. [314] with the permission of The Korean Institute of Metals and Materials)

the geometries and material parameters for a perfect metamaterial absorber with improvements in absorber thickness, bandwidth, and polarization insensitive and oblique incidence properties.

The metamaterial absorber designed by simulation is manufactured using the conventional PCB manufacturing process. Using a copper-plated FR4 as the substrate, a designed pattern can be produced through the processes of mask making by CAD and printing, photo resist (PR) coating, ultraviolet (UV) exposure, developing, and final etching and cleaning. Figure 6.32a shows an example of metamaterial absorber composed of grid conductor on FR4 board grounded by copper film [314]. The size of the specimen is 50 cm × 50 cm and contains 5 × 5 unit cells. For the dimensions of unit cell (p = 100 mm, l = 73 mm, w_1 = 9.0 mm, w_2 = 4.5 mm) shown in Fig. 6.32b, the reflection loss is −10 dB at f_1 = 0.88 GHz and −18 dB at f_2 = 2.48 GHz, which is very consistent with simulation result, as shown in Fig. 6.32c.

A frequency selective surface (FSS) is a periodic surface with identical two-dimensional arrays of elements arranged on a dielectric substrate. Depending on the pattern or geometry of the conductive elements, it exhibits a bandpass or band-stop characteristics for a specific frequency band, and can be treated as one type of metamaterials. The resistive FSS, that requires a relatively large sheet resistance (10 ~ 200 Ω/sq), is produced by screen printing the conductive ink on FR4 substrate. The experimental tools required for screen printing include a screen-printing machine, a printed board, a mask frame with a printing pattern, inks, and squeeze (Fig. 6.33).

In general, a polymer film or a dielectric plate is used as the substrate. The mask frame is manufactured in the form of a pattern mesh so that the printing ink can pass through it. Printing ink is a paste of conductive particles mixed with a solvent (binder) with an appropriate viscosity (Fig. 6.33b). Carbon black is usually used for conductive particles, and metal particles or flakes are sometimes added to lower the resistance. A mask pattern is placed atop a substrate. Ink is placed on top of the screen, and a squeeze is used to push the ink through the holes in the mesh (Fig. 6.33c). Characteristics of screen mesh (geometrical and physical properties, elasticity, tension, etc.), selection of printing ink (type and size of particles, viscosity, solid content), and printing process (speed of squeegee, angle, pressure) should be controlled to produce FSS with uniform sheet resistance.

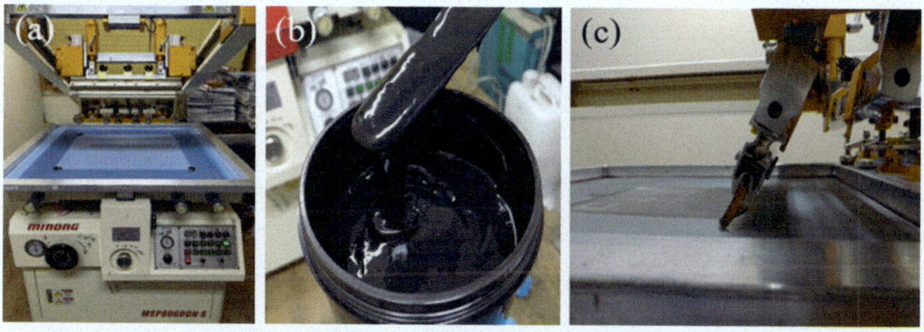

Fig. 6.33 **a** Screen-printing machine, **b** carbon black paste, and **c** squeeze

Figure 6.34a shows a prototype FSS absorber fabricated by screen printing. Carbon paste was printed on a 0.15 mm thick FR4 substrate. The surface resistance measured by the four-prove method was about 200 Ω/sq. The spacer is a polymer composite of carbonyl iron particles (thickness = 2.5 mm). The back side was terminated by copper. The size of test sample is 50 cm × 50 cm. The dimensions of the unit cell and square loop conductor are period $p = 10$ mm, conductor length $d = 8$ mm, and conductor width $w = 0.5$ mm. The reflection loss measured by the free-space method (shown in Fig. 6.34b) is almost identical to the simulation result, as depicted in Fig. 6.34c [117].

Since the size of the resonator becomes smaller as the resonance of a metamaterial moves to a higher frequency, it requires the higher precision fabrication techniques. In THz range, metamaterial resonators are on the order of tens of micrometers. For these sizes, photolithography is the most effective method of fabrication [297, 299, 315]. At higher frequencies surpassing the capability of photolithography, such as in the infrared and visible range, the metamaterial absorbers are best fabricated using e-beam lithography and focused ion beam (FIB) [306, 316]. These methods are capable of making structures with sizes that are on the order of nanometers. However, compared to photolithography, these methods are a much slower batch process and usually used to fabricate the small-area samples. An optical metamaterial absorber with a unit cell period of 998 nm and a pixel size of 67 nm was fabricated using e-beam lithography and demonstrated a high absorptivity greater than 90% over a broad optical wavelength range between 1.95 and 4.80 μm [308].

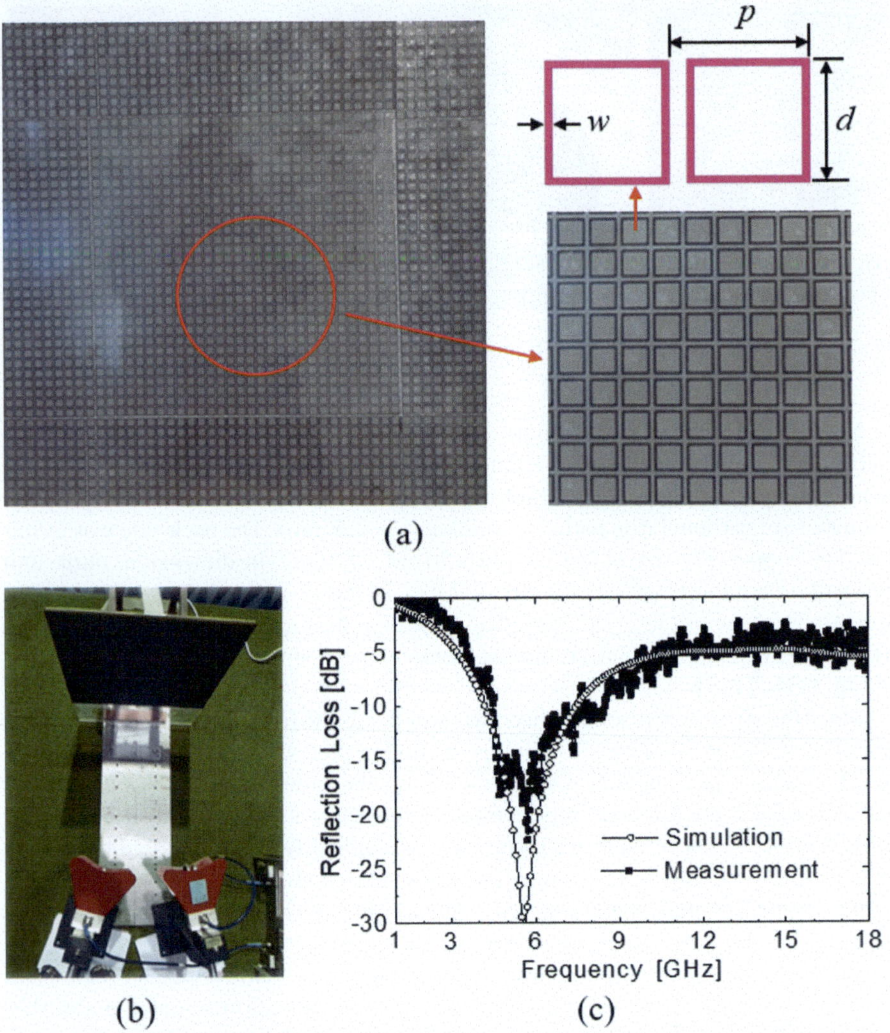

Fig. 6.34 **a** A test sample of square loop FSS absorber fabricated by screen printing, **b** measurement by free-space method, and **c** reflection loss (reproduced from Ref. [117] with the permission of The Korean Institute of Metals and Materials)

References

1. F. Qin, C. Brosseau, J. Appl. Phys. **111**, 061301 (2012)
2. A.R. Oganov, R.J. Hemley, R.M. Hazen, A.P. Jones, Rev. Mineral. Geochem. **75**, 47–77 (2013)
3. L. Liu, P. He, K. Zhou, T. Chen, J. Mater. Sci. **49**, 4379–4386 (2014)
4. N.J. Tang, Y. Yang, K.J. Lin, W. Zhong, C.T. Au, Y.W. Du, J. Phys. Chem. C **112**, 19316 (2008)
5. D.L. Zhao, Z.M. Shen, Mater. Lett. **62**, 3704–3706 (2008)
6. A.K. Geim, K.S. Novoselov, Nature Mater. **6**, 183–191 (2007)

7. K.S. Novoselov, A.K. Geim, S.V. Morozov, D. Jiang, M.I. Katsnelson, I.V. Grigorieva, S.V. Dubonos, A.A. Firsov, Nature **438**, 197–200 (2005)
8. A.H. Castro Neto, F. Guinea, N.M.R. Peres, K.S. Novoselov, A.K. Geim, Rev. Mod. Phys. **81**, 109 (2009)
9. S. Iijima, Nature **354**, 56–58 (1991)
10. S. Iijima, P.M. Ajayan, T. Ichihashi, Phys. Rev. Lett. **69**, 3100 (1992)
11. S.K. Kwon, J.M. Ahn, G.H. Kim, C.H. Chun, J.S. Hwang, J.H. Lee, Polym. Eng. Sci. **42**, 2165 (2002)
12. K.-K. Choi, S.-B. Cho, K.-I. Kwon, S.-S. Kim, J.-M. Kim, J. Kor. Ceram. Soc. **29**, 293 (1992)
13. J.-H. Oh, K.-S. Oh, C.-G. Kim, C.-S. Hong, Compos. B **35**, 49 (2004)
14. P.M. Ajayan, T.W. Ebbesen, Rep. Prog. Phys. **60**, 1025–1062 (1997)
15. Q.Z. Zhao, M.B. Nardelli, J. Bernholc, Phys. Rev. B **65**, 144105 (2002)
16. N. Saifuddin, A.Z. Raziah, A.R. Junizah, J. Chem. **2013**, 676815 (2012)
17. P.A. Martins-Júnior, C.E. Alcântara, R.R. Resende, A.J. Ferreira, J. Dent. Res. **92**(7), 575 (2013)
18. F.H. Gojny, M.H.G. Wichmann, U. Kopke, B. Fiedler, K. Schulte, Compos. Sci. Technol. **64**, 2363 (2004)
19. Y. Zhou, F. Pervin, L. Lewis, S. Jeelani, Mater. Sci. Eng. A **452–453**, 657 (2007)
20. A. Saib et al., in *Proceedings of the 2005 European Microwave Conference* (Paris, France, 2005), pp. 1–4
21. M. Kallumottakkal, M.I. Hussein, M.Z. Iqbal, Front. Mater. **8**, 633079 (2021)
22. P. Savi, M. Miscuglio, M. Giorcelli, A. Tagliaferro, Prog. Electromag. Res. Lett. **44**, 63 (2014)
23. P. Savi, M. Giorcelli, S. Quaranta, Appl. Sci. **9**, 851 (2019)
24. H. Wang, D. Zhu, W. Zhou, L. Fou, J. Alloys Compd. **648**, 313 (2015)
25. D.-L. Zhao, X. Li, Z.-M. Shen, J. Alloys Compd. **471**, 457 (2009)
26. R.C. Che, C.Y. Zhi, C.Y. Liang, X.G. Zhou, Appl. Phys. Lett. **88**, 033105 (2006)
27. X.J. Lee, B.Y.Z. Hiew, K.C. Lai, L.Y. Lee, S. Gan, S. Thangalazhy-Gopakumar, J. Taiwan Inst. Chem. Eng. **98**, 163 (2019)
28. Electronic properties of graphene on Wikipedia. https://en.wikipedia.org/wiki/Electronic_properties_of_graphene
29. C.-Y. Chen, N.-W. Pu, Y.-M. Liu, L.-H. Chen, C.-H. Wu, T.-Y. Cheng, Compos. Part B Eng. **135**, 119 (2018)
30. M. Bozzi, L. Pierantoni, S. Bellucci, Radio Eng. **24**, 661 (2015)
31. Y. Bai, B. Zhong, Y. Yu, M. Wang, J. Zhang, B. Zhang, NPJ 2D mater. Appl. **3**, 32 (2019)
32. J. Li et al., Chem. Eng. J. **425**, 131558 (2021)
33. P. Bhattacharya, C.K. Das, S.S. Kalra, J. Mater. Sci. Res. **1**, 126 (2012)
34. J.S. Li, H. Huang, Y.J. Zhou, C.Y. Zhang, Z.T. Li, J. Mater. Res. **32**, 1213 (2017)
35. L. Hualiang, Y. Guo, Z. Yang, Y. Cheng, L.P. Wang, B. Zhang, Y. Zhao, Z.J. Xu, G. Ji, J. Mater. Chem. C **5**, 491 (2017)
36. C.-Y. Chen, N.-W. Pu, Y.-M. Liu, S.-Y. Huang, C.-H. Wu, M.-D. Ger, Compos. Part B Eng. **114**, 395 (2017)
37. V.K. Singh, A. Shukla, M.K. Patra, L. Saini, R.K. Jani, S.R. Vadera, Carbon **50**, 2202 (2012)
38. M.J. McAllister et al., Chem. Mater. **19**, 4396 (2007)
39. S. Stankovich et al., Nature **442**, 282 (2006)
40. S. Stankovich et al., Carbon **45**, 1558 (2007)
41. S. Syama, P.V. Mohanan, Nano-Micro Lett. **11**, 6 (2019)
42. A. Asif Iqbal, N. Sakib, A.K.M. Parvez, D.M. Nuruzzaman, Materialia **12**, 100815 (2020)
43. V.K. Singh, A. Shukla, M.K. Patra, L. Saini, R.K. Jani, S.R. Vadera, N. Kumar, Carbon **50**, 2202 (2012)
44. Z. Du, X. Chen, Y. Zhang, X. Que, P. Liu, X. Zhang, Materials **13**, 3065 (2020)
45. Q. Fan, L. Zhang, H. Xing, H. Wang, X. Ji, J. Mater. Sci. Mater. Electron. **31**, 3005 (2020)
46. W. Feng, Y. Wang, J. Chen, L. Wang, L. Guo, J. Ouyang, Carbon **108**, 52 (2016)
47. M. Naguib, V.N. Mochalin, M.W. Barsoum, Y. Gogotsi, Adv. Mater. **26**, 992 (2011)
48. C.E. Shuck et al., Adv. Eng. Mater. **22**, 1901241 (2020)

49. MXenes on Wikipedia. https://en.wikipedia.org/wiki/MXenes
50. J. Halim et al., Appl. Surf. Sci. **362**, 406 (2016)
51. N.R. Hemanth, T. Kim, B. Kim, A.H. Jadhav, K. Lee, N.K. Chaudhari, Mater. Chem. Front. **5**, 3298 (2021)
52. A.N. Enyashin, A.L. Ivanovskii, J. Phys. Chem. C **117**, 13637 (2013)
53. Y. Qing, W. Zhou, F. Luo, D. Zhu, Ceram. Int. **42**, 16412 (2016)
54. W. Feng, H. Luo, Y. Wang, S. Zeng, L. Deng, X. Zhou, RSC Adv. **8**, 2398 (2018)
55. H. Luo, W. Feng, C. Liao, L. Deng, S. Liu, H. Zhang, J. Appl. Phys. **123**, 104103 (2018)
56. J. Smit, H.P.J. Wijn, *Ferrites*, (Philips Technical Library, Eindhoven, 1959), pp. 202–211
57. T. Tsutaoka, J. Appl. Phys. **93**, 2789 (2003)
58. D.-H. Han, *Dissertation* (MS Thesis) (Chungbuk National University, 1994)
59. S.S. Kim, S.B. Cho, K.K. Choi, K.I. Gueon, J.M. Kim, K.S. Churn, IEEE Trans. Magn. **27**, 5462 (1991)
60. H. Su, H. Zhang, X. Tang, X. Xiang, J. Magn. Magn. Mater. **283**, 157 (2004)
61. J. Murbe, J. Topper, J. Electrocer. **15**, 215 (2005)
62. X. Shen, Y. Wang, X. Yang, L. Lu, L. Huang, J. Mater. Sci. Mater. Electron. **21**, 630 (2010)
63. S. Yan, W. Liu, Z. Chen, Y. Nie, X. Wang, Z. Feng, J. Appl. Phys. **115**, 17A529 (2014)
64. M.V. Cabanas, J.M. Gonzalez-Calbet, J. Fodriguez-Carvajal, M. Vallet-Regi, J. Solid State Chem. **111**, 229 (1994)
65. J.L. Snoek, Phys. **14**, 207 (1948)
66. G.H. Jonker, H.P.J. Wijn, P.B. Braun, Philips Tech. Rev. **18**, 145 (1956–57)
67. M. Sugimoto, in *Properties of Ferroxplana-type Hexagonal Ferrites*, ed. by E. P. Wohlfarth, *Ferromagnetic Materials*, vol. 3 (North-Holland Pub., Amsterdam, 1982), pp. 428–433
68. Q.A Pankhurst, D.H. Jones, A.H. Morrish, X.Z. Zhou, A.R. Corradi, in *Proceedings ICF-5* (1989), pp. 323–327
69. B. Sugg, H. Vincent, J. Mag. Mag. Mater. **139**, 364 (1995)
70. E. Brando, H. Vincent, O. Dubrinfant, A. Fourrier-Lamer, R. Lebourgeois, J. Phys. IV **7**, C1-421 (1997)
71. E. Brando, H. Vincent, J. Rodriguez-Carjaval, J. Phys. IV **7**, C1-303 (1997)
72. A. Tauber, J.A. Kohn, R.O. Savage, J. Appl. Phys. **34**, 1265–1267 (1963)
73. D. Autissier, A. Podemski, C. Jacquiod, J. Phys. IV **7**, C1-409 (1997)
74. H.S, Shin, J. Kor. Cerm. Soc. **49**, 205 (2012)
75. J. Smit, H.P.J. Wijn, *Ferrites*, (Philips Technical Library, Eindhoven, 1959), pp. 46–54
76. R.C. Pullar, S.G. Appleton, A.K. Bhattacharya, J. Magn. Magn. Mater. **186**, 326 (1998)
77. I. Nedkov, A. Petkov, V. Karpov, IEEE Trans. Magn. **26**, 1483 (1990)
78. E. Neckenburger, H. Severin, J.K. Vogel, G. Winkler, Z. Angew, Phys. **18**, 65 (1964)
79. H.-S. Cho, S.-S. Kim, IEEE Trans. Magn. **35**, 3151 (1999)
80. S. Sugimoto, S. Kondo, K. Okayama, H. Hakamura, D. Book, T. Kagotani, M. Homma, H. Ota, M. Kimura, R. Sato, IEEE Trans. Magn. **35**, 3154 (1999)
81. Y.-J. Kim, S.-S. Kim, J. Electrocer. **24**, 314 (2010)
82. Y.-J. Kim, *Dissertation* (MS Thesis) (Chungbuk National University, 2001)
83. Y.-J. Kim, S.-S. Kim, IEEE Trans. Magn. **38**, 3108 (2002)
84. H.-I. Hsiang, H.-H. Duh, J. Mater. Sci. **36**, 2081 (2001)
85. A. Paoluzi, F. Licci, O. Moze, G. Turilli, A. Deriu, G. Albanese, E. Calabrese, J. Appl. Phys. **63**, 5074 (1988)
86. H. Graetsh, H. Haberey, R. Leckebusch, M. Rosenberg, K. Sahl, IEEE Trans. Magn. **20**, 495 (1984)
87. H.-S. Cho, S.-S. Kim, Ceram. Inter. **45**, 9406 (2019)
88. Z.W. Li, L. Chen, Y. Wu, C.K. Ong, J. Appl. Phys. **96**, 534 (2004)
89. G. Albanese, M. Carbucicchio, J. Appl. Phys. **11**, 81 (1976)
90. H.-S. Cho, *Dissertation* (Ph.D. Thesis) (Chungbuk National University, 2019)
91. Y.P. Wu, C.K. Ong, Z.W. Li, L. Chen, G.O. Lin, S.J. Wang, J. Appl. Phys. **97**, 063909 (2005)
92. Z.W. Li, Y.P. Wu, G.Q. Lin, J. Appl. Phys. **102**, 083908 (2007)
93. Z.W. Li, Y.P. Wu, G.Q. Lin, L.B. Kong, J. Phys. D Appl. Phys. **42**, 095007 (2009)

94. Z.W. Li, Y.P. Wu, G.Q. Lin, L.B. Kong, IEEE Trans. Magn. **45**, 670 (2009)
95. W. Jing, Z. Hong, B. Shuxin, C. Ke, Z. Changrui, J. Magn. Magn. Mater. **312**, 310 (2007)
96. L. Deng, L. Ding, K. Zhou, S. Huang, Z. Hu, B. Yang, J. Magn. Magn. Mater. **323**, 1895 (2011)
97. F. Guo, G. Ji, J. Xu, H. Zou, S. Gan, X. Xu, J. Magn. Magn. Mater. **324**, 1209 (2012)
98. N. Fuchikami, J. Phys. Soc. Jpn. **20**, 760 (1965)
99. G. Albanese, A. Deriu, S. Rinaldi, J. Phys. C Solid State Phys. **9**, 1313 (1976)
100. H.J. Kwon, J.Y. Shin, J.H. Oh, J. Appl. Phys. **75**, 6109 (1994)
101. M. Matsumoto, Y. Miyata, J. Appl. Phys. **79**, 5486 (1996)
102. I.G. Chen, S.H. Hsu, Y.H. Chang, Appl. Phys. **87**, 6247 (2000)
103. L.B. Kong, Z.W. Li, L. Liu, R. Huang, M. Abshinova, Z.H. Yang, C.B. Tang, P.K. Tan, C.R. Deng, S. Matitsine, Inter. Mater. Rev. **58**, 203 (2013)
104. J.Y. Shin, J.H. Oh, IEEE Trans. Magn. **29**, 3437 (1993)
105. Y. Bai, J. Zhou, Z. Gui, L. Li, Mater. Sci. Eng. B **103**, 115 (2003)
106. Z.W. Li, M.J. Chua, Z.H. Yang, J. Magn. Magn. Mater. **382**, 134 (2015)
107. R.A. Braden, I. Gordon, R.L. Harvey, IEEE Trans. Magn. **2**, 43 (1966)
108. T. Nakamura, K. Hatakeyama, IEEE Trans. Magn. **36**, 3415 (2000)
109. M.J. Park, S.-S. Kim, Electron. Mater. Lett. **12**, 610 (2016)
110. M.J. Park, *Dissertation* (MS Thesis) (Chungbuk National University, 2000)
111. Z.W. Li, Y.P. Wu, G.Q. Lin, L.F. Chen, J. Magn. Magn. Mater. **310**, 145 (2007)
112. X. Wang, L. Li, S. Su, Z. Gui, Z. Yue, J. Zhou, J. Magn. Magn. Mater. **234**, 255 (2001)
113. X. Wang et al., J. Eur. Ceram. Soc. **23**, 715 (2003)
114. T. Nakamura, E. Hankui, J. Magn. Magn. Mater. **257**, 158 (2003)
115. Z.W. Li, G.Q. Lin, L.F. Chen, Y.P. Wu, C.K. Ong, J. Appl. Phys. **99**, 063905 (2006)
116. BASF web-stie. https://www.basf.com/global/en.html
117. H.-S. Cho, S.-S. Kim, Kor. J. Met. Mater. **57**, 741 (2019)
118. S.-H. Lee, D.-H. Lee, J.-R. Yoon, Bull. Korean Inst. Electr. Electron. Mater. Eng. **27**, 3 (2014)
119. R.M. Walser, W. Kang, IEEE Trans. Magn. **34**, 1144 (1998)
120. F.S. Wen, L. Qiao, D. Zhou, W.L. Zuo, H.B. Yi, F.S. Li, Chin. Phys. Lett. **17**, 2263 (2008)
121. R. Han, L. Qiao, T. Wang, F.S. Li, J. Alloys Compd. **509**, 2734 (2011)
122. F. Wen, W. Zuo, H. Yi, N. Wang, L. Qiao, F. Li, Phys. B **404**, 3567 (2009)
123. R.B. Yang, W.F. Liang, J. Appl. Phys. **109**, 07A311 (2011)
124. S.-S. Kim, S.-T. Kim, Y.-C. Yoon, K.-S. Lee, J. Appl. Phys. **97**, 10F905 (2005)
125. P.H. Zhou, L.J. Deng, J.L. Xie, D.F. Liang, J. Alloys Compd. **448**, 303 (2008)
126. J. Wei, Z. Zhang, B. Wang, T. Wang, F. Li, J. Appl. Phys. **108**, 123908 (2010)
127. L.J. Deng, P.H. Zhou, J.L. Xie, L. Zhang, J. Appl. Phys. **101**, 103916 (2007)
128. R. Gong, X. Wang, W. Cheng, X. Shen, Mater. Lett. **62**, 266 (2008)
129. L. Qiao, F. Wen, J. Wei, J. Wang, F. Li, J. Appl. Phys. **103**, 063903 (2008)
130. S.T. Kim, H.S. Cho, S.-S. Kim, IEEE Trans. Magn. **41**, 3562 (2005)
131. K.-S. Lee, Y.-C. Yun, S.-W. Kim, S.-S. Kim, J. Appl. Phys. **103**, 07E504 (2008)
132. S. Yoshida, M. Sato, E. Sugawara, Y. Shimada, J. Appl. Phys. **85**, 4636 (1999)
133. M. Yamaguchi, K. Maruta, H. Ono, IEEE Trans. Magn. **41**, 3565 (2005)
134. S.-S. Kim, Res. Chem. Intermed. **40**, 2553 (2014)
135. S. Yoshida, H. Ono, S. Ando, F. Tsuda, T. Itoh, Y. Shimada, M. Yamaguchi, K.I. Arai, S. Ohnuma, T. Matsumoto, IEEE Trans. Magn. **37**, 2401 (2001)
136. Y. Shimada, M. Yamaguchi, S. Ohnuma, T. Itoh, W.D. Li, S. Ikeda, K.H. Kim, H. Nagura, IEEE Trans. Magn. **39**, 3052 (2003)
137. S. Ohnuma, H. Nagura, H. Fugimori, T. Masumoto, IEEE Trans. Magn. **40**, 2712 (2004)
138. K.H. Kim, M. Yamaguchi, IEEE Trans. Magn. **40**, 2847 (2004)
139. H. Ono, T. Ito, S. Yoshida, Y. Takase, O. Hashimoto, Y. Shimada, IEEE Trans. Mang. **40**, 2853 (2004)
140. N. Matsushita, T. Nakamura, M. Abe, J. Appl. Phys. **93**, 7133 (2003)
141. K. Kondo, T. Chiba, H. Ono, S. Yoshida, Y. Shimada, N. Matsushita, M. Abe, J. Magn. Magn. Mater. **301**, 107 (2006)

142. J. Smit, H.P.J. Wijn, *Ferrites* (Philips Technical Library, Eindhoven, 1959) pp. 278–283
143. S.-S. Kim, IEEE Trans. Magn. **48**, 3490 (2012)
144. M. Matumoto, Y. Miyata, J. Appl. Phys. **79**, 5486 (1996)
145. L.P. Pereverzeva, Y.M. Povlavko, Y.S. You, *ISAF'96, Proceedings of the Tenth IEEE International Symposium on Applications of Ferroelectrics*, vol. 2 (1996), p. 841
146. S.L. Swartz, T.R. Shrout, W.A. Schulze, L.E. Cross, J. Am. Ceram. Soc. **67**, 311 (1984)
147. A.A. Bokov, Z.-G. Ye, J. Mater. Sci. **41**, 31 (2006)
148. S.-S. Kim, Y.C. Yoon, K.H. Kim, J. Electrocer. **10**, 95 (2003)
149. H.C. Wang, W.A. Schulze, J. Am. Ceram. Soc. **73**, 825 (1990)
150. R.E. Newnham, S.J. Jang, M. Xu, F. Jones, Microwaves: theory and application in materials processing, in *Ceramic Transactions*, vol. 21, ed. by D.E. Clark, F.D. Gac (The American Ceramic Society, 1991), pp. 51–67
151. Y.M. Poplavko, Y.S. You, J. Kor. Phys. Soc. **32**, S1737 (1998)
152. Y.-J. Lee, Y.-C. Yun, S.-S. Kim, Kor. J. Mater. Res. **12**, 160 (2002)
153. S.M. Abbas, A.K. Dixit, R. Chatterjee, T.C. Goel, Mater. Sci. Eng. B **123**, 167 (2005)
154. X.D. Chen, G.Q. Wang, Y.P. Duan, S.H. Liu, J. Alloys Compd. **453**, 433 (2008)
155. G.Q. Wang, X.D. Chen, Y.P. Duan, S.H. Liu, J. Alloys Compd. **454**, 340 (2008)
156. H.J. Yang, J. Yuan, Y. Li, Solid State Commun. **163**, 1 (2013)
157. Q. Li, X.W. Yin, W.Y. Duan, J. Alloys Compd. **565**, 66 (2013)
158. G.R. Fisher, P. Barnes, Philos. Mag. **61**, 217 (1990)
159. T. Muranaka, Y. Kikuchi, T. Yoshizawa, N. Shirakawa, J. Akimitsu, Sci. Technol. Adv. Mater. **9**, 044204 (2008)
160. K. Yoshida, J. Ceram. Soc. Jpn. **118**, 82 (2010)
161. R. Naslain, Compos. Sci. Technol. **64**, 155 (2004)
162. R.R. Naslain, R.J-F. Paller, J.L. Lamon, Int. J. Appl. Ceram. Tec. **7**, 263 (2010)
163. A.R. Bunsell, A. Piant, J. Mater. Sci. **41**, 823 (2006)
164. H.T. Liu, H. Tian, J. Eur. Ceram. Soc. **32**, 2505 (2012)
165. B. Zhang, J. Li, J. Sun, S. Zhang, H. Zhai, Z. Du, J. Eur. Ceram. Soc. **22**, 93 (2002)
166. H.J. Yang, J. Yuan, Y. Li, Z.L. Hou, H.B. Jin, X.Y. Fang, M.S. Cao, Solid State Commun. **163**, 1 (2013)
167. Y. Takeuchi, T. Abe, T. Kageyama, H. Sakai, in *Proceedings of 2005 Particle Accelerator Conference* (Knoxville, TN, 2005), pp. 1195–1197
168. Y.-K. Dou, J.-B. Li, X.-Y. Fang, H.-B. Jin, M.-S. Cao, Appl. Phys. Lett. **104**, 052102 (2014)
169. C. Liu, D. Yu, D.W. Kirk, Y. Xu, RSC Adv. **7**, 595–605 (2017)
170. R. Wu, K. Zhou, C.Y. Yue, J. Wei, Y. Pan, Prog. Mater. Sci. **72**, 1 (2015)
171. S.C. Chiu, H.C. Yu, Y.Y. Li, J. Phys. Chem. C **114**, 1947 (2010)
172. R.B. Wu, K. Zhou, Z.H. Yang, X.K. Qian, J. Wei, L. Liu, Cryst. Eng. Comm. **15**, 570 (2013)
173. R.G. Chaudhuri, S. Paria, Chem. Rev. **112**, 2373 (2012)
174. Y. Bhattacharjee, S. Bose, A.C.S. Appl, Nano Mater. **4**, 949 (2021)
175. X. Liu et al., Carbon **95**, 870 (2015)
176. Y.-J. Chen, G. Xiao, T.-S. Wang, Q.-Y. Ouyang, L.-H. Qi, Y. Ma, P. Gao, C.-L. Zhu, M.-S. Cao, H.-B. Jin, J. Phys. Chem. C **115**, 13603 (2011)
177. X. Meng, W. Yang, G. Han, Y.S. Yu, Ma, W. Liu, Z. Zhang, J. Magn. Magn. Mater. **502**, 166518 (2020)
178. J. Pan, X. Sun, T. Wang, Z. Zhu, Y. He, W. Xia, J. He, Appl. Surf. Sci. **457**, 271 (2018)
179. Z. Li, E. Yang, X. Qi, R. Xie, T. Jing, S. Qin, C. Deng, W. Zhong, J. Colloid Interface Sci. **565**, 227 (2020)
180. Y. Liu, Y. Lin, H. Yang, J. Alloys Compd. **805**, 130 (2019)
181. Y. Huang, W. Xing, J. Fan, J. Dai, Q. Liu, F. Hu, G. Xu, J. Magn. Magn. Mater. **502**, 166543 (2020)
182. X. Wang, T. Zhu, S. Chang, Y. Lu, W. Mi, W. Wang, A.C.S. Appl, Mater. Interfaces **12**, 11252 (2020)
183. L. Long, E. Yang, X. Qi, R. Xie, Z. Bai, S. Qin, W. Zhong, J. Mater. Chem. C **7**, 8975 (2019)

184. H. Wang, Y. Dai, W. Gong, D. Geng, S. Ma, D. Li, W. Liu, Z. Zhang, Appl. Phys. Lett. **102**, 223113 (2013)
185. P. Liu, S. Gao, Y. Wang, Y. Huang, Y. Wang, J. Luo, A.C.S. Appl, Mater. Interfaces **11**, 25624 (2019)
186. D. Guo, H. Yuan, X. Wang, C. Zhu, Y. Chen, A.C.S. Appl, Mater. Interfaces **12**, 9628 (2020)
187. J. Jiang, D. Li, D. Geng, J. An, J. He, W. Liu, Z. Zhang, Nanoscale **6**, 3967 (2014)
188. F. Zhang, W. Zhang, W. Zhu, B. Cheng, H. Qiu, S. Qi, Appl. Surf. Sci. **463**, 182 (2019)
189. R. Kuchi, T. Latif, S.W. Lee, V. Dongquoc, P.C. Van, D. Kim, J.-R. Jeong, RSC Adv. **10**, 1172 (2020)
190. X.-J. Zhang, S.-W. Wang, G.-S. Wang, Z. Li, A.-P. Guo, J.-Q. Zhu, D.-P. Liu, P.-G. Yin, RSC Adv. **7**, 22454 (2017)
191. T. Zhu, W. Shen, X. Wang, Y.-F. Song, W. Wang, Chem. Eng. J. **378**, 122159 (2019)
192. S. Dong, W. Zhang, X. Zhang, P. Hu, J. Han, Chem. Eng. J. **354**, 767 (2018)
193. B. Wei, J. Zhou, Z. Yao, A.A. Haidry, K. Qian, H. Lin, X. Guo, W. Chen, Appl. Surf. Sci. **508**, 145261 (2020)
194. J. Liu, J. Xu, R. Che, H. Chen, M. Liu, Z. Liu, Chemistry **19**, 6746 (2013)
195. M. Yu, C. Liang, M. Liu, X. Liu, K. Yuan, H. Cao, R. Che, J. Mater. Chem. C **2**, 7275 (2014)
196. R. Qiang, Y. Du, Y. Wang, N. Wang, C. Tian, J. Ma, P. Xu, X. Han, Carbon **98**, 599 (2016)
197. M. Zhang, J. Qiu, Z. Xin, X. Sun, Materials **14**, 4610 (2021)
198. W. Ma, P. He, T. Wang, J. Xu, X. Liu, Q. Zhuang, Z.-K. Cui, S. Lin, Chem. Eng. J. **420**, 129875 (2021)
199. W.L. Gindrup, US patent No. 5,786,785, (1998)
200. W.L. Gindrup. US patent No. 4,624,798, (1986)
201. S. Shukla, S. Seal, J. Akesson, R. Oder, R. Carter, Z. Rahman, Appl. Surf. Sci. **181**, 35 (2001)
202. S. Shukla, S. Seal, Z. Rahaman, K. Scammon, Mater. Lett. **57**, 151 (2002)
203. W.-J. Kim, S.-S. Kim, Appl. Surf. Sci. **329**, 219 (2015)
204. W. Wang, Q. Li, Y. Li, H. Xu, J.P. Zhai, J. Phys. D Appl. Phys. **42**, 215306 (2009)
205. S.-T. Kim, S.-S. Kim, IEEE Trans. Magn. **48**, 3494 (2012)
206. S.-S. Kim, S.T. Kim, J.M. Ahn, K.H. Kim, J. Magn. Magn. Mater. **271**, 39 (2004)
207. S.-T. Kim, S.-S. Kim, J. Alloys Compd. **687**, 22 (2016)
208. A.X. Zeng, W.H. Xiong, J. Xu, Surf. Coat. Technol. **197**, 142 (2005)
209. A.X. Zeng, W.H. Xiong, J. Xu, Mater. Lett. **59**, 524 (2005)
210. G.H. Mu, X.F. Pan, H.G. Shen, M.Y. Gu, Mater. Sci. Eng. A **4454–46**, 563 (2007)
211. A.X. Zeng, W.H. Xiong, C.F. Wang, Q.H. Zhou, J. Wuhan. Univ. Technol. Mater. Sci. **9**, 129 (2006)
212. N. Ranjbar, C. Kuenzel, Fuel **207**, 1 (2017)
213. H.P.J. Wijn, *Magnetic Properties of Metals: d-Elements, Alloys and Compounds, Part of Book Series Data Science and Technology* (Springer, Berlin, 1991), pp.22–62
214. Z.W. Liu, L.X. Phua, Y. Liu, C.K. Ong, J. Appl. Phys. **100**, 093902 (2006)
215. S.-T. Kim, S.-S. Kim, J. Appl. Phys. **115**, 17A528 (2014)
216. S. Zhou, Q. Zhang, H. Liu, X. Gong, J. Huang, Mater. Chem. Phys. **134**, 224 (2012)
217. B. Lu, X.L. Dong, H. Huang, X.F. Zhang, X.G. Zhu, J.P. Lei, J.P. Sun, J. Magn. Magn. Mater. **320**, 1106 (2008)
218. J. Wei, J. Liu, S. Li, J. Magn. Magn. Mater. **312**, 414 (2007)
219. W. Fu, S. Liu, W. Fan, H. Yang, X. Pang, J. Xu, G. Zou, J. Magn. Magn. Mater. **316**, 54 (2007)
220. G. Mu, N. Chen, X. Pan, K. Yang, M. Gu, Appl. Phys. Lett. **91**, 043110 (2007)
221. J. Wang, J. Wang, B. Zhang, Y. Sun, W. Chen, T. Wang, J. Magn. Magn. Mater. **401**, 209 (2016)
222. R.A. Stonier, Stealth aircraft and technology from world war II to the Gulf, part II: applications and design. SAMPE J. **27**, 9–17 (1991)
223. C. Wang, M. Chen, H. Lei, K. Yao, H. Li, W. Wen, D. Fang, Compos. B Eng. **123**, 19 (2017)
224. D. Munalli, G. Dimitrakis, D. Chronopoulos, S. Greedy, A. Long, Compos. B Eng. **173**, 106906 (2019)
225. S.-Y. Kim, S.-S. Kim, Polym. Polym. Compos. **26**, 105 (2018)

226. W.I. Lee, G.S. Springer, J. Comp. Mater. **18**, 357 (1984)
227. S.-S. Kim, Kor. J. Mater. Res. **8**, 161 (1998)
228. S.H. Kim, Y.G. Park, S.-S. Kim, Phys. Stat. Sol. (C) **4**, 4602 (2007)
229. S.-S. Kim, Key Eng. Mater. **858**, 140 (2020)
230. W.S. Chin, D.G. Lee, Compo. Struct. **77**, 457 (2007)
231. S.-E. Lee, J.-H. Kang, C.-G. Kim, Compo. Struct. **76**, 397 (2006)
232. P.C. Kim, D.G. Lee, Compo. Struct. **87**, 161 (2009)
233. J.-H. Oh, K.-S. Oh, C.-G. Kim, C.-S. Hong, Compos.: Part B **35**, 49 (2004)
234. K.-Y. Park, J.-H. Han, J.-B. Kim, S.-K. Lee, J. Comp. Mater. **45**, 2773 (2011)
235. A.R. Busnell, A. Poant, J. Mater. Sci. **41**, 823 (2006)
236. A.G. Simon, R. Bunsell, J. Mater. Sci. **19**, 3658 (1986)
237. M. Sugimoto, T. Shimoo, K. Okamura, T. Seguchi, J. Am. Ceram. Soc. **78**, 1849 (1995)
238. T. Ishikawa, S. Kajii, T. Hisayuki, Y. Kohtoku, Ceram. Eng. Sci. Proc. **19**, 283 (1998)
239. R. Mo, X. Yin, M. Li, F. Ye, X. Fan, L. Cheng, J. Amer. Ceram. Soc. **103**, 4352 (2020)
240. R. Mo, X.W. Yin, F. Ye, X.F. Liu, L.F. Cheng, L.T. Zhang, Ceram. Int. **43**, 8922 (2017)
241. X.F. Liu, L.T. Zhang, X.W. Yin, F. Ye, Y.S. Liu, L.F. Cheng, Ceram. Int. **43**, 3267 (2017)
242. D.H. Ding, W.C. Zhou, B. Zhang, F. Luo, D.M. Zhu, J. Mater. Sci. **46**, 2709 (2011)
243. F. Ye, L.T. Zhang, X.W. Yin, Y.S. Liu, L.F. Cheng, J. Mater. Sci. Technol. **29**, 55 (2013)
244. X.W. Yin, L. Kong, L.T. Zhang, L.F. Cheng, N. Travitzky, P. Greil, Int. Mater. Rev. **59**, 326 (2014)
245. T. Ishikawa, Compos. Sci. Technol. **51**, 135 (1994)
246. D. Kim, W.-J. Kim, J.Y. Park, Oxid. Met. **80**, 389 (2013)
247. R.R. Naslain, R.J.-F. Paller, J.L. Lamon, Int. J. Appl. Ceram. Tec. **7**, 263 (2010)
248. P. Wang, F. Liu, H. Wang, H. Li, Y. Gou, J. Mater. Sci. Tech. **35**, 2743 (2019)
249. R.J. Kerans, R.S. Hay, T.A. Parthasarathy, M.K. Cinibulk, J. Am. Ceram. Soc. **85**, 2599 (2002)
250. Y. Mu, W.C. Zhou, F. Luo, Ceram. Int. **40**, 3411 (2014)
251. H.T. Liu, H.F. Cheng, J. Wang, J. Alloys Compd. **491**, 248 (2010)
252. H. Tian, H.T. Liu, H.F. Cheng, Ceram. Int. **40**, 9009 (2014)
253. D. Ding, Y. Shi, Z. Wu, W. Zhou, F. Luo, J. Chen, Carbon **60**, 552 (2013)
254. Y. Mu, Y. Mu, W. Zhou, Y. Hu, H. Wang, F. Luo, D. Ding, Y. Qing, J. Europ. Ceram. Soc. **35**, 2991 (2015)
255. T. Hao, H.-T. Liu, H.-F. Cheng, Comp. Sci. Tech. **90**, 202 (2014)
256. W.J. Padilla, D.N. Basov, D.R. Smith, Mater. Today **9**, 27 (2006)
257. Y. Liu, X. Zhang, Chem. Soc. Rev. **40**, 2494 (2011)
258. V.G. Veselago, Sov. Phys. Usp. **10**, 509 (1968)
259. J.B. Pendry, A.J. Holden, W.J. Stewart, I. Youngs, Phys. Rev. Lett. **76**, 4773 (1996)
260. J.B. Pendry, A.J. Holden, D.J. Robbins, W.J. Stewart, IEEE Trans. Microw. Theory Tech. **47**, 2075 (1999)
261. D.R. Smith, W.J. Padilla, D.C. Vier, S.C. Nemat-Nasser, S. Schultz, Phys. Rev. Lett. **84**, 4184 (2000)
262. R.A. Shelby, D.R. Smith, S. Schultz, Science **292**, 77 (2001)
263. M. Bayindir, K. Aydin, E. Ozbay, P. Markoš, C.M. Soukoulis, Appl. Phys. Lett. **81**, 120 (2002)
264. I.V. Shadrivov, D.A. Powell, S.K. Morrison, Y.S. Kivshar, G.N. Milford, Appl. Phys. Lett. **90**, 201919 (2007)
265. T. Weiland, R. Schuhmann, R.B. Greegor, C.G. Parazzoli, A.M. Vetter, D.R. Smith, D.C. Vier, S. Schultz, J. Appl. Phys. **90**, 5419 (2001)
266. C. Caloz, Mater. Today **12**, 12 (2009)
267. J.B. Pendry, Phys. Rev. Lett. **85**, 3966 (2000)
268. A. Grbic, G.V. Eleftheriades, Phys. Rev. Lett. **92**, 117403 (2004)
269. N. Fang, H. Lee, C. Sun, X. Zhang, Science **308**, 534 (2005)
270. X. Zhang, Z. Liu, Nat. Mater. **7**, 435 (2008)
271. J.B. Pendry, D. Schurig, D.R. Smith, Science **312**, 1780 (2006)
272. D. Schurig, J.J. Mock, B.J. Justice, S.A. Cummer, J.B. Pendry, A.F. Starr, D.R. Smith, Science **314**, 977 (2006)

273. N.A. Zharova, I.V. Shadrivov, Y.S. Kivshar, Opt. Express **16**, 4615 (2008)
274. S. Zhang et al., Phys. Rev. Lett. **95**, 137404 (2005)
275. J. Zhou, T. Koschny, M. Kafesaki, E.N. Economou, J.B. Pendry, C.M. Soukoulis, Phys. Rev. Lett. **95**, 223902 (2005)
276. A.N. Grigorenko, A.K. Geim, H.F. Gleeson, Y. Zhang, A.A. Firsov, I.Y. Khrushchev, J. Petrovic, Nature **438**, 335 (2005)
277. C. Caloz, A. Sanada, T. Itoh, IEEE Trans. Microw. Theory Tech. **52**, 980 (2004)
278. M.C.K. Wiltshire, J.B. Pendry, I.R. Young, D.J. Larkman, D.J. Gilderdale, J.V. Hajnal, Science **291**, 849 (2001)
279. N.I. Landy, S. Sajuyigbe, J.J. Mock, D.R. Smith, W.J. Padilla, Phys. Rev. Lett. **100**, 207402 (2008)
280. C. Hu, X. Li, Q. Feng, X. Chen, X. Luo, Opt. Express **18**, 6598 (2010)
281. R.F. Huang, Z.W. Li, L.B. Kong, L. Liu, S. Matitsine, Prog. In Electromag. Res. B **14**, 407 (2009)
282. V.D. Lam, J.B. Kim, S.J. Lee, Y.P. Lee, Opt. Express **15**, 16651 (2007)
283. C.M. Watts, X. Liu, W.J. Padilla, Adv. Mat. **24**, OP98 (2012)
284. F. Costa, S. Gonovesi, A. Monorchio, IEEE Trans. Antennas Propag. **61**, 1201 (2013)
285. R.L. Fante, M.T. McCormack, IEEE Trans. Antennas Propag. **36**, 1443 (1988)
286. E.F. Knott, J.F. Shaeffer, M.T. Tuley, *Radar Cross Section*, 2nd edn. (Artech House, Boston, 1993), pp.240–247
287. Y.I. Abdulkarim, A. Mohanty, O.P. Acharya, B. Appasani, M.S. Khan, S.K. Mohapatra, F.F. Muhammadsharif, J. Dong, Front. Phys. **10**, 893791 (2022)
288. W. Xin, Z. Binzhen, W. Wanjun, W. Junlin, D. Junping, IEEE Photon. J. **9**, 1 (2017)
289. P. Garg, P. Jain, AEU-Int. J. Electron. Commun. **116**, 153063 (2020)
290. X.-W. Dai, S.-Y. Lin, S.-W. Mao, G.O. Luo, AEU–Int. J. Electron. Commun. **119**, 153179 (2020)
291. M. Bakır, M. Karaaslan, E. Unal, O. Akgol, C. Sabah, Opto-Electronics Rev. **25**, 318 (2017)
292. J.W. Park et al., Opt. Express **21**, 9691 (2013)
293. H.Y. Zheng et al., Opt. Express **20**, 24002 (2012)
294. S. Ghosh, S. Bhattacharyya, Y. Kaiprath, J. Appl. Phys. **115**, 104503 (2014)
295. N. Wang et al., IEEE Photon. J. **7**, 1 (2015)
296. H. Tao, N.I. Landy, C.M. Bingham, X. Zhang, R.D. Averitt, W.J. Padilla, Opt. Express **16**, 7181 (2008)
297. H. Tao, C.M. Bingham, A.C. Strikwerda, D. Pilon, D. Shrekenhamer, N.I. Landy, K. Fan, X. Zhang, W.J. Padilla, R.D. Averitt, Phys. Rev. B **78**, 241103(R) (2008)
298. A.K. Azad, A.J. Taylor, E. Smirnova, J.F. O'Hara, Appl. Phys. Lett. **92**, 011119 (2008)
299. N.I. Landy, C.M. Bingham, T. Tyler, N. Jokerst, D.R. Smith, W.J. Padilla, Phys. Rev. B **79**, 125104 (2009)
300. H. Zhou, F. Ding, Y. Jin, S.L. He, Prog. In Electromag. Res. **119**, 449 (2011)
301. K. Iwaszczuk, A.C. Strikwerda, K. Fan, X. Zhang, R.D. Averitt, P.U. Jepsen, Opt. Express **20**, 635 (2012)
302. C.M. Bingham, H. Tao, X. Liu, R.D. Averitt, X. Zhang, W.J. Padilla, Opt. Express **16**, 18565 (2008)
303. P. Singh, S.K. Amer, L. Chao, M.N. Afsar, S. Sonkusale, Prog. In Electromagn. Res. **142**, 625 (2013)
304. X. Liu, T. Starr, A.F. Starr, W.J. Padilla, Phys. Rev. Lett. **104**, 207403 (2010)
305. J. Hao, J. Wang, X. Liu, W.J. Padilla, L. Zhou, M. Qiu, Appl. Phys. Lett. **96**, 251104 (2010)
306. N. Liu, M. Mesch, T. Weiss, M. Hentschel, H. Giessen, Nano Lett. **10**, 2342 (2010)
307. J. Yang, C. Xu, S. Qu, H. Ma, J. Wang, Y. Pang, J. Adv. Dielect. **08**, 1850007 (2018)
308. J.A. Bossard, L. Lin, S. Yun, L. Liu, D.H. Werner, T.S. Mayer, ACS Nano **8**, 1517 (2014)
309. J. Grant, I.J.H. McCrindle, D.R.S. Cumming, Opt. Express **24**, 3451 (2016)
310. Y.L. Liao, Y. Zhao, Sci. Rep. **10**, 1480 (2020)
311. Y. Wang et al., Results Phys. **16**, 102951 (2020)
312. F. Qin et al., Nanomaterials **10**, 207 (2020)

313. Z. Li, L. Stan, D.A. Czaplewski, X. Yang, J. Gao, Opt. Express **26**, 5616 (2018)
314. Y.-H. Ryu, S.-S. Kim, Kor. J. Met. Mater. **58**, 131 (2020)
315. Y. Cheng, H. Yang, J. Appl. Phys. **108**, 034906 (2010)
316. Z.H. Jiang, S. Yun, F. Toor, D.H. Werner, T.S. Mayer, ACS Nano **5**, 4641 (2011)

Chapter 7
Measurement

In this last chapter, the methods and principles for measuring the high-frequency properties (complex permittivity ε_r and permeability μ_r) and EM wave absorption are described. Explanation is given for the conversion principle from measured *S* parameters to ε_r and μ_r, with the attachment of associated computer program in Appendix. Two commonly used methods, waveguide and free-space methods, are described.

7.1 Introduction

One of the important points in development of EM wave absorbers is the precise measurement of material parameters (complex permittivity ε_r, complex permeability μ_r) and reflection loss in microwave frequencies. There are two commonly used methods for measuring the material parameters. The first method is to use waveguides such as a coaxial airline or a rectangular waveguide. From the reflection and transmission coefficients measured in the transmission line partially filled with a material, ε_r and μ_r can be derived using the theoretical equations [1, 2]. The measurement frequency range is very wide, from 500 MHz to 110 GHz. It has the advantages that the measuring fixture is easy to install and measurement is possible even in gaseous or liquid state. On the other hand, it is difficult to measure precisely for low-loss materials and waveguides cannot be used for low-frequency measurements below 500 MHz [3, 4]. The second method is free-space test. It is a method of using free space as a transmission line, placing a test sample in space and obtaining ε_r and μ_r from the measured reflection and transmission coefficients. This method allows the measurement of material parameters in wideband frequency range. Since the test sample is installed in free space, it does not need to be machined in the form of a measurement fixture, and it can be measured in the state of as-received sheet or plate.

S. Kim, *Fundamentals and Principles of Electromagnetic Wave Absorbers*, Springer Series in Advanced Microelectronics 66, https://doi.org/10.1007/978-981-97-6477-8_7

7.2 Waveguide Method

7.2.1 S Parameters

Consider a transmission line partially filled with a sample of thickness d and material parameters ε_r and μ_r, as shown in Fig. 7.1. Due to the difference of characteristic impedance between air Z_0 and sample Z_1, wave reflection occurs at both faces of the sample. At the front surface, the reflection coefficient Γ is given by

$$\Gamma = \frac{Z_1 - Z_0}{Z_1 + Z_0} \tag{7.1}$$

and, at the back side, the reflection coefficient becomes

$$-\Gamma = \frac{Z_0 - Z_1}{Z_0 + Z_1} \tag{7.2}$$

For the incident wave of field V_0, the muti-reflection inside the sample yields the total field of reflected waves R and the total field of the transmitted waves T, expressed as

$$R = \Gamma V_0 - \Gamma\left(1 - \Gamma^2\right)V_0 e^{-2\gamma d} - \Gamma^3\left(1 - \Gamma^2\right)V_0 e^{-4\gamma d} - \cdots$$

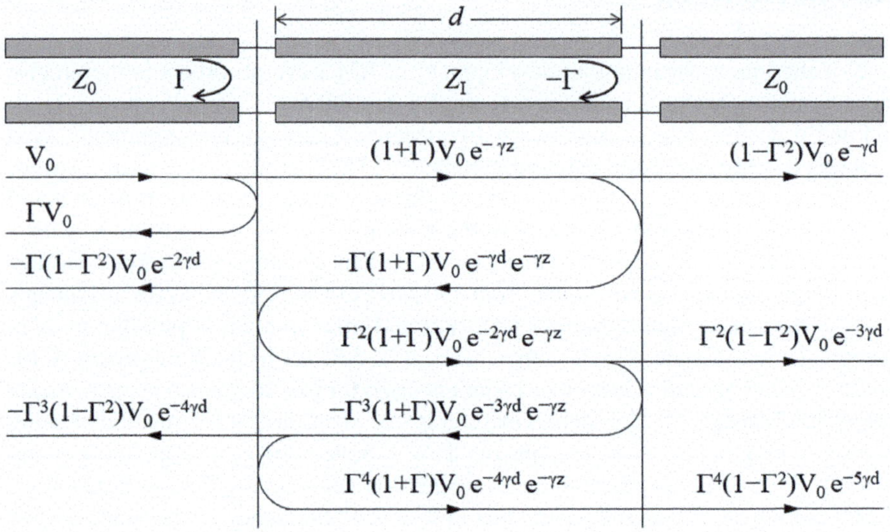

Fig. 7.1 Reflection and transmission of EM waves in transmission line filled with a sample of thickness d

$$= \frac{V_0\Gamma\left(1 - e^{-2\gamma d}\right)}{1 - \Gamma^2 e^{-2\gamma d}} \tag{7.3}$$

$$T = \left(1 - \Gamma^2\right)V_0 e^{-\gamma d} + \Gamma^2\left(1 - \Gamma^2\right)V_0 e^{-3\gamma d} + \Gamma^4\left(1 - \Gamma^2\right)V_0 e^{-5\gamma d} + \cdots$$
$$= \frac{\left(1 - \Gamma^2\right)V_0 e^{-2\gamma d}}{1 - e^{-2\gamma d}} \tag{7.4}$$

where γ is the propagation constant of sample medium.

For 2-port network, the S parameters indicating the reflection and transmission are given by

$$S_{11} = \frac{R}{V_0} = \frac{\Gamma\left(1 - e^{-2\gamma d}\right)}{1 - \Gamma^2 e^{-2\gamma d}} \tag{7.5}$$

$$S_{21} = \frac{T}{V_0} = \frac{\left(1 - \Gamma^2\right)e^{-2\gamma d}}{1 - e^{-2\gamma d}} \tag{7.6}$$

If we define $P = e^{-\gamma d}$, Γ and P can be expressed as a function of measured parameters S_{11} and S_{21}:

$$\Gamma = \frac{-\left(S_{21}^2 - S_{11}^2 - 1\right) \pm \sqrt{\left(S_{21}^2 - S_{11}^2 - 1\right)^2 - 4S_{11}^2}}{2S_{11}} \tag{7.7}$$

$$P = \frac{S_{11} + S_{21} - \Gamma}{1 - \Gamma(S_{11} + S_{21})} \tag{7.8}$$

Using the relations $\Gamma = (Z_1 - Z_0)/(Z_1 + Z_0) = \left(\sqrt{\mu_r/\varepsilon_r} - 1\right)/\left(\sqrt{\mu_r/\varepsilon_r} + 1\right)$ and $P = \exp\left[-j(\omega/c)\sqrt{\mu_r\varepsilon_r}d\right]$ where ω is the angular frequency and c is the velocity of light, we get

$$\frac{\mu_r}{\varepsilon_r} = \left(\frac{1 + \Gamma}{1 - \Gamma}\right)^2 \tag{7.9}$$

$$\mu_r\varepsilon_r = -\left[\frac{c}{\omega d}\ln\left(\frac{1}{P}\right)\right]^2 \tag{7.10}$$

When using a rectangular waveguide, μ_r and ε_r are given by

$$\mu_r = \frac{1 + \Gamma}{1 - \Gamma}\frac{1}{\Lambda}\sqrt{\frac{\lambda_0^2\lambda_c^2}{\lambda_0^2 - \lambda_c^2}} \tag{7.11}$$

$$\varepsilon_r = \frac{\lambda_0^2}{\mu_r}\left(\frac{1}{\Lambda^2} + \frac{1}{\lambda_c^2}\right) \tag{7.12}$$

where

$$\frac{1}{\Lambda^2} = -\left[\frac{1}{2\pi d}\ln\left(\frac{1}{P}\right)\right]^2 \tag{7.13}$$

and λ_0: wavelength in free space, λ_c: wavelength at cutoff frequency in waveguide. For the coaxial airline with TEM propagation mode, $\lambda_c = \infty$. The material parameters ε_r and μ_r can be computed from the measured S_{11} and S_{21}. An example of the calculation of ε_r and μ_r from the *S* parameters is shown in Appendix B and a MATLAB conversion program is presented in Appendix C.

Rectangular waveguides and coaxial airlines are commonly used for *S*-parameters measurement (Fig. 7.2). Coaxial airlines cover a broad frequency range (DC to 18 GHz for 7 mm airline and DC to 26.5 GHz for 3.5 mm airline), and the frequency coverage is banded for rectangular waveguides. Table 7.1 shows the commonly used rectangular waveguides and their inner size, frequency range used, and cutoff frequencies. Each waveguide has a cutoff frequency and will not pass the waves of greater wavelength. Since this method requires the sample in the form of a waveguide, the gap between waveguide wall and sample causes the measurement error, which is more pronounced in high-permittivity materials. Therefore, the precise machining is required to minimize the random errors.

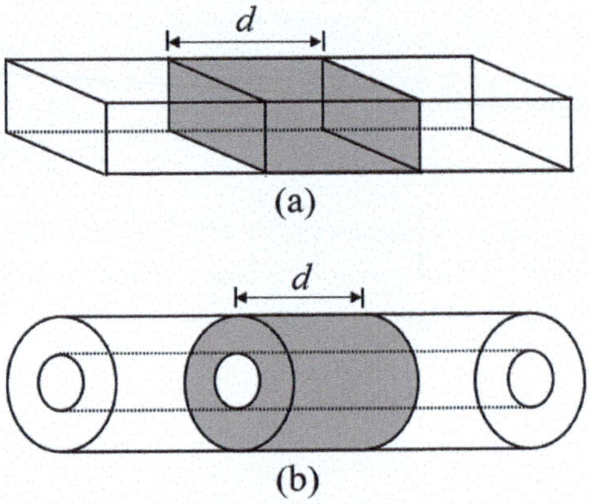

Fig. 7.2 Transmission lines filled with a sample of thickness *d*: **a** rectangular waveguide and **b** coaxial airline

Table 7.1 Some commonly used rectangular waveguides and their characteristics

Types	Size (mm^2)	Frequency range (GHz)	Cutoff frequency (GHz)
WR-137	34.85 × 15.8	5.85–8.2	4.304
WR-90	22.86 × 10.16	8.2–12.4	6.562
WR-62	15.8 × 7.9	12.4–18	9.494
WR-42	10.7 × 4.3	18–26.5	14.058
WR-28	7.11 × 3.56	26.5–40	21.097

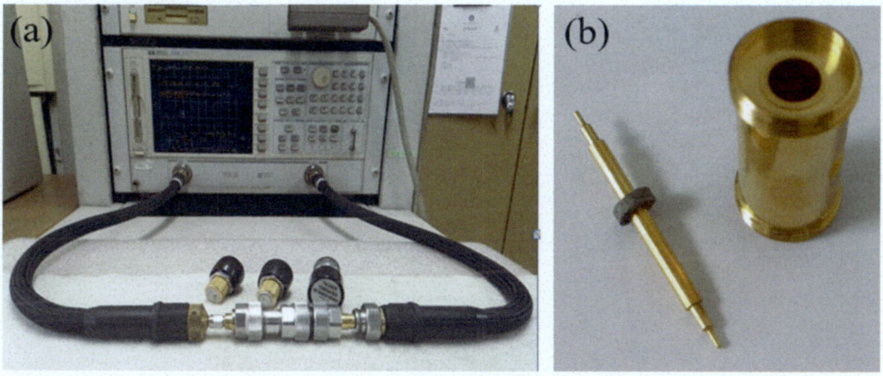

Fig. 7.3 A measurement system consisting of **a** vector network analyzer and **b** coaxial airline sample holder

A measurement system using the waveguide method consists of a vector network analyzer (VNA), a coaxial airline or waveguide section, software to calculate ε_r and μ_r, and an external computer. Figure 7.3 shows a measurement setup with a coaxial airline sample holder. A vector network analyzer consists of a signal source, a receiver, and a display. For calibration of the VNA with waveguides, which removes systematic errors from the measurement system, the TRL (Thru-Reflect-Line) method is used. After calibration, S_{11} and S_{22} are lower than − 50 dB and S_{21} are close to 0 dB. After connecting with sample holder, S_{11} must be less than − 30 dB and S_{21} must be within − 0.1 dB. The measured data includes the magnitude and phase of S parameters at discrete frequencies over the measurement range. A procedure for performing the measurement of material constants using VNA and waveguide is given in Appendix D.

7.2.2 Short-Circuited Line Method

In contrast to the 2-port microwave network method, a short-circuited line method is used for the measurement of permittivity of non-magnetic materials ($\mu_r = 1$). Figure 7.4 shows the wave reflection in 1-port measurement method in which the back side of sample is shorted and then S_{11} is only measured. A test sample of thickness d is placed near the shorted end of transmission line. For the incident wave of field V_0, the multi-reflection inside the sample yields the total field of reflected waves R, expressed as

$$R = \Gamma V_0 - \left(1 - \Gamma^2\right)V_0 e^{-2\gamma d} - \Gamma\left(1 - \Gamma^2\right)V_0 e^{-4\gamma d} - \cdots \quad (7.14)$$

Then S_{11} is given by

Fig. 7.4 Reflection of EM waves in short-circuited transmission line filled with a sample of thickness *d*

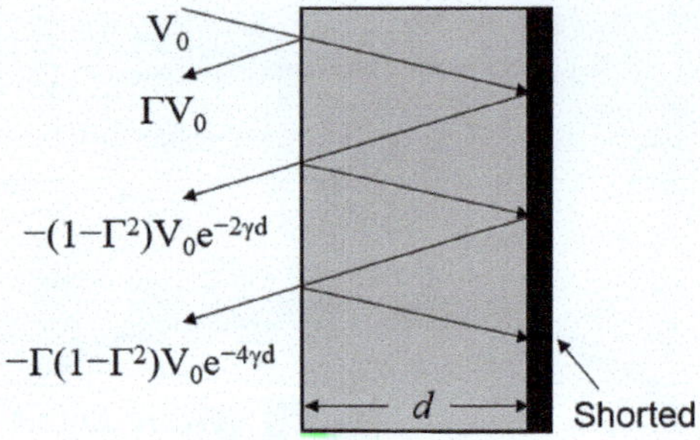

$$S_{11} = \frac{R}{V_0} = \Gamma - \frac{(1 - \Gamma^2)e^{-2\gamma d}}{1 - \Gamma e^{-2\gamma d}} \tag{7.15}$$

Following the procedure described earlier, we can determine the complex permittivity [5]:

$$\varepsilon_r = \left(\frac{1 - \Gamma}{1 + \Gamma}\right)^2 \tag{7.16}$$

7.3 Free-Space Method

Free space can be treated as one of transmission line. The reflection and transmission coefficients measured by antennas can be used to obtain the complex permittivity and complex permeability of a material. In the free-space measurement, several methods are proposed to obtain the material information under test. The first is two-samples method where the material parameters can be obtained from the reflection coefficients measured for two thicknesses *d* and 2*d*. The second is two-terminations method measuring the reflective coefficients under two different termination conditions (for example, open-circuited and short-circuited). The third is *S* parameters method, which uses the reflection and transmission coefficients of measurement sample, as explained in Sect. 7.2.1. In the methods above, it is possible to calculate the complex permittivity and permeability from two measured parameters with changing the thickness, termination conditions, or *S*-parameters.

When measuring the reflection coefficient in free space, the measurement errors must be eliminated from the coupling between transmission antenna and receiving antenna and the unnecessary waves from the periphery of measuring sample and the surrounding walls. For this demand, it is recommended to measure in an anechoic chamber and install EM wave absorbers in vicinity of samples. For frequencies above X-band, antenna with good directivity is recommended. The size of test sample is large enough to avoid the diffraction at the sample edges, especially when measuring at low frequencies. For example, a test sample in the form of a square plate must

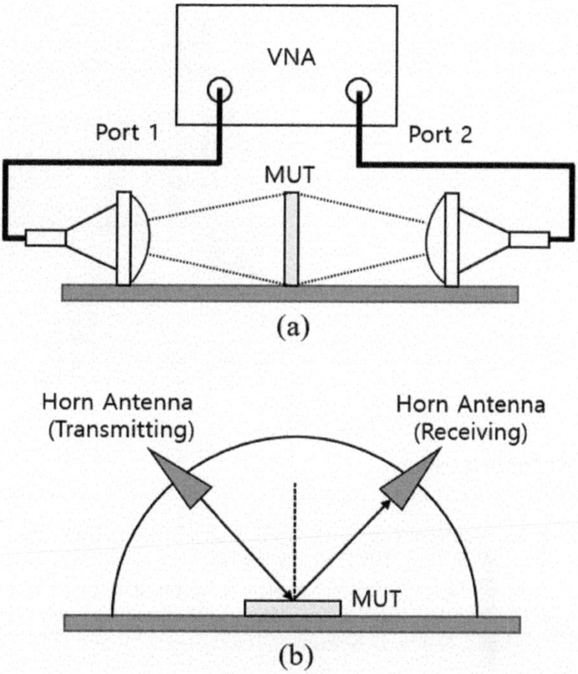

Fig. 7.5 Free-space method: **a** *S*-parameter configuration and **b** arch test

have at least one side length of measurement wavelength. The spot-focusing antenna is effective for minimizing the diffraction effect and reducing the sample size [3].

The free-space method is a non-contacting and non-destructive method and can be applied to materials to be tested under high temperatures. This method has two typical measurement setups: an *S*-parameter configuration with a beam focusing or not, and the NRL arch, as shown in Fig. 7.5. The measurement principle is the same as the transmission line method, which replaces the waveguide with free space using antennas. Figure 7.6 shows the free-space measurement setup for reflection loss in anechoic chamber.

Fig. 7.6 Free-space measurement setup for reflection loss in anechoic chamber

References

1. A.M. Nicolson, G.F. Ross, IEEE Trans. Instrum. Meas. **19**, 377 (1970)
2. W.B. Weir, Proc. IEEE **62**, 33 (1974)
3. D.K. Ghodgaonkar, V.V. Varadan, V.K. Varadan, IEEE Trans. Instrum. Meas. **37**, 789 (1989)
4. C.C. Courtney, IEEE Trans. Microw. Theory Techn. **46**, 517 (1998)
5. D. Houtz, D. Gu, D.K. Walker, IEEE Trans. Microw. Theory Techn. **64**, 3820 (2016)

Appendix A
HFSS Simulation of Metamaterial Absorbers

A design process is presented by using HFSS simulation tool for a sample of metamaterial absorber composed of cross dipoles on the grounded FR4 substrate. 3D design modeling, creating boundaries and port excitation, analysis setup, and results display (*S* parameter, *E* field, and current distribution) are sequentially included.

1. **Design Review**
   The metamaterial absorber (MA) is composed of cross strip (CS) on the top layer and ground plane (GP) on the bottom layer separated by dielectric substrate, as shown in Fig. A.1a. The structure is illuminated normally by TEM wave in z-axis with electric field (E) and magnetic field (H) polarized in x-axis and y-axis, respectively. The dimension of unit cell is shown in Fig. A.1b where $p = 2.0$ mm, $d = 1.0$ mm, and $w = 0.1$ mm. The substrate material is FR4 with dielectric permittivity $\varepsilon_r = 4.1 - j0.03$ and its thickness is $t_{FR4} = 0.2$ mm. The metal on the top and bottom layer is copper with electrical conductivity $\sigma = 5.8 \times 10^7$ S/m and thickness $t_m = 0.35$ μm. Frequency range of simulation is 75–80 GHz.
2. **Ansoft Desktop Terms**
   The HFSS Desktop has several optional panels, which are shown in Fig. A.2.

   - A ***Project Manager*** which contains a design tree which lists the structure of the project.
   - A ***Message Manager*** that displays any errors or warnings before a simulation.
   - A ***Property Window*** that displays the model parameters or attributes.
   - A ***Progress Window*** that displays the solution progress.
   - A ***3D Modeler Window*** which contains the model and model tree for the active design.

S. Kim, *Fundamentals and Principles of Electromagnetic Wave Absorbers*, Springer Series in Advanced Microelectronics 66, https://doi.org/10.1007/978-981-97-6477-8

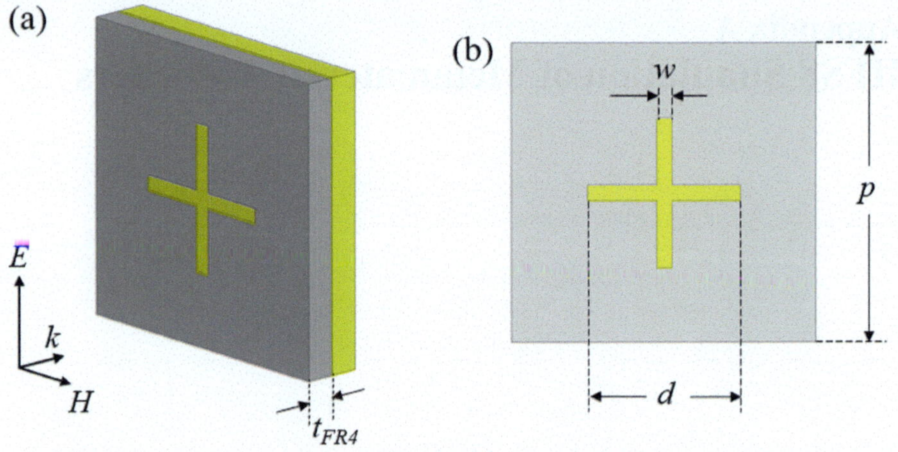

Fig. A.1 Metamaterial absorbers of a cross dipole: **a** schematic structure, **b** dimension of unit cell

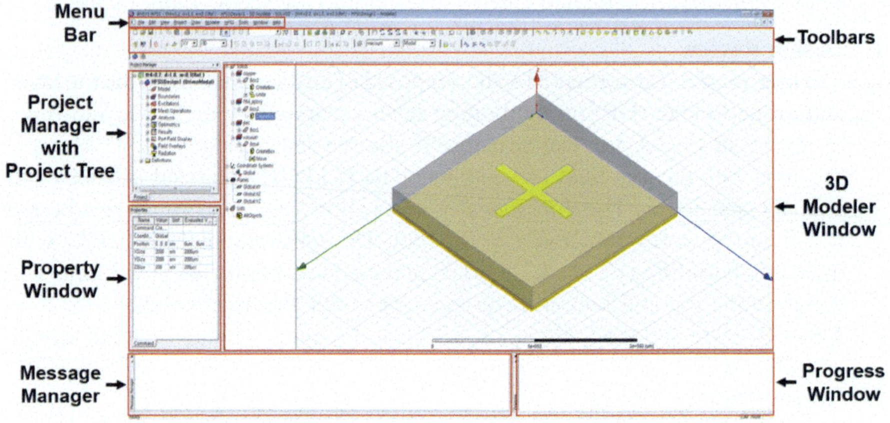

Fig. A.2 HFSS Desktop optional panels

3. **Getting Started**
   To access HFSS, select the menu item ***Start > Program > ANSYS Electromagnetics 15.0***.
4. **Opening a New Project**
   To open a new project, select the menu item ***File > New***.
   From the Project menu, select Insert HFSS Design.
5. **Set Solution Type**
   To set the solution type, select the menu item ***HFSS > Solution Type***.
   In Solution Type Window, choose ***Driven Terminal***, and click the OK button.

6. **Creating the 3D Model**

   6.1. Set the units

   - To set the units, select the menu item ***Modeler > Units Modeler > Units***.
   - Select units in μm and click the OK button.

   6.2. Create the ground plane

   - Select the menu item ***Draw > Box***.
   - Using the coordinate entry fields, enter the box position.
     X: 0, *Y*: 0, *Z*: 0,
     Dimension: *X*size: 2000 μm, *Y*size: 2000 μm, *Z*size: – 40 μm.
     Here, to easily edit the FSS units used in the analysis, it is recommended that the ground plane be designed in the negative direction of the *z*-axis.
   - Select the menu item ***Properties > Box > Material*** and select a material copper (Cu).

   6.3. Create the substrate

   - Select the menu item ***Draw > Box***.
   - Using the coordinate entry fields, enter the box position and enter the dimension of the box.
     Position: *X*: 0, *Y*: 0, *Z*: 0,
     Dimension: *X*size: 2000 μm, *Y*size: 2000 μm, *Z*size: 200 μm.
   - Select the menu item ***Properties > Box > Material*** and add a new material 'FR4' and enter its dielectric properties ($\varepsilon_r = 4.1$, loss tangent = 0.007).

   6.4. Create the cross diploe on FR4

   - Select the menu item ***Draw > Box***.
   - Using the coordinate entry fields, create the cross dipole.
   - Select the menu item ***Properties > Box > Material*** and select a material copper (Cu).

   6.5. Complete the model design (Fig. A.3)

7. **Creating 3D Air Box**

   7.1. Create the 3D air box for port excitation

   - Select the menu item ***Draw > Box***.
   - Using the coordinate entry fields, enter the box position.
     Position: *X*: 0, *Y*: 0, *Z*: – 750 μm,
     Dimension: *X*size: 2000 μm, *Y*size: 2000 μm, *Z*size: 1500 μm.
     The height of air box in *z*-axis should be adjusted to be greater than the size of the FSS (approximately 10 ~ 15 times the size of the FSS).
   - Select the menu item ***Properties > Box > Material*** and enter material (vacuum) with transparency (0.8).

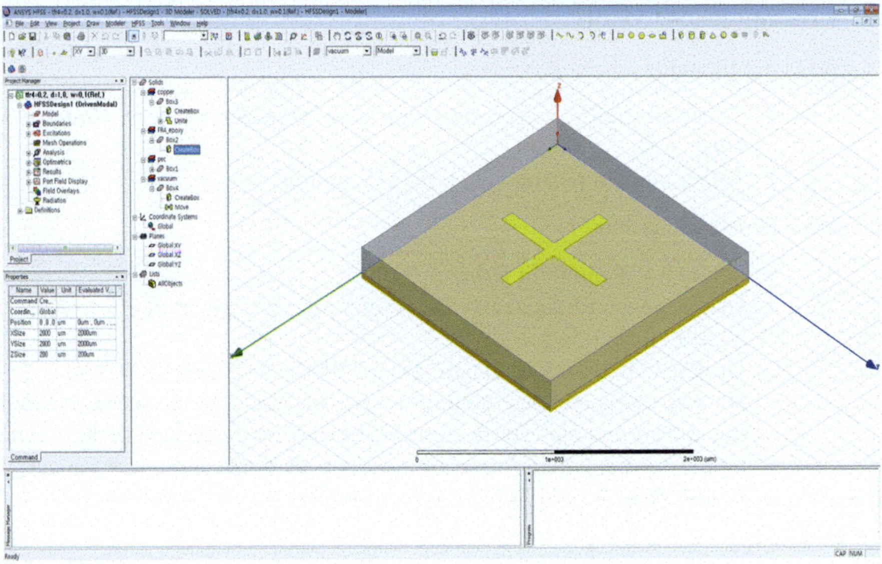

Fig. A.3 3D model for a metamaterial absorber (cross dipole/FR4/ground plane)

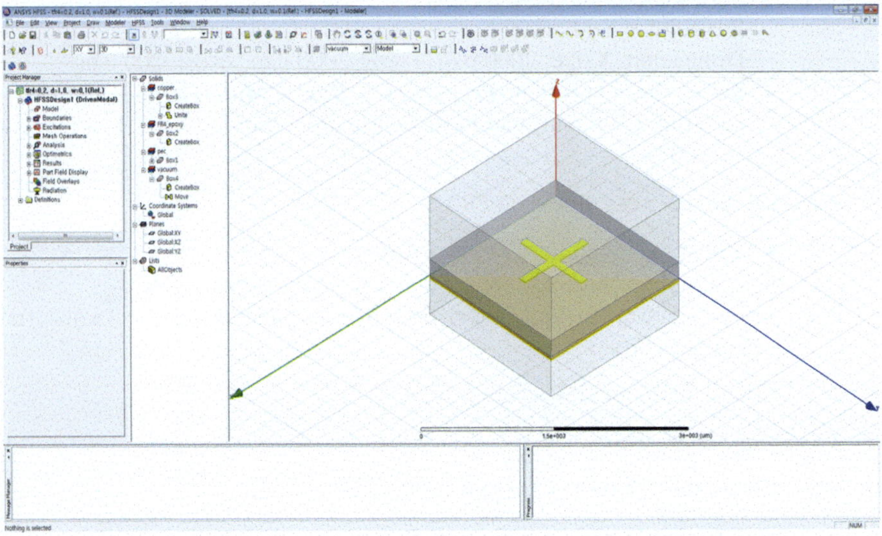

Fig. A.4 3D air box design

7.2. Complete the air box design (Fig. A.4)

8. **Creating Boundaries and Excitation**

When modeling the infinitely large repeating array structures such as metamaterials, Master-Slave boundary condition is useful instead of the PEC/PMC boundary condition. Master-Slave boundaries must exist in pairs on opposite sides. The Floquet port can only be granted under the condition of the Master-Slave boundary.

8.1. Create the first Master boundary

- To select the Master boundary face, select the menu item ***Edit > Select > Faces***.
  In the case of the existing object, select the menu item ***Edit > Select > By Name***.
- To create the Master boundary, select the menu item ***HFSS > Boundaries > Assign > Master***.
- Select the face to which we want to grant the Master boundary.
- Complete Master boundary dialog window name: Master1.

8.2. In a same way, create the second Master boundary in *y*-axis of the air box: Master2

8.3. Create the first Slave boundary

- To create the Slave boundary, select the menu item ***HFSS > Boundaries > Assign > Slave***.
- Select the symmetrical face facing the face that we specified as the Master boundary as the Slave boundary.
- Complete the General Data Tab name: Slave1.

8.4. In a same way, create the second Slave boundary in *y*-axis of the air box: Slave2
Figure A.5 shows the assignment of Master–Slave boundaries.

8.5. Create the Floquet port

- To select the Floquet port face, select the menu item ***Edit > Select > Faces***.
- To create Floquet port: select the menu item ***HFSS > Excitation > Assign > Floquet Port***.
  Figure A.6 shows the assignment of Floquet port.

9. **Creating an Analysis Setup**

- To create an Analysis Setup: select the menu item ***HFSS > Analysis Setup > Add Solution Setup***.
- Solution Setup Window: click the General tab
  - Solution frequency: 77 GHz
  - Maximum number of passes: 10
  - Maximum Delta S: 0.01.

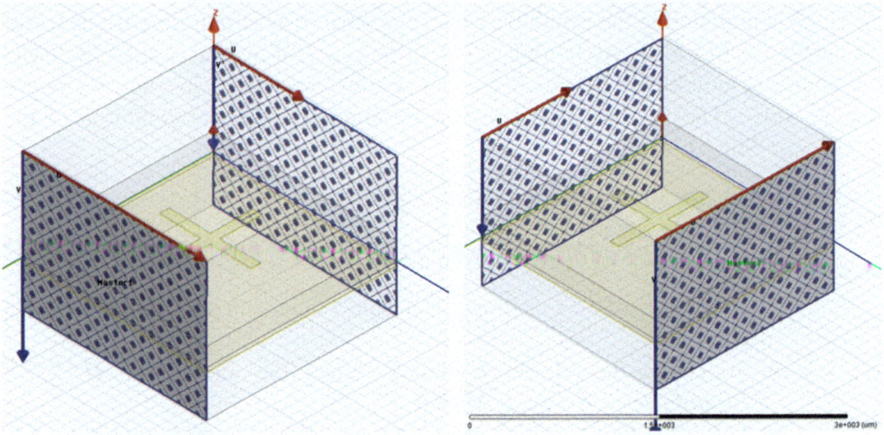

(a) (b)

Fig. A.5 Boundary conditions in HFSS: **a** Master1-Slave1, **b** Master2-Slave2

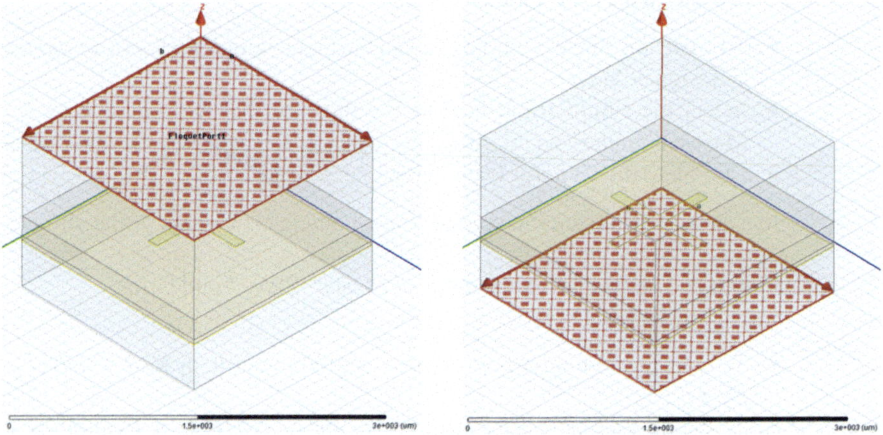

(a) (b)

Fig. A.6 Floquet port excitation: **a** port 1, **b** port 2

- Click the OK button.

10. **Creating a Frequency Sweep**

- To create an Analysis Setup, select the menu item ***HFSS > Analysis Setup > Add Frequency Sweep***.
- Edit Sweep Window

- Sweep type: Interpolating
- Frequency setup type: Linear Step
- Start: 75 GHz
- Stop: 80 GHz
- Step: 0.01 GHz.

- Click the OK button.

11. **Analyze**
To start the solution process, select the menu item ***HFSS > Analyze All***.
We can see the solution progress in ***Progress Window***.
12. **Post-processing**
Solution data are *S* parameter, *E* field, and current distributions.

12.1. *S* parameter plot

- To create a report, select the menu item ***HFSS > Results > Create Terminal Solution Data Report > Rectangular Plot***.
- New Report Window:
  - Solution: **Setup1: Sweep1**
  - Domain: **Sweep**
  - Category: **Terminal *S* Parameter**
  - Quantity: S_{11}
  - Function: **dB**
  - Click **New Report** button
  - Display and view graph (Fig. A.7)
  - Click **Close** button.

12.2. Field plot (*E* field)

- To create an *E* field plot, select the menu item ***HFSS > Results > Fields > E > Mag_E***.
- Select frequency and phase, check quantity and click [done].
- Display and view graph (Fig. A.8).

12.3. Field plot (current density)

- To create a current density, select the menu item ***HFSS > Results > Fields > E > Vector_E***.
- Select frequency and phase, check quantity and click [done].
- Display and view graph (Fig. A.9).

To analyze the electric field strength and the current distribution at the same time, select [***Complex Mag_E***] and analyze it.

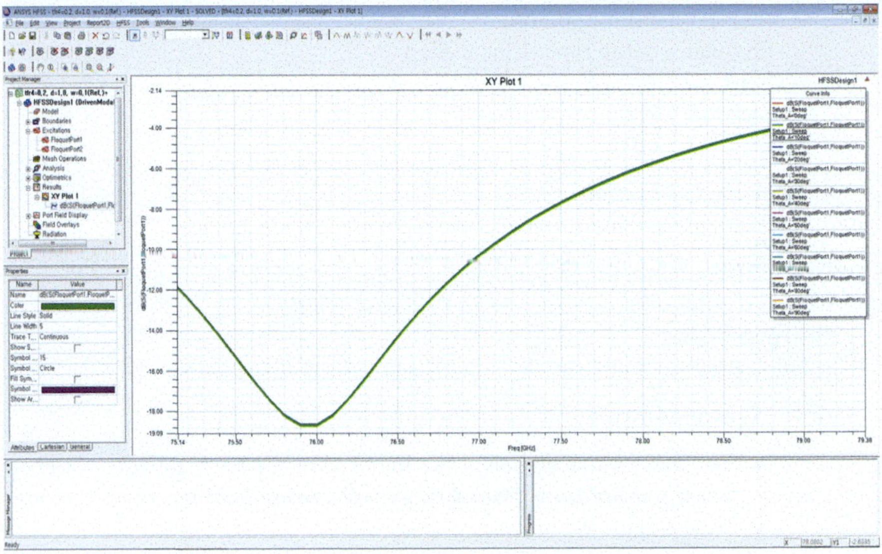

Fig. A.7 S_{11} parameter result

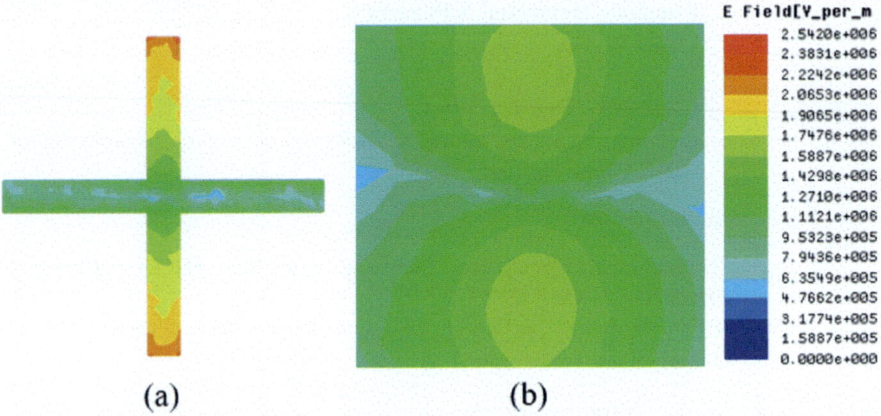

Fig. A.8 *E* field distribution in **a** cross dipole and **b** ground plane

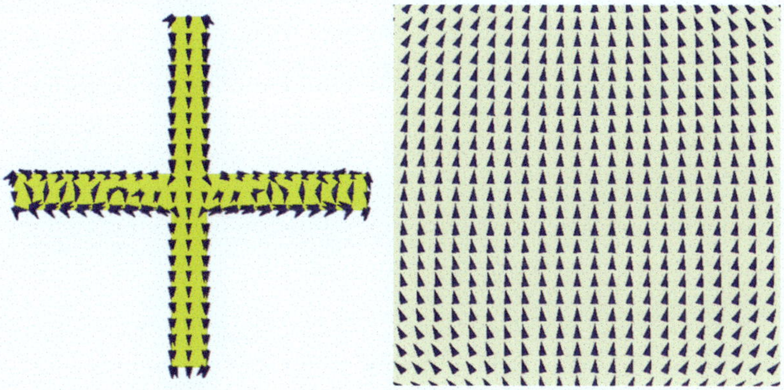

(a) (b)

Fig. A.9 Current density distribution in **a** cross dipole and **b** ground plane

Appendix B
Calculation Example of μ_r and ε_r from S Parameters

Following is an example for the calculation of complex permeability μ_r and permittivity ε_r from the measured S parameters, which are given by

$$S_{11}(\omega) = 0.652\angle 170^\circ = -0.6421 + j0.1132$$

$$S_{21}(\omega) = 0.405\angle -150^\circ = -0.3507 - j0.2025$$

The measurement frequency is $f_0 = 10$ GHz. A sample is inserted into the X-band sample holder (WR-90 waveguide). The length of sample is $d = 2$ mm, and the cutoff frequency for X-band waveguide is $f_c = 6.557$ GHz.

First calculate K,

$$K = \frac{\{S_{11}^2(\omega) - S_{21}^2(\omega)\} + 1}{2S_{11}(\omega)} = -1.126 + j0.051 \quad \text{(B.1)}$$

Then, we obtain reflection coefficient Γ,

$$\Gamma = K \pm \sqrt{K^2 - 1} = (-1.126 + j0.051) \pm \sqrt{(-1.126 + j0.051)^2 - 1} \quad \text{(B.2)}$$

Because $|\Gamma| \leq 1$,

$$\Gamma = (-1.126 + j0.051) + (0.526 - j0.109) = -0.6 - j0.058$$

Using Eq. (7.8), we can calculate

$$P = \frac{\{S_{11}(\omega) + S_{21}(\omega)\} - \Gamma}{1 - \{S_{11}(\omega) + S_{21}(\omega)\}\Gamma} = -0.410 - j0.201 = 0.456\angle -154^\circ \quad \text{(B.3)}$$

S. Kim, *Fundamentals and Principles of Electromagnetic Wave Absorbers*, Springer Series in Advanced Microelectronics 66, https://doi.org/10.1007/978-981-97-6477-8

$$\begin{aligned}\ln\left(\frac{1}{P}\right) &= \ln(2.192\angle 154^\circ) \\ &= \ln(2.192\angle 2.685\,\text{rad}) = \ln(2.192) + j(2.685 + 2\pi n) \end{aligned} \tag{B.4}$$

where, $n = 0, 1, 2, 3, \ldots$, the integer of (d/λ_g). This n can be determined from the analysis of group delay λ_g. For the case $d/\lambda_g < 1$, the parameter n is equal to 0, then we get

$$\ln\left(\frac{1}{P}\right) = 0.785 + j2.685$$

From Eq. (7.13), we obtain

$$\begin{aligned}\frac{1}{\Lambda^2} &= -\left[\frac{1}{2\pi d}\ln\left(\frac{1}{P}\right)\right]^2 = -\left[\frac{1}{2\pi \times 0.2}(0.785 + j2.685)\right]^2 \\ &= -(-4.179 + j2.672) = -4.960\angle 147^\circ \end{aligned} \tag{B.5}$$

and

$$\frac{1}{\Lambda} = \pm j(2.227\angle 73.5^\circ) = \pm j(0.625 + j2.138)$$

Because $\mathrm{Re}\left(\frac{1}{\Lambda}\right) \geq 0$, we obtain

$$\frac{1}{\Lambda} = 2.138 - j0.625 = 2.227\angle - 16.3^\circ$$

$$\frac{1+\Gamma}{1-\Gamma} = 0.248 - j0.045 = 0.252\angle - 10.3^\circ$$

Using Eq. (7.11) with values of $f_0 = 10$ GHz ($\lambda_0 = 3.0$ cm) and $f_c = 6.557$ GHz ($\lambda_c = 4.575$ cm), we get

$$\begin{aligned}\mu_r &= \frac{1+\Gamma}{\Lambda(1-\Gamma)\sqrt{\frac{1}{\lambda_0^2} - \frac{1}{\lambda_c^2}}} = \frac{2.227\angle - 16.3^\circ \times 0.252\angle - 10.3^\circ}{\sqrt{\frac{1}{3^2} - \frac{1}{4.575^2}}} \\ &= 8.908\angle - 26.6^\circ = 8.0 - j4.0 \end{aligned} \tag{B.6}$$

Thus

$$\mu_r' = 8.0 \text{ and } \mu_r'' = 4.0$$

Using Eq. (7.12), we obtain

$$\varepsilon_r = \frac{\left(\frac{1}{\Lambda^2} + \frac{1}{\lambda_c^2}\right)\lambda_0^2}{\mu_r} = \frac{\left((2.138 - j0.625)^2 + \frac{1}{4.575^2}\right) \times 3^2}{8.0 - j4.0} = 5.0 - j0.5 \qquad \text{(B.7)}$$

Thus

$$\varepsilon_r' = 5.0 \text{ and } \varepsilon_r'' = 0.5$$

Appendix C
MATLAB Program for Calculation of μ_r and ε_r from *S* Parameters Measured Using Coaxial Airline

```
clear all

% ------------Data input--------------------
sta=input('Input the start frequency(GHz) : ');
ed=input('Input the end frequency(GHZ) : ');
num=input('Input the number of point : ');
d=input('Input the material thickness d(mm) : ');
%L=input ('Input the coaxial air line length L(cm) : ');

% ------------ S parameter transform --------------

L=3;
f0=sta:(ed-sta)./(num-1):ed;
lamda=3./(f0.*10);
lamda=lamda';
theta=((L.*0.01-d.*0.001)./lamda)*2.*pi;

load rcs1s11.dat;
load rcs1s21.dat;
load rcs1s22.dat;

s11=rcs1s11(:,1)+j.*rcs1s11(:,2);
s22=rcs1s22(:,1)+j.*rcs1s22(:,2);
s21=rcs1s21(:,1)+j.*rcs1s21(:,2);
s11a=abs(s11);
s22a=abs(s22);
s21a=abs(s21);

s11b=angle(s11);
s22b=angle(s22);
s21b=angle(s21);

for h=1:num
sh=s11b(h);
```


S. Kim, *Fundamentals and Principles of Electromagnetic Wave Absorbers*, Springer Series in Advanced Microelectronics 66, https://doi.org/10.1007/978-981-97-6477-8

```
if( sh<0) s11b(h)=sh;
elseif(sh>0) s11b(h)=sh-2.*pi;
end
end

flag=0;
s11p(1)=s11b(1);
for h=2:num
if(s11b(h)>s11b(h-1)+3) flag=flag+1;
else flag=flag;
end
s11p(h)=s11b(h)-2.*flag.*pi;
end
s11p=s11p';

for h=1:num
sh=s22b(h);
if( sh<0) s22b(h)=sh;
elseif(sh>0) s22b(h)=sh-2.*pi;
end
end

flag=0;
s22p(1)=s22b(1);
for h=2:num
if(s22b(h)>s22b(h-1)+3) flag=flag+1;
else flag=flag;
end
s22p(h)=s22b(h)-2.*flag.*pi;
end
s22p=s22p';

for h=1:num
sh=s21b(h);
if( sh<0) s21b(h)=sh;
elseif(sh>0) s21b(h)=sh-2.*pi;
end
end

flag=0;
s21p(1)=s21b(1);
for h=2:num
if(s21b(h)>s21b(h-1)+3) flag=flag+1;
else flag=flag;
end
s21p(h)=s21b(h)-2.*flag.*pi;
end
s21p=s21p';
s11ta=sqrt(s11a.*s22a);
s11tp=(s11p+s22p)./2 + theta;
s21ta=s21a;
s21tp=s21p+theta;

s11=s11ta.*cos(s11tp)+j.*s11ta.*sin(s11tp);
s21=s21ta.*cos(s21tp)+j.*s21ta.*sin(s21tp);
```

```
%-----------------ur, er calculation------------
f0=f0';

k=(s11.^2-s21.^2+1)./(2.*s11);
gam1=k+sqrt(k.^2-1);
gam2=k-sqrt(k.^2-1);

for h=1:num
ga=abs(gam1(h));
if (ga <1.001) gam(h)=gam1(h);
else gam(h)=gam2(h);
end
end

gam=gam.';

t=(s11+s21-gam)./(1-(s11+s21).*gam);
ta=abs(t);
tang=angle(t);
lnt=log(ta)+j.*tang;
x=(1+gam)./(1-gam);
y=j.*(0.3./(f0.*2.*pi.*d.*0.001)).*lnt;
ur=x.*y;
er=y./x;
```

Appendix D
Measurement Procedure for μ_r and ε_r

A general measurement procedure for ε_r and μ_r using the transmission/reflection line method is shown in Fig. D.1.

The followings are descriptions on each process:

(1) Set up the vector network analyzer (VNA) including checking cable connection and running operation program.
(2) Set the frequency range, for example 8.2–12.4 GHz for X-band waveguide, and number of points, usually 201 or 401.
(3) Calibrate the system using the calibration kit available. Full 2-port calibration with TRL method is conducted in coaxial airline or waveguide measurements. Reflection is conducted at both port 1 and port 2 with short-circuited, open-circuited, and matched-load terminations. Conduct the transmission forward and reverse after connecting port 1 and port 2 with 'trans thru' and 'match thru'. After calibration, S_{11} and S_{22} are lower than $-$ 50 dB and S_{21} are very close to 0 dB. After connecting with sample holder, $S_{11} \leq -30$ dB and $S_{21} \geq -0.1$ dB.
(4) The material under test (MUT) is placed in a sample holder. A precise machining of MUT is needed in order to minimize the air gaps between sample and holder. An optimum sample thickness is close to $\lambda/4$.
(5) Reflection coefficient (S_{11}, S_{22}) and transmission coefficient (S_{21}) are measured.
(6) The complex permittivity and permeability are calculated from the S parameters with input of sample thickness and cutoff frequencies using the external programs (presented in Appendix C).
(7) Display and save the data.

S. Kim, *Fundamentals and Principles of Electromagnetic Wave Absorbers*, Springer Series in Advanced Microelectronics 66, https://doi.org/10.1007/978-981-97-6477-8

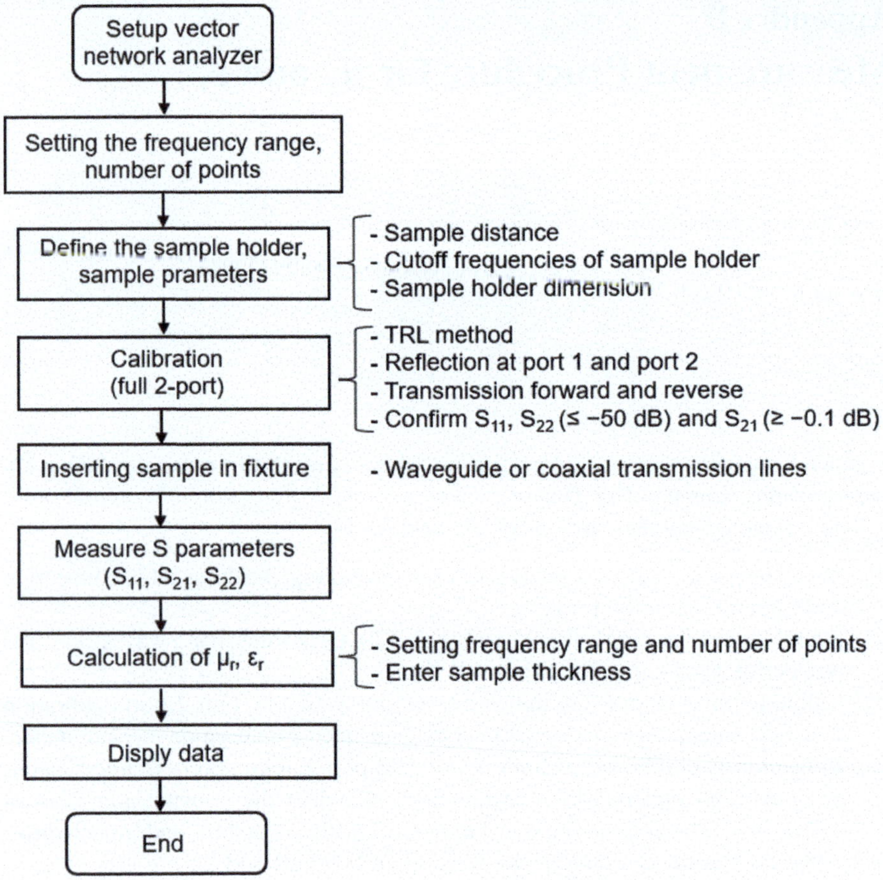

Fig. D.1 A procedure for performing the material constants measurement

Index

S. Kim, *Fundamentals and Principles of Electromagnetic Wave Absorbers*, Springer Series in Advanced Microelectronics 66, https://doi.org/10.1007/978-981-97-6477-8

I

J

L

M

SPRINGER NATURE

GPSR Compliance

The European Union's (EU) General Product Safety Regulation (GPSR) is a set of rules that requires consumer products to be safe and our obligations to ensure this.

If you have any concerns about our products, you can contact us on ProductSafety@springernature.com

In case Publisher is established outside the EU, the EU authorized representative is:

Springer Nature Customer Service Center GmbH
Europaplatz 3
69115 Heidelberg, Germany

The manufacturer's authorised representative in the EU is Springer Nature Customer Service Centre GmbH, Europaplatz 3, 69115 Heidelberg, Germany. If you have any concerns regarding our products, please contact ProductSafety@springernature.com

Printed and bound by CPI Group (UK) Ltd, Croydon, CR0 4YY
01/10/2026
02225962-0005